THE BIOLOGY
OF *EUGLENA*

Volume III
Physiology

CONTRIBUTORS TO THIS VOLUME

G. BENJAMIN BOUCK

EUGENE C. BOVEE

GIULIANO COLOMBETTI

BODO DIEHN

G. DUBERTRET

LELAND N. EDMUNDS, JR.

ELLIS S. KEMPNER

GORDON F. LEEDALE

M. LEFORT-TRAN

FRANCESCO LENCI

JEROME A. SCHIFF

STEVEN D. SCHWARTZBACH

THE BIOLOGY
OF *EUGLENA*

Edited by DENNIS E. BUETOW

DEPARTMENT OF PHYSIOLOGY AND BIOPHYSICS
UNIVERSITY OF ILLINOIS
URBANA, ILLINOIS

Volume III

Physiology

1982

ACADEMIC PRESS
A Subsidiary of Harcourt Brace Jovanovich, Publishers
New York   London
Paris   San Diego   San Francisco   São Paulo   Sydney   Tokyo   Toronto

ACADEMIC PRESS, INC.
111 Fifth Avenue, New York, New York 10003

United Kingdom Edition published by
ACADEMIC PRESS, INC. (LONDON) LTD.
24/28 Oval Road, London NW1   7DX

Library of Congress Cataloging in Publication Data

Buetow, Dennis E., Date.
    The biology of Euglena.

    Includes bibliographies and indexes.
    CONTENTS: v. 1. General biology and ultra-
structure--v. 2. Biochemistry--v. 3. Physiology.
I. Euglena.   II. Title.
QL368.F5B8   589.4'4   68-14645
ISBN 0-12-139903-6 (v. 3)   AACR2

To the memory
of
Theodore Louis Jahn

CONTENTS

Chapter 1. **Ultrastructure**

GORDON F. LEEDALE

Chapter 2. **Flagella and the Cell Surface**

G. BENJAMIN BOUCK

LIST OF CONTRIBUTORS

Numbers in parentheses indicate the pages on which the authors' contributions begin.

G. BENJAMIN BOUCK (29), Department of Biological Sciences, University of Illinois, Chicago, Illinois 60680

EUGENE C. BOVEE (143), Department of Physiology and Cell Biology, University of Kansas, Lawrence, Kansas 66045

GIULIANO COLOMBETTI (169), Consiglio Nazionale delle Ricerche, Istituto di Biofisica, 56100 Pisa, Italy

BODO DIEHN (169), Department of Zoology, Michigan State University, East Lansing, Michigan 48824

G. DUBERTRET (253), Laboratoire de Cytophysiologie de la Photosynthese, CNRS, 91190 Gif-sur-Yvette, France

LELAND N. EDMUNDS, JR.* (53), Department of Biology, State University of New York, Stony Brook, New York 11794

ELLIS S. KEMPNER (197), Laboratory of Physical Biology, National Institute of Arthritis, Metabolism and Digestive Diseases, National Institutes of Health, Bethesda, Maryland 20205

GORDON F. LEEDALE (1), Department of Plant Sciences, University of Leeds, Leeds LS2 9JT, England

M. LEFORT-TRAN (253), Laboratoire de Cytophysiologie de la Photosynthese, CNRS, 91190 Gif-sur-Yvette, France

FRANCESCO LENCI (169), Consiglio Nazionale delle Ricerche, Istituto di Biofisica, 56100 Pisa, Italy

JEROME A. SCHIFF (313), Institute for Photobiology of Cells and Organelles, Brandeis University, Waltham, Massachusetts 02254

STEVEN D. SCHWARTZBACH (313), School of Life Sciences, University of Nebraska, Lincoln, Nebraska 68588

*Present address: Department of Anatomical Sciences, School of Medicine, Health Sciences Center, State University of New York, Stony Brook, New York 11794.

PREFACE

Volumes I and II of "The Biology of *Euglena*" were published in 1968. These early volumes collated the very widely scattered literature on the genus *Euglena* and covered its biology comprehensively from taxonomy and ecology to biochemistry. Since 1968, the literature on *Euglena* has greatly expanded. This is the first of two new volumes on the biology of *Euglena*. Both update the earlier volumes and cover new areas of investigation on *Euglena*. Volume III covers "Physiology" and Volume IV will cover "Subcellular Organelles and Molecular Biology."

As before, a wide variety of biological experimentation continues to be performed with *Euglena*. Indeed, the amount of new literature cited in Volumes III and IV will show that members of the genus *Euglena,* and especially *Euglena gracilis,* are among the most widely used and researched eukaryotic microorganisms in biology. This is the result, of course, of the unique taxonomic position held by this genus which shows both animal-like and plant-like characteristics. In particular, *Euglena gracilis* is easily grown in large quantities on relatively simple media, readily adapts to a variety of nutritional and environmental conditions, and contains "conditional" chloroplasts which develop and fix CO_2 only in the light.

The efforts of all the contributors to this volume are certainly appreciated.

DENNIS E. BUETOW

CONTENTS OF PREVIOUS VOLUMES

CHAPTER 1

ULTRASTRUCTURE

Gordon F. Leedale

I. Introduction

The ultrastructure of *Euglena* has been reviewed on several previous occasions, either within the context of a general survey of euglenoid flagellates (Leedale, 1967a, 1967b, 1971) or in relation to studies on particular species of the genus, especially *E. spirogyra* (Leedale *et al.*, 1965) and *E. gracilis* (see Buetow, 1968). The present account aims to summarize the essential features of organelle distribution and structure in *Euglena,* including observations made during the last decade. The reader should refer to the accounts cited above for more detailed information, but it is hoped that the following compilation, though brief, is sufficiently self-contained to stand as a basic guide to *Euglena* cytology

THE BIOLOGY OF *EUGLENA*, VOL. 3

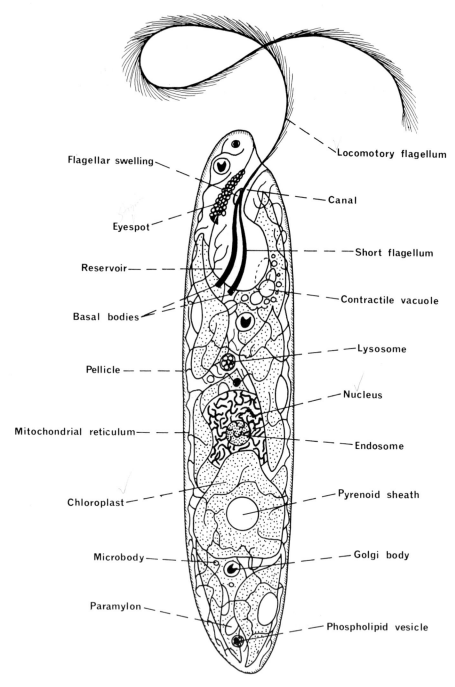

Flagellar swelling

Eyespot

Reservoir

Basal bodies

Pellicle

Mitochondrial reticulum

Chloroplast

Microbody

Paramylon

Locomotory flagellum

Canal

Short flagellum

Contractile vacuole

Lysosome

Nucleus

Endosome

Pyrenoid sheath

Golgi body

Phospholipid vesicle

Fig. 1. Euglena cell of the *E. gracilis* type, showing the main organelles.

and ultrastructure. Previously unpublished micrographs have been used, where appropriate, in order to increase the range of illustrations available for study.

II. The *Euglena* Cell

Distribution of organelles in an idealized cell of the *E. gracilis* type is illustrated in Fig. 1. The actual size, shape, chloroplast distribution, nuclear location and flagellar emergence of *E. gracilis* are shown in Fig. 2, and details of the cell anterior are clarified in Fig. 3. A living cell of *E. spirogyra* is illustrated in Fig. 4 for comparison. Comments on individual organelles will be found in the subsequent pages.

Cells of *E. gracilis* are approximately 50 μm long × 10 μm wide and cigar-shaped (Fig. 2); cells of *E. spirogyra* are said to range from 45 to 250 μm in length but characteristically are 100–150 μm long × 15 μm wide (Fig. 4). Other species range in length from ca. 10 μm (*E. minuta*) to more than 500 μm (*E. ehrenbergii, E. oxyuris*) and in shape from spherical (*E. inflata*), through fusiform (*E. viridis* and many others), to narrowly elongate (*E. acus*), always with some degree of spiralling. Sizes and shapes are documented by Gojdics (1953), Huber-Pestalozzi (1955), and Pringsheim (1956) and in many earlier publications cited by these authors; Pringsheim (1956) makes pertinent comments on size ranges and changes of shape in relation to identification of species.

III. Pellicle and Associated Structures

The pellicle of *Euglena* (Fig. 5) consists of helically disposed strips of elaborate cross-sectional shapes that overlap and articulate along their edges (Figs. 6 and 8; see also 22) and interdigitate at both ends of the cell. The strips are proteinaceous (Barras and Stone, 1965; Hofmann and Bouck, 1976) and intracellular, lying immediately beneath the plasmalemma (Figs. 6–8, 11, and 14; see also 22). Associated with the strips are species-specific arrangements of microtubules and muciferous bodies (Figs. 6–11; see also 22).

Ultrastructure of the pellicular complex was first described in detail by Leedale (1964) for *E. spirogyra* (Figs. 5 and 6) and has been summarized several times since (e.g., Leedale, 1967a; Buetow, 1968). The cross-sectional shape of the pellicular strips varies from species to species (Figs. 6–8, 11; see also 22) but all exhibit some form of groove–ridge articulation by means of which adjacent strips can slide against one another; the groove–ridge shape remains intact under various disruptive chemical treatments (Silverman and Hikida, 1976). Thickness of strips is related to cell mobility: Rigid species have thick pellicles (Figs. 6 and 8; see also 22); highly mobile species have thin pellicles (Figs. 7, 11, and 14). On

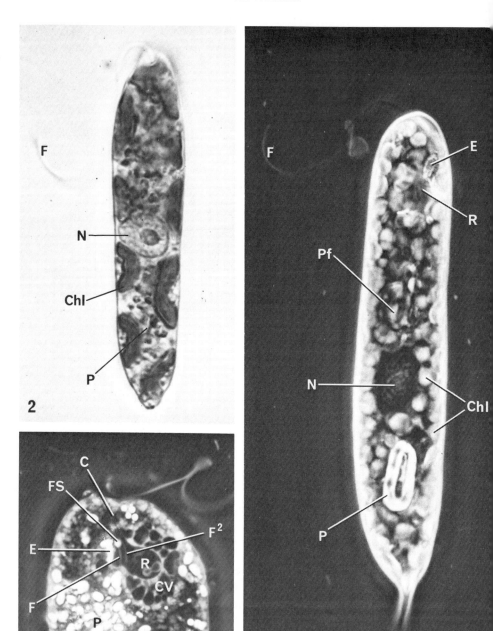

the other hand, distribution and numbers of pellicular microtubules (Figs. 6, 7, and 11; see also 22) do not seem to be related to cell elasticity.

Freeze-etch studies of the pellicle of *E. gracilis* reveal particulate and striated layers (Holt and Stern, 1970; Schwelitz *et al.*, 1970) which can be correlated with plasmalemmar and pellicular strip ultrastructure as seen in sections. Miller and Miller (1978) demonstrate non-complementarity of fracture faces in what they term the "cell membrane" of *E. gracilis* and claim that this plus unusually high protein content is evidence for a unique type of membrane; however, these authors seem not to appreciate the nature of the plasmalemma-pellicle combination in *Euglena*. Guttman and Ziegler (1974) describe some degree of continuity between pellicular and canal microtubules; Hofmann and Bouck (1976) suggest that microtubule distribution in *E. gracilis* is important in relation to the development of new pellicular strips during cell growth and division. Formation of daughter pellicular complexes between existing ones (Leedale, 1968) is shown by immunological techniques to occur by intussusceptive growth (Hofmann and Bouck, 1976), new strips originating by intercalation between the old strips within the canal. Relative ages of strips are indicated in some species by patterns of external ornamentation (Fig. 5).

The surface of all *Euglena* cells is coated with mucilage that is secreted from the muciferous bodies (Figs. 6-8, 10, 11, and 14; see also 22) via mucilage canals (Figs. 6 and 8-10) to the outside. In some species the mucilage forms an extensive layer (e.g., *E. viridis*, Fig. 14) which may eventually become a cyst; in others it is reduced to a few strands (e.g., swimming *E. gracilis*, Rosowski, 1977) or serves as a template for extracellular ornamentation (e.g., the warts of *E. spirogyra*, Fig. 5); in the colonial euglenoid *Colacium*, the mucilage is essential to the epizoic habit of this genus (Rosowski and Willey, 1977); in *Trachelomonas*, it is involved in formation of extracellular envelopes (Leedale, 1975a). The muciferous bodies are part of an extensive subpellicular system of endoplasmic reduction (ER), arranged helically, parallel to the pellicular strips (Fig. 10). Some species of *Euglena* additionally have large ejectile mucocysts

---

Fig. 2. Osmium-fixed cell of *E. gracilis*, showing characteristic shape, locomotory flagellum (F) emerging from the subapical canal opening, 9-10 large chloroplasts (Chl), central nucleus (N), and paramylon (P). ×2000.

Fig. 3. Anterior end of living cell of *E. gracilis*, flattened to show the two flagella (F, F²) inserted on the wall of the reservoir (R), flagellar swelling (FS) opposite the eyespot (E), canal (C), contractile vacuolar region (CV), and small paramylon granules (P). ×1500.

Fig. 4. Living cell of *E. spirogyra*, showing numerous discoid chloroplasts (Chl), locomotory flagellum (F), eyespot (E), reservoir (R), nucleus (N), large paramylon granule (P), and position of forming paramylon granule (Pf). ×1500.

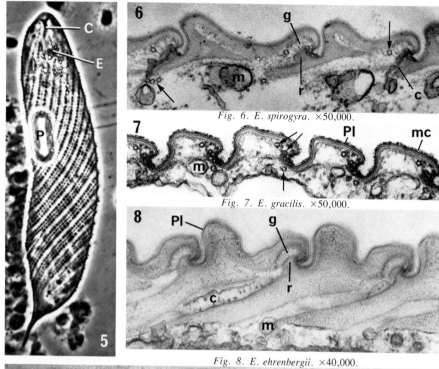

Fig. 6. *E. spirogyra.* ×50,000.

Fig. 7. *E. gracilis.* ×50,000.

Fig. 8. *E. ehrenbergii.* ×40,000.

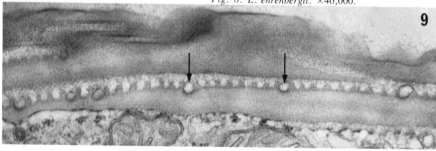

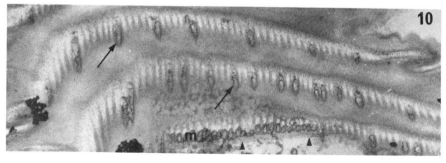

(e.g., *E. splendens;* Hausmann and Mignot, 1977) or pellicle pores with dense contents (e.g., *E. granulata;* Arnott and Walne, 1967). Secretion of Golgi-produced mucilage into the reservoir during cyst formation (Triemer, 1980) is described in Section VII. The types of cysts are reviewed by Buetow (1968).

The structure, chemistry, growth, and physiology of the pellicle are discussed in detail in Chapter 2 of this volume.

IV. Organelles of the Cell Anterior

A. CANAL AND RESERVOIR

Cells of *Euglena* have an anterior invagination consisting of a tubular canal and bulbous reservoir (Figs. 1 and 3). The canal has pellicular material and microtubules underlying the plasmalemma (Figs. 5 and 11), with further helical swirls of microtubules, presumably skeletal in function, in the adjacent cytoplasmic matrix (Leedale, 1967a). The reservoir membrane is not underlain by pellicle (Figs. 12–14; see also 22) but is still lined by species-specific patterns of microtubules (Figs. 14 and 22). The invagination microtubules and a ribosome-free region of cytoplasmic matrix are often delimited from the rest of the cell by a cylindrical sheath of ER (Fig. 22).

The canal is a rigid tube (Fig. 5) through which the locomotory flagellum passes to emerge from the apical or subapical (ventral; see Leedale, 1967a) canal opening (Fig. 2). The reservoir membrane provides the main site in the cell for exo- and endocytosis; the contractile vacuole empties into the reservoir (see Section IV,D) but Kivic and Vesk (1974) have demonstrated experimentally that pinocytosis also occurs. This is thought to be a mechanism for recovery of large molecules lost during contractile vacuolar discharge. Images consistent with exo-

---

Fig. 5. Pellicle of *E. spirogyra,* with rows of ferric hydroxide-impregnated warts of mucilage spiralling round the cell; since the rows are laid down on individual pellicular strips and the size of warts is related to mucilage production and the iron content of the growth medium, the rows indicate the alternation of older and younger strips. C, Pellicular lining of canal; E, eyespot droplets; P, paramylon granule. ×750.

Figs. 6–8. Pellicles of *Euglena* spp. sectioned transversely to the pellicular strips. c, Muciferous canal; g, groove of one strip; r, ridge of next strip; m, muciferous body; mc, mucilage; Pl, plasmalemma; arrows, pellicular microtubules.

Fig. 9. Pellicle of *E. ehrenbergii* with strips in oblique longitudinal section, showing muciferous canals (arrows) passing between teeth on the flange of one strip. ×40,000.

Fig. 10. Oblique section of pellicle of *E. ehrenbergii,* showing strips, teeth, muciferous canals (arrows), muciferous bodies (m), and microtubules (arrowheads). ×25,000.

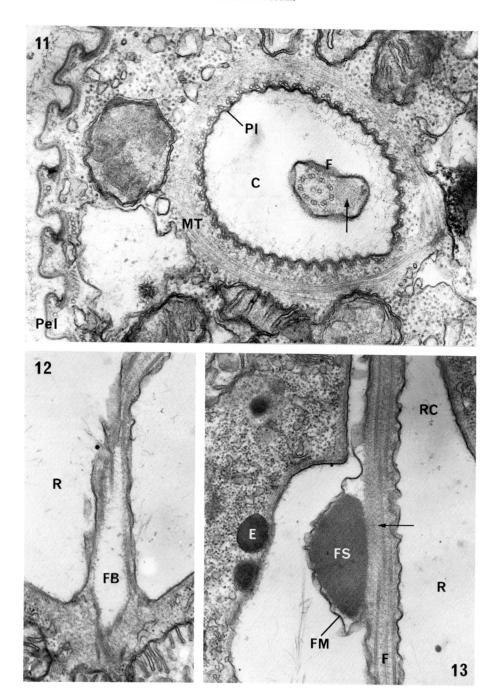

and endocytosis appear in freeze-etch preparations of reservoir membrane (Miller and Miller, 1978).

B. FLAGELLA

Euglena has two flagella (Figs. 3 and 14) inserted on the dorsal wall of the reservoir; one emerges from the canal as the organelle of swimming (Fig. 2), and the other ends within the reservoir (Fig. 3). In a few species of *Euglena* both flagella are nonemergent (e.g., *E. obtusa*) and the organism undergoes creeping euglenoid movements. The flagellar axoneme is the characteristic eukaryotic 9 + 2 microtubular configuration, alongside of which runs a paraflagellar rod (Figs. 11, 13, and 22). A unilateral array of fine hairs (each ca. 3 μm long) and an extensive coating of accessory material is borne on the emergent flagellum (Leedale, 1967a). Recent studies by Hofmann and Bouck (1976) show that the longer hairs are attached to the paraflagellar rod, whereas 1.5 μm fibrils are the centrifugally arranged portions of structural complexes attached parallel to and outside of the flagellar membrane; these units are in near register laterally and overlap longitudinally by one-half of a unit length (see Chapter 2).

The flagellar bases are swollen and appear hollow in section (Fig. 12); details of the transitional zone are described by Piccini and Mammi (1978). Three microtubular roots run from the bases, with different numbers of component microtubules in different species (and genera: 5-8-5 in *Menoidium bibacillatum* and 6-8-2 in *Rhabdomonas costata*, Leedale and Hibberd, 1974; 6-9-4 in *Eutreptiella* cf. *braarudii*, Moestrup, 1978). One root passes over the eyespot (Moestrup, 1978), whereas another runs into the cell and may represent what was earlier termed a rhizoplast by light microscopists. In *E. gracilis* the root configuration is probably 5-7-3, with a striated fibril also involved (Piccinni and Mammi, 1978); in *E. acus* the largest root has at least 12 microtubules plus dense material (see Fig. 22). According to Moestrup (1978), the ultrastructure of this root in *Eutreptiella* is similar to the multilayered structure found in many green algae and the spermatozoids of bryophytes and higher plants.

Fig. 11. Transverse section (T.S.) of canal (C) of *E. viridis;* the locomotory flagellum (F) in T.S. shows the 9 + 2 axoneme and paraflagellar rod (arrow) side by side; paired microtubules and pellicular material underlie the canal plasmalemma (Pl), with a layer of helical microtubules (MT) beneath these; structures of the pellicular complex (Pel) are seen at left. ×50,000.

Fig. 12. Longitudinal section (L.S.) of hollow flagellar base (FB) of *E. gracilis.* R, reservoir. ×30,000.

Fig. 13. Continuation of L.S. of same flagellum as in Fig. 12, to show flagellar swelling (FS) inside the flagellar membrane (FM), the 9 + 2 axoneme (F), and paraflagellar rod (arrow). E, eyespot droplet; R, reservoir; RC, reservoir–canal transition. ×30,000.

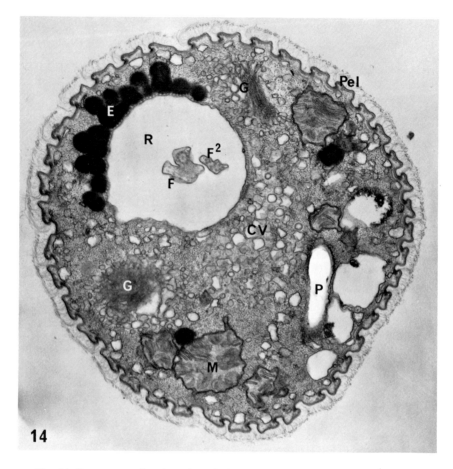

Fig. 14. Transverse section of anterior end of *E. viridis;* eyespot droplets (E) ensheath the neck of the reservoir (R) within which the two flagellar bases (F, F²) are sectioned. A single layer of microtubules underlies the reservoir membrane; the contractile vacuolar region (CV) is flanked by two Golgi bodies (G). M, mitochondrial profiles; P, paramylon; Pel, pellicle, with copious external mucilage appearing as a fibrous layer. ×15,000.

The latest information on appendages, membrane, axoneme, and growth of the flagellum in *Euglena* is presented in Chapter 2 of this volume; movement is discussed in Chapter 4.

C. FLAGELLAR SWELLING AND EYESPOT

It has been known for some time (Leedale, 1967a; Walne and Arnott, 1967; Walne, 1971; Kivic and Vesk, 1972) that the flagellar swelling of *Euglena* is a crystalline structure within the flagellar membrane (Fig. 13). Piccinni and

Mammi (1978) describe it as a crystal with a monoclinic or slightly distorted hexagonal unit cell ($a = 8.9$ nm; $b = 7.7$ nm; $c = 8.3$ nm; $\beta = 110°$). The swelling has a complex structural relationship with the paraflagellar rod (Fig. 13), discussed most recently with reference to the genus *Trachelomonas* by West *et al.* (1980). The swelling contains flavins but no other pigments (Benedetti *et al.*, 1976).

The eyespot in *Euglena* consists of a group of extraplastidial orange-red droplets grouped on the dorsal side of the canal or canal–reservoir transition (Fig. 14), opposite the flagellar swelling (Figs. 3 and 13). Individual droplets are membrane-bound but there is no common membrane; a layer of microtubules lies between the eyespot and the reservoir–canal membrane (Fig. 14). Detailed studies on eyespot structure, pigments, and relationship with other organelles are reviewed by Walne (1980); eyespot droplets contain flavins, lipids, and numerous carotenoid pigments. Recent experimental studies on the function of eyespot and flagellar swelling in photoreception and phototaxis are discussed in Chapter 5 of this volume.

D. Contractile Vacuole

The contractile vacuolar region in *Euglena* curves around the ventral and posterior surfaces of the reservoir (Fig. 3), diametrically opposite to the eyespot (Fig. 14). Discharge of the contractile vacuole occurs every 20–30 seconds by confluence of vacuolar and reservoir membranes. After discharge, numerous accessory vacuoles (Fig. 14) coalesce to form larger ones (Fig. 3) and these eventually fuse to produce the single spherical contractile vacuole, which then empties into the reservoir. Alveolate (coated) vesicles, 100 nm in diameter, occur among the accessory vacuoles (Leedale *et al.*, 1965) and are thought to be concerned in phase segregation, retaining solutes in the cytoplasm and secreting water (and, possibly, waste products) into the vacuole. Golgi bodies in the region are apparently involved in the osmoregulatory activity, although direct origin of alveolate vesicles from Golgi cisternae, known for green algae, has not been observed in *Euglena* (Leedale, 1967a). Recovery of the reservoir to its normal shape after each systole is probably accomplished by means of its lining of microtubules (Figs. 13, 14; see also 22).

Further observations on the structure and function of the contractile vacuole, including the possible existence of some form of permanent vacuole that refills at each cycle, are discussed by Leedale *et al.* (1965), Leedale (1967a,b), and Buetow (1968).

V. Chloroplasts and Pyrenoids

Chloroplasts in *Euglena* spp. (see Leedale, 1967a. Figs. 146–157) range from small pyrenoid-less discs (Fig. 4) to large plates or complexes with naked or

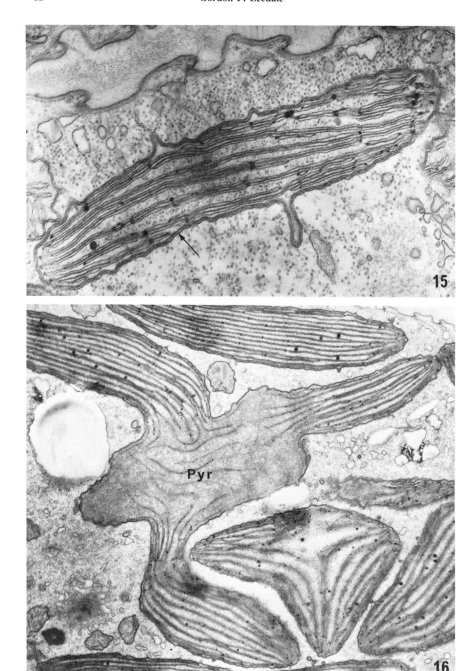

paramylon-sheathed pyrenoids (Figs. 2 and 16). Electron microscopy reveals a three-membraned chloroplast envelope (Fig. 15), photosynthetic lamellae composed typically of three thylakoids (Figs. 15 and 16), although patterns involving 2–12 may occur, a 70 S ribosomal matrix or stroma, infrequent plastoglobuli, and DNA (see this treatise, Volume IV, Chapter 6). Pyrenoids have finely granular stroma penetrated by lamellae reduced to one or two thylakoids (Figs. 16 and 17).

The three-membraned envelope has recently been put forward by Gibbs (1978) as evidence that the chloroplasts are evolved by reduction from endosymbiotic green algae: The inner two membranes are interpreted as the chloroplast envelope proper, and the outer one as the plasmalemma of the original symbiont. Contrary to earlier reports (Leedale, 1968), the outer membrane is not ER (i.e., it does not bear ribosomes, it does not connect to the nuclear envelope or cytoplasmic ER, and there are no periplastidial vesicles between the chloroplast membranes). In written communication with Gibbs, the author has suggested that the endosymbiosis might more probably have involved green *chloroplasts* rather than green algae. Phagotrophic euglenoids such as *Peranema* can "cut" their way into algal cells (e.g., *Cladophora*) and ingest the chloroplasts and other contents (Leedale, 1967a); if digestion of the chloroplasts had failed to occur in a colorless ancestral euglenoid, the transient food vacuoles could have become functional euglena chloroplasts. The two inner membranes would then be the original envelope of the green algal chloroplast and the third (outer) membrane would be the phagocytotic membrane (i.e., cytosome plasmalemma) of the host euglenoid. The phylogenetic import of this is discussed in Section XII.

More research time has been devoted to the chloroplasts of *E. gracilis* than to any other component of the *Euglena* cell. Chapters 7 and 8 of this volume and Chapters 5, 6, and portions of Chapters 1 and 7–9 of Volume IV review chloroplast structure, development, chemistry, and metabolism. The reader is referred to these chapters for the most recent information.

VI. Mitochondria

The mitochondrial system of *E. gracilis* is a single reticulum (Leedale and Buetow, 1970) that ramifies throughout the cell (Fig. 1). Three-dimensional reconstructions from serial sections show the chondriome to be a permanent

---

Fig. 15. Section of chloroplast of *E. viridis* to show three-membraned envelope (arrow), two- and three-thylakoid lamellae, and granular stroma. ×40,000.

Fig. 16. Sectioned pyrenoid (Pyr) of *E. viridis,* with radiating chloroplast ribbons showing typical lamellate organization. ×12,500.

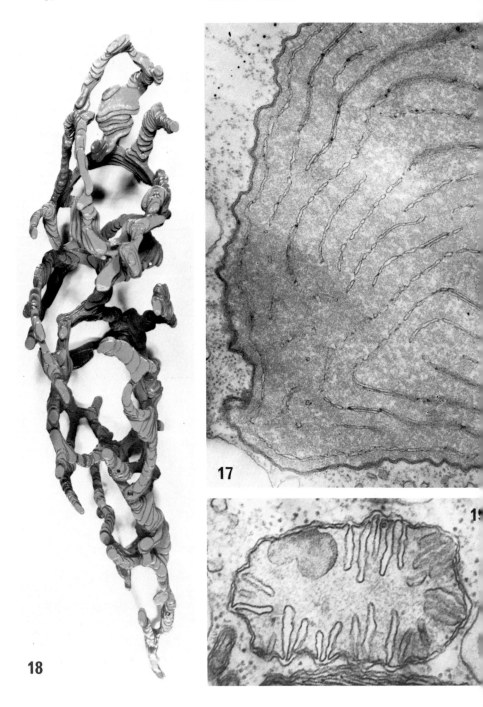

17

18

network of threads 0.4–0.6 μm thick (Fig. 18) that grows and divides with the cell, always constituting about 6% of the cell volume (Pellegrini, 1980a). Under carbon starvation (Leedale and Buetow, 1970) or during bleaching (Pellegrini, 1980b), the morphology, volume, and component thickness of the reticulum changes; in heterotrophic cells (dark-grown on sodium acetate) the chondriome becomes a fenestrated shell occupying 15–16% of the cell volume. The appearance of the mitochondrial system in living cells suggests it is also a reticulum in other species of *Euglena*, although this has not been shown by serial sectioning.

The elements of the mitochondrion show a cristate organization in section (Fig. 19). The outer membrane is strongly undulated; the cristae are ovoid or circular, with characteristic constriction at the base. The most recent information on mitochondrial morphology, ultrastructure, isolation, physiology, and function will be found in Pellegrini (1980a, 1980b) and in Volume IV, Chapter 3 of this treatise.

VII. Golgi Bodies

Species of *Euglena* contain several to many Golgi bodies (dictyosomes), the number increasing with cell size. They are distributed throughout the organism but there is always a concentration in the region around the reservoir (Fig. 14), presumably in relation to exo- and endocytosis, secretion of new pellicle, osmoregulation, and other functions associated with the cell anterior. Each Golgi body is a large structure, consisting of 15–30 cisternae more than 1 μm in diameter (Fig. 20); vesicles 25–80 nm in diameter are proliferated in large numbers from the fenestrated edges of the cisternae.

The Golgi bodies are involved in metabolite conversion and secretory activity, as in other eukaryotes. They contain acid phosphatase (Brandes *et al.*, 1964; Sommer and Blum, 1965) and have been implicated in the formation of lysosomes under conditions of carbon starvation (Malkoff and Buetow, 1964).

Recently, Triemer (1980) has reported on the role of the Golgi apparatus in mucilage production and cyst formation in *E. gracilis*. When dark-grown cells are placed in minus-nitrogen media for 48–72 hours in the dark, the Golgi cisternae decrease in number, dilate, and fill with mucilage, which is then se-

Fig. 17. Detail of pyrenoid of *E. viridis* to show one- and two-thylakoid lamellae within dense stroma. ×35,000.

Fig. 18. Three-dimensional reconstruction of the single mitochondrial reticulum of a young cell of *E. gracilis* (courtesy of Dr. M. Pellegrini).

Fig. 19. Section of mitochondrion of *E. spirogyra* to show typical cristae. ×50,000.

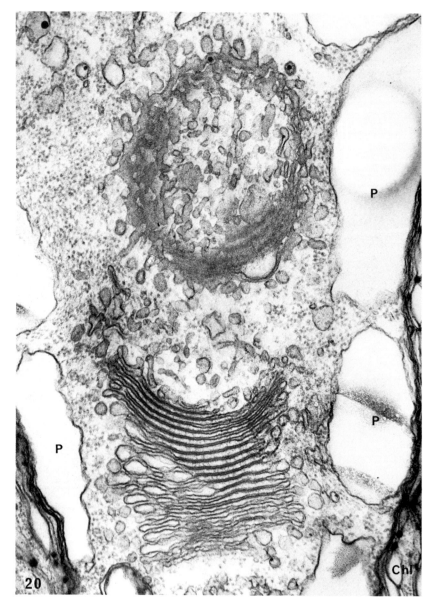

Fig. 20. Two Golgi bodies of *E. gracilis,* the lower one in vertical section, the upper one in transverse section; paramylon granules (P) adjacent to chloroplasts (Chl) are bounded by single membranes. ×45,000.

creted into the reservoir. From here the polysaccharide material is passed to the cell exterior, the rotation of the encysting cell ensuring that a layer of even thickness is deposited over the cell surface. It is interesting to note that as the Golgi bodies become more active, the number of cisternae per organelle decreases. This is also the case during cell division when Golgi bodies are actively producing new pellicular material for sequestration to the growing strips (Leedale, 1968).

VIII. Lysosomes, Autophagic Vacuoles, and Microbodies

Carbon-starved cells of streptomycin-bleached *E. gracilis* form large cytolysomes or autophagic vacuoles (Brandes *et al.*, 1964; Malkoff and Buetow, 1964; Leedale and Buetow, 1976) within which mitochondrial elements, chloroplast remnants, other membranes, and granular cytoplasm (Fig. 21) are degraded to provide the starving cells with materials for basic metabolism. Systems of layered ER, apparently proliferated from the nuclear envelope, are implicated in the formation and supply of membranes to the cytolysomes. Accompanying senescence phenomena include a drastic decrease in cell size, paramylon digestion, coarsening of the mitochondrial reticulum, thickening of the chromosomes, shedding of the external flagellum, and loss of the flagellar swelling, all relative to a progressive decrease in per cell levels of carbohydrate, protein, RNA, and DNA. The reduced cell retains viability for several weeks and rejuvenates if carbon is supplied. With prolonged starvation, some cells undergo lysis and the carbon thereby released into the medium allows recovery of other, still viable, cells; thus the population can survive starvation for several months.

Changes in chromosome texture and DNA content in relation to this process are described by Leedale (1975b) and further observations on aging cells of *Euglena* are recorded by Gomez *et al.* (1974). Current information on the existence in *Euglena* of small (less than 1 μm diameter) single-membraned inclusions with granular and enzymatic contents (peroxidase, catalase, and oxidases), usually referred to as microbodies (or, more specifically, as peroxisomes), are discussed in this treatise in Volume IV, Chapter 4.

IX. Nucleus

The cytology and ultrastructure of the nucleus in *Euglena* were reviewed by Leedale in 1968 with reference to interphase, mitosis, amitosis, and meiosis. A two-membraned nuclear envelope surrounds permanently condensed chromosomes, evenly distributed in granular nucleoplasm around a central endosome (nucleolus). The perinuclear space (between the two nuclear membranes) averages 20 nm in width but in some species (e.g., *E. ehrenbergii*) is swollen and

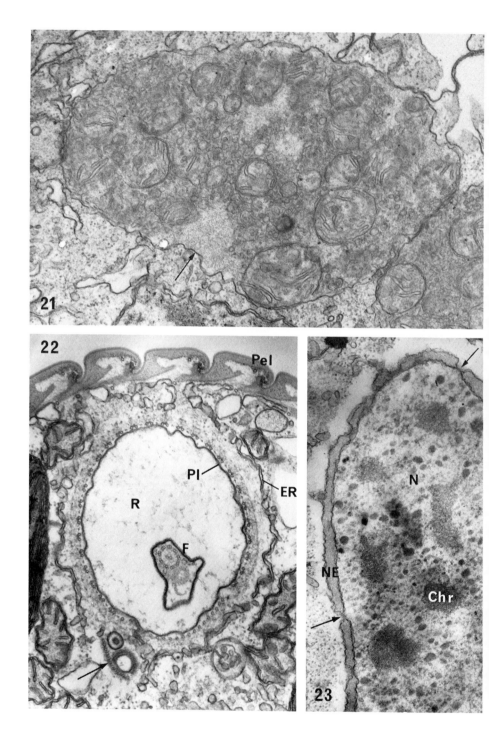

filled with material of an unknown nature; this makes the nuclear pores particularly distinct (Fig. 23). Extensions of the outer nuclear membrane link up with elements of tubular (''smooth'') ER (Leedale, 1968) but not with the chloroplasts (Gibbs, 1978). The filamentous construction of the chromatin varies during the cell cycle but the condensed chromosomes never lose their individuality. The 10-nm thickness of the chromonema reported by O'Donnell (1965) has been confirmed by study of isolated *E. gracilis* chromosomes spread on water (Haapala and Soyer, 1975). The latter authors demonstrate axial and chromomere fibrils as in higher eukaryotes and interpret *Euglena* chromosomes as circular uninemic structures; Moyne *et al.* (1975) describe fibrils of DNA linking chromosomes to one another in *E. gracilis*. Most of the other studies on *Euglena* nuclei in the last decade have been concerned with variations in the structure of the endosome and chromosomes through the cell cycle (e.g., Chaley *et al.*, 1977), variations due to carbon starvation (Leedale, 1975b) and vitamin B_{12}-deficiency, nuclear isolation (Buetow, 1978) and template activity, and the biochemistry of nuclear DNA and RNA and nucleoproteins (including histones). These areas of research are discussed in Volume IV, Chapter 2.

The ultrastructure of mitosis in *Euglena* was described for the first time by Leedale (1968) and can be characterized as follows:

(1) The nuclear envelope remains intact throughout nuclear division.

(2) The endosome elongates along the division axis and divides into two daughter endosomes.

(3) By means of a sequence of movements that differ in different species (Leedale, 1958, 1967a, 1968), the chromosomes come to lie along the division axis parallel to the dividing endosome. In *E. gracilis,* these are daughter chromosomes (chromatids), the separation and segregation of the longitudinally replicated parent chromosomes having occurred during orientation; in other species the parallel threads are chromosomes that replicate into pairs of chromatids after orientation (*E. spirogyra*) or previously formed pairs of chromatids (*E. viridis*), with segregation occurring subsequently in both cases.

Fig. 21. Autophagic vacuole (cytolysome) from carbon-starved cell of *E. gracilis;* mitochondrial profiles and other cytoplasmic components are undergoing breakdown within the organelle, which is bounded by a single membrane (arrow). ×30,000.

Fig. 22. Transverse section of the reservoir–canal transition in *E. acus;* microtubules and fine matrix underlie the reservoir–canal plasmalemma (Pl), delimited by a sheath of endoplasmic reticulum (ER); a large microtubular flagellar root is seen in cross section (arrow). F, locomotory flagellum in T.S.; R, reservoir–canal; Pel, pellicle. ×30,000.

Fig. 23. Nucleus (N) of *E. ehrenbergii,* showing nuclear envelope (NE) with expanded perinuclear space and nuclear pores (arrows). Chr, chromosome. ×30,000.

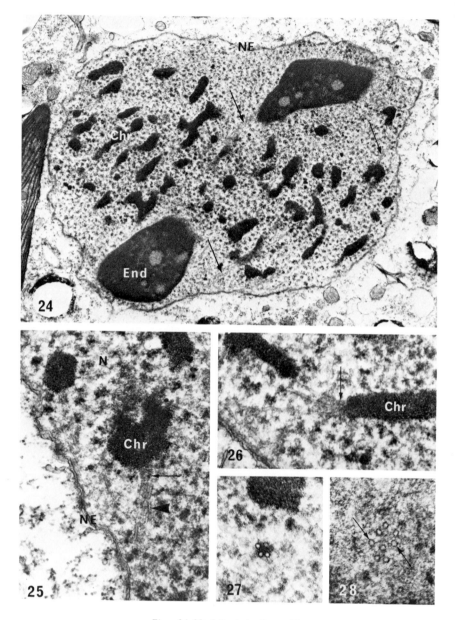

Figs. 24–28. Mitosis in *E. gracilis*.

(4) Separation, segregation, and poleward movements of chromatids are staggered such that some reach the poles while others are still at the equator.

(5) Microtubules appear in the nucleoplasm during prophase and form bundles between the chromosomes and alongside the dividing endosome during metaphase and anaphase. These nucleoplasmic microtubules end near the poles within the nuclear envelope; there are no cytoplasmic microtubules associated with the dividing nucleus.

Most of these features are illustrated in the longitudinal section of mitotic metaphase shown in Fig. 24 and are described in similar terms by Pickett-Heaps and Weik (1977) and Gillott and Triemer (1978) and in Volume IV, Chapter 2 of this treatise. The major new observation in these recent accounts is the demonstration of localized centromeres (kinetochores) on the mitotic chromsomes of *E. gracilis*. Pickett-Heaps and Weik (1977) record and illustrate several microtubules inserted into the chromosomal substance, described as a "distinct, layered kinetochore," in metaphase; they could not find such structures in mitotic stages of *Phacus*. Gillott and Triemer (1978) similarly record "distinct trilayered kinetochores" on *E. gracilis* chromosomes during mitosis.

These reports have prompted a reinvestigation of mitosis* in *E. gracilis* by the present author, resulting in micrographs of metaphases and anaphases in which centromeres are apparent (including one micrograph actually published in 1971, reproduced here as Fig. 24). In longitudinal section, several microtubules are seen ending in a matrix that is slightly less electron-dense than the main chromosomal substance; this region of microtubule insertion (the centromere)

Fig. 24. Section of nucleus in mitotic metaphase; the ends of the dividing endosome (End) are included but not the thinning central region; the nuclear envelope (NE) is intact; numerous chromosomes (Chr) are sectioned in approximate L.S. along the division axis; bundles of microtubules (arrows) run roughly parallel to the division axis. ×18,000.

Fig. 25. Detail from Fig. 24, showing chromosomes (Chr) with median centromeres; chromosomal microtubules (arrowhead) attach to centromere matrix (arrow). N, nucleus; NE, nuclear envelope. ×45,000.

Fig. 26. Chromosome (Chr) with terminal centromere (arrow). ×45,000.

Fig. 27. Transverse section of centromere, showing four microtubules at the periphery of dense matrix. ×72,000.

Fig. 28. Transverse section of bundle of eight microtubules in nucleoplasm (minispindle, see text); some of the tubules (arrowed) show one side arm directed towards the center of the bundle. ×72,000.

*This topic is covered in detail in Volume IV, Chapter 2.

may appear median (Fig. 25) or terminal (Fig. 26); in neither case does the matrix have a layered substructure. Immediately adjacent to this centromere matrix, the chromosomal microtubules are strictly parallel for a short distance (ca. 100 nm) before diverging away from the chromosome; the matrix and parallel microtubular region are best seen in Fig. 25 and the diverging microtubules in Fig. 26. The bundles of microtubules approach or run parallel to the nuclear envelope (Figs. 25 and 26) but no obvious attachments have been seen. More significantly, transverse sections of the centromere show four microtubules spaced around the periphery of a block of dense material (Fig. 27), an arrangement that can be compared with the bundles of 8–12 microtubules reported in the nucleoplasm by Leedale (1968) and shown in Fig. 28. Pickett-Heaps and Weik (1977) "gained the strong impression" that a few chromosomal microtubules are interspersed with other longer tubules traversing the spindle (i.e., the continuous microtubules) and that each chromatid can be envisaged as having its own miniature subspindle. Present observations support this idea and suggest that four chromosomal microtubules associate with 4–8 continuous ones in each minispindle. Such an arrangement could be expected to result in the individual poleward chromosomal migration that actually occurs in *E. gracilis* and would help to explain the long duration of anaphase (Leedale, 1959, 1967a). The four centromeric microtubules (Fig. 27) appear denser and narrower in diameter than do the 8–12 microtubules of the mini-spindle (Fig. 28), although the latter group is presumed to contain the chromosomal microtubules. Whether this is a genuine structural difference or a reflection of different fixation (the centromeric matrix perhaps preventing any swelling of the microtubules) is unknown. Side arms can be seen extending perpendicularly from some of the microtubules toward the centre of the mini-spindle (Fig. 28), possibly representing crossbridges that are involved in the sliding mechanism of microtubules in spindles and flagella (see Dustin, 1978).

Pickett-Heaps and Weik (1977) also provide new information on the persistence of interzonal microtubules within nuclear envelope material between the daughter nuclei at telophase. They suggest that activity of this region in late anaphase results in the final partitioning of the endosome.

X. Endoplasmic Reticulum, Ribosomes, and Microtubules

As indicated at several points in the present account and by Leedale (1967a), the ER in *Euglena* constitutes the peripheral complex of muciferous bodies, various tubular ("smooth") and vesiculate ("rough") components in the cell, a canal sheath, etc. Additionally, elements of the tubular ER are associated with or directly connected to the Golgi bodies (Leedale, 1967a), the nuclear envelope, the membranes of the paramylon granules, and other structures. However, there is apparently no connection between the ER and the chloroplast membranes (Gibbs, 1978). Ribosomes and polyribosomes of typical eukaryotic size and chemistry occur in the cytoplasmic matrix; chloroplast and mitochondrial ribosomes are of the smaller "prokaryotic" type, as in these organelles in other eukaryotes. Ribosomal RNA, ribosomes, and polyribosomes are the subject of Chapter 7 in Volume IV of the present treatise.

Microtubules in *Euglena* occur in species-specific patterns in relation to the pellicle, canal, reservoir, and flagellar roots; microtubules of the flagellar axoneme are discussed in the next chapter; spindle microtubules are all nucleoplasmic, both chromosomal and continuous types now being known.

XI. Paramylon and Other Cell Inclusions

Paramylon, the characteristic reserve carbohydrate of *Euglena* and other euglenoids, is a β-1,3-glucan occurring as extraplastidial granules with helical substructure, both free in the cytoplasmic matrix and as sheaths around pyrenoids (but still outside the chloroplast envelope). The granules are membrane-bound (Fig. 20), and continuity exists between the membrane and elements of the ER (Leedale, 1967a). Shape, size, structure, location, and chemistry of paramylon granules were reviewed in detail by Barras and Stone (1968).

Vacuoles and vesicles with assorted contents (and often unknown function) have been recorded by many authors. Leedale *et al.* (1965) described phospholipid vesicles that are 0.5–2 μm-diameter single-membraned organelles con-

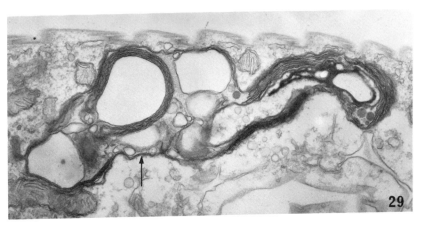

Fig. 29. Membranelle (degenerate chloroplast) of *E. gracilis* bleached by treatment with leucomycin (kitasamycin); various patterns of membranes, lipid globules, and stroma lie within a three-membraned envelope (arrow). ×22,500.

taining brown or orange pigments, lipids, myelin figures, and proliferations of unit membranes. These vesicles have been equated with metachromatic granules (volutin), commonly described in algal cells, and with lipofuscin granules of aging mammalian cells (Leedale, 1967a). Degenerate chloroplasts resulting from treatment of *Euglena* with various drugs (e.g., Siegesmund *et al.*, 1962) have often been reported. These structures, referred to as "membranelles" by Leedale and Buetow (1976), have the three membranes of the chloroplast envelope and various patterns of internal membranes (Fig. 29). Although apochlorotic and therefore photosynthetically functionless, the membranelles replicate and persist in colorless strains of *E. gracilis*.

XII. Final Comments

Euglena displays the structural complexity characteristic of protists in which all functions and activities of the organism are manifest in a single cell. Without the possibility of differentiation or division of labor, the *Euglena* cell is structurally equipped for maintenance and change of shape, swimming, photoreception, osmoregulation, exo- and endocytosis, photosynthesis, respiration, metabolite production and conversion, autolysis, protein synthesis, genetic continuity, reproduction, as well as the myriad other syntheses, metabolisms, responses, and activities, the physiology and chemistry of which are discussed in this four-volume treatise.

The features of diversity upon which the taxonomy of the genus *Euglena* is based (Leedale, 1967a) include cell size and shape; lengths and activity of flagella; numbers, size, and shape of chloroplasts; pyrenoids; cell (pellicular) rigidity; muciferous bodies; and the form of paramylon granules. Features of *Euglena* cytology and ultrastructure that are unique to or characteristic of all euglenoids (Leedale, 1967a) include the mode of insertion of flagella into an anterior invagination of the cell; the chemistry, construction, and helical symmetry of the pellicle; the structure, appendages, bases, and roots of the flagella; the crystalline flagellar swelling and extraplastidial eyespot; the form and functioning of the contractile vacuole; the basal constriction of the mitochondrial cristae; the large Golgi bodies; paramylon; and the structure and division of the nucleus.

The taxonomic isolation of euglenoids made evident by this formidable array of clear-cut diagnostic features is discussed by Leedale (1967a, 1978), as is the phylogeny within the group. Familiarity with the concept of chloroplast origin by symbiosis now makes it seem more logical to propose schemes of phylogeny in which the ancestral euglenoid line is colorless (Leedale, 1978) rather than green (Leedale, 1967a). Comments on the significance of the three membranes of the chloroplast envelope in *Euglena* are directly relevant here and both forms of symbiosis proposed above (see Section V), green alga or green algal chloroplasts, would explain the presence in *Euglena* of chlorophyll *b* and other photosynthetic pigments typical of Chlorophyta and land plants, without the need to suggest any close relationship between these groups and the euglenoid flagellates. Other ultrastructural features that might serve as indicators of evolution and phylogeny in the euglenoids include flagellar appendages, flagellar roots (especially the possible presence of a multilayered structure), and centromeres (which indicate that the nuclear division is not as different from classical mitosis as thought earlier).

References

Arnott, H. J., and Walne, P. L. (1967). Observations on the fine structure of the pellicle pores of *Euglena granulata*. *Protoplasma* **64**, 330–344.

Barras, D. R., and Stone, B. A. (1965). The chemical composition of the pellicle of *Euglena gracilis* var. *bacillaris*. *Biochem. J.* **97**, 14–15.

Barras, D. R., and Stone, B. A. (1968). Carbohydrate composition and metabolism in *Euglena*. *In* "The Biology of *Euglena*" (D. E. Buetow, ed.), Vol. II, pp. 149–191. Academic Press, New York.

Benedetti, P. A., Bianchini, G., Checucci, A., Ferrara, R., Grassi, S., and Percival, D. (1976). Spectroscopic properties and related functions of the stigma measured in living cells of *Euglena gracilis*. *Arch. Mikrobiol.* **111**, 73–76.

Brandes, D., Buetow, D. E., Bertini, F., and Malkoff, D. B. (1964). Role of lysosomes in cellular lytic processes. I. Effect of carbon starvation in *Euglena gracilis*. *Exp. Mol. Pathol.* **3**, 583–609.

Buetow, D. E. (1968). Morphology and ultrastructure of *Euglena*. *In* "The Biology of *Euglena*" (D. E. Buetow, ed.), Vol. I, pp. 109–184. Academic Press, New York.

Buetow, D. E. (1978). Nuclei and chromatin from *Euglena gracilis*. *In* "Handbook of Phycological Methods. Physiological and Biochemical Methods" (J. A. Hellebust and J. S. Craigie, eds.), pp. 15–23. Cambridge Univ. Press, London and New York.

Chaley, N., Lord, A., and Fontaine, F. G. (1977). A light and electron microscope study of nuclear structure throughout the cell cycle in the euglenoid *Astasia longa* (Jahn). *J. Cell Sci.* **27**, 23–45.

Dustin, P. (1978). "Microtubules." Springer-Verlag, Berlin and New York.

Gibbs, S. P. (1978). The chloroplasts of *Euglena* may have evolved from symbiotic green algae. *Can. J. Bot.* **56**, 2883–2889.

Gillott, M. A., and Triemer, R. E. (1978). The ultrastructure of cell division in *Euglena gracilis*. *J. Cell Sci.* **31**, 25–35.

Gojdics, M. (1953). "The Genus *Euglena*." Univ. of Wisconsin Press, Madison.

Gomez, M. P., Harris, J. B., and Walne, P. L. (1974). Studies of *Euglena gracilis* in ageing cultures. II. Ultrastructure. *Br. Phycol. J.* **9**, 175–193.

Guttman, H. N., and Ziegler, H. (1974). Clarification of structures related to function in *E. gracilis*. *Cytobiol.* **9**, 10–22.

Haapala, O. K., and Soyer, M.-O. (1975). Organization of chromosome fibrils in *Euglena gracilis*. *Hereditas* **80**, 185–194.

Hausmann, K., and Mignot, J.-P. (1977). Untersuchungen an den Mucocysten von *Euglena splendens* Dangeard 1901. *Protistologica* **13**, 213–217.

Hofmann, C., and Bouck, G. B. (1976). Immunological and structural evidence for patterned intussusceptive surface growth in a unicellular organism. *J. Cell Biol.* **69**, 693-715.

Holt, S. C., and Stern, A. I, (1970). The effect of 3-(3,4-dichlorophenyl)-1,1-dimethylurea on chloroplast development and maintenance in *Euglena gracilis*. I. Ultrastructural characterization of light-grown cells by the techniques of thin sectioning and freeze-etching. *Plant Physiol.* **45**, 475-483.

Huber-Pestalozzi, G. (1955). Das Phytoplankton des Süsswassers. 4. Euglenophyceen. *In* "Die Binnengewasser" (A. Thienemann, ed.), Vol. 16, pp. 1-606. Schweizerbart'sche Verlagsbuchhandlung, Stuttgart.

Kivic, P. A., and Vesk, M. (1972). Structure and function of the euglenoid eyespot. *J. Exp. Bot.* **23**, 1070-1075.

Kivic, P. A., and Vesk, M. (1974). Pinocytotic uptake of protein from the reservoir of *Euglena*. *Arch. Mikrobiol.* **96**, 155-159.

Leedale, G. F. (1958). Nuclear structure and mitosis in the Euglenineae. *Arch. Mikrobiol.* **32**, 32-64.

Leedale, G. F. (1959). The time-scale of mitosis in the Euglenineae. *Arch. Mikrobiol.* **32**, 352-360.

Leedale, G. F. (1964). Pellicle structure in *Euglena*. *Br. Phycol. Bull.* **2**, 291-306.

Leedale, G. F. (1967a). "Euglenoid Flagellates." Prentice-Hall, Englewood Cliffs, New Jersey.

Leedale, G. F. (1967b). Euglenida/Euglenophyta. *Ann. Rev. Microbiol.* **21**, 31-48.

Leedale, G. F. (1968). The nucleus in *Euglena*. *In* "The Biology of *Euglena*" (D. E. Buetow, ed.), Vol. I, pp. 185-242. Academic Press, New York.

Leedale, G. F. (1971). The euglenoids. *In* "Oxford Biology Readers" (J. J. Head and O. E. Lowenstein, eds.), No. 5, pp. 1-16. Oxford Univ. Press, London and New York.

Leedale, G. F. (1975a). Envelope formation and structure in the euglenoid genus *Trachelomonas*. *Br. Phycol. J.* **10**, 17-41.

Leedale, G. F. (1975b). Preliminary observations on nuclear cytology and ultrastructure in carbon-starved streptomycin-bleached *Euglena gracilis*. *Colloq. Internat. CNRS* **240**, 285-290.

Leedale, G. F. (1978). Phylogenetic criteria in euglenoid flagellates. *BioSystems* **10**, 183-187.

Leedale, G. F., and Buetow, D. E. (1970). Observations on the mitochondrial reticulum in living *Euglena gracilis*. *Cytobiologie* **1**,195-202.

Leedale, G. F., and Buetow, D. E. (1976). Observations on cytolysome formation and other cytological phenomena in carbon-starved *Euglena gracilis*. *J. Microsc. Biol. Cell.* **25**, 149-154.

Leedale, G. F., and Hibberd, D. J. (1974). Observations on the cytology and fine structure of the euglenoid genera *Menoidium* Perty and *Rhabdomonas* Fresenius. *Arch. Protistenkd.* **116**, 319-345.

Leedale, G. F., Meeuse, B. J. D., and Pringsheim, E. G. (1965). Structure and physiology of *Euglena spirogyra*. I and II. *Arch. Mikrobiol.* **50**, 68-102.

Malkoff, D. B., and Buetow, D. E. (1964). Ultrastructural changes during carbon starvation in *Euglena gracilis*. *Exp. Cell Res.* **35**, 58-68.

Miller, K. R., and Miller, G. J. (1978). Organization of the cell membrane in *Euglena*. *Protoplasma* **95**, 11-24.

Moestrup, Ø. (1978). On the phyogenetic validity of the flagellar apparatus in green algae and other chlorophyll *a* and *b* containing plants. *BioSystems* **10**, 117-144.

Moyne, G., Bertaux, O., and Puvion, E. (1975). The nucleus of *Euglena*. I. An ultracytochemical study of the nucleic acids and nucleoproteins of synchronized *Euglena gracilis* Z. *J. Ultrastruct. Res.* **52**, 362-376.

O'Donnell, E. H. J. (1965). Nucleolus and chromosomes in *Euglena gracilis*. *Cytologia* **30**, 118-154.

Pellegrini, M. (1980). Three-dimensional reconstruction of organelles in *Euglena gracilis* Z. I.

Qualitative and quantitative changes of chloroplasts and mitochondrial reticulum in synchronous photoautotrophic culture. *J. Cell Sci.* **43**, 137–166.

Pellegrini, M. (1980b). Three-dimensional reconstruction of organelles in *Euglena gracilis* Z. II. Qualitative and quantitative changes of chloroplasts and mitochondrial reticulum in synchronous cultures during bleaching. *J. Cell Sci.* **46**, 313–340.

Piccinni, E., and Mammi, M. (1978). Motor apparatus of *Euglena gracilis:* ultrastructure of the basal portion of the flagellum and the paraflagellar body. *Boll. Zool.* **45**, 405–414.

Pickett-Heaps, J. D., and Weik, K. (1977). Cell division in *Euglena* and *Phacus.* I. Mitosis. *In* "Mechanisms of Control of Cell Division," (L. Rost and E. M. Gifford, eds.), pp. 308–336. Dowden, Hutchinson and Ross, Stroudsburg, Pennsylvania.

Pringsheim, E. G. (1956). Contributions toward a monograph of the genus *Euglena. Nova Acta Leopoldina* **18**, 1–168.

Rosowski, J. R. (1977). Development of mucilaginous surfaces in euglenoids. II. Flagellated, creeping and palmelloid cells of *Euglena. J. Phycol.* **13**, 323–328.

Rosowski, J. R., and Willey, R. L. (1977). Development of mucilaginous surfaces in euglenoids. I. Stalk morphology of *Colacium mucronatum. J. Phycol* **13**, 16–21.

Schwelitz, F., Evans, W. R., and Mollenhauer, H. H. (1970). The fine structure of the pellicle of *Euglena gracilis* as revealed by freeze-etching. *Protoplasma* **69**, 341–349.

Siegesmund, K. A., Rosen, W. G., and Gawlik, S. R. (1962). Effects of darkness and of streptomycin on the fine structure of *Euglena gracilis. Am. J. Bot.* **49**, 137–145.

Silverman, H., and Hikida, R. S. (1976). Pellicle complex of *Euglena gracilis:* characterization by disruptive treatments. *Protoplasma* **87**, 237–252.

Sommer, J. R., and Blum, J. J. (1965). Cytochemical localization of acid phosphatase in *Euglena gracilis. J. Cell Biol.* **24**, 235–251.

Triemer, R. E. (1980). Role of Golgi apparatus in mucilage production and cyst formation in *Euglena gracilis* (Euglenophyceae). *J. Phycol.* **16**, 46–52.

Walne, P. L. (1971). Comparative ultrastructure of eyespots in selected euglenoid flagellates. *In* "Contributions in Phycology" (B. C. Parker and R. M. Brown, eds.), pp. 107–120. Allen Press, Lawrence, Kansas.

Walne, P. L. (1980). Euglenoid flagellates. *In* "Phytoflagellates" (E. Cox, ed.), 165–212. Elsevier, Amsterdam.

Walne, P. L., and Arnott, H. J. (1967). The comparative ultrastructure and possible function of eyespots: *Euglena granulata* and *Chlamydomonas eugametos. Planta* **77**, 325–353.

West, L. K., Walne, P. L., and Rosowski, J. R. (1980). *Trachelomonas hispida* var. *coronata* (Euglenophyceae): I. Ultrastructure of cytoskeletal and flagellar systems. *J. Phycol.* **16**, 489–497.

CHAPTER 2

FLAGELLA AND THE CELL SURFACE

G. Benjamin Bouck

I. Introduction

Concepts of surface organization and behavior have taken spectacular new directions since this subject was last considered in Volumes I and II of this series. In particular, the demonstration of surface motility in fluid membranes, of transmembrane glycoproteins, and of intracellular mediation of the movements of surface molecules are all now well-established properties of many cells. Many of these properties have been resolved through the use of cross-linking reagents, by detergent dissociation of hydrophobic membrane moieties, and by antibody labeling methods that have permitted detailed analyses of membrane and membrane-related events. Some of these methods have in recent years also been applied to the *Euglena* surface and have led to a number of interesting and unexpected findings, for example:

1. The dramatic paracrystalline arrays of intramembrane particles revealed by freeze-etching was not anticipated and is apparently unique among eukaryotic membranes.

THE BIOLOGY OF *EUGLENA*, VOL. 3

2. The discovery by use of tagged antibodies of large, differentiated membrane domains is unusual and potentially experimentally useful in approaching the general problem of the control of surface synthesis and organization.

3. Detergent and biochemical dissection of the flagellar surface indicates that its surface is even more complex than originally thought and challenges any interpretation of organization and development.

Some of the foregoing and other properties of euglenoid surfaces will be detailed in this chapter, with particular emphasis placed on findings published since 1968. Earlier studies have been well-reviewed by Buetow (1968) and Leedale (1967).

II. The Cell Surface

A. EXTERNAL SURFACE

1. *General Features**

The concept of the euglenoid surface as a distinct "small skin" or pellicle is perhaps at present misleading in considering its separate parts. It is well-known that the surface of most forms is relatively rigid and is folded into characteristic ridges or grooves on the exposed surfaces (i.e., directly visible to the viewer), but that, within the invaginated canal and resevoir, the ridges lose their identity and the surfaces are smooth. The surface is thought by some to consist of interlocking strips that are oriented in screw or helical symmetry relative to the long axis of the cell. The number of ridges and grooves decreases at the anterior and posterior ends of the cell by a more or less abrupt initiation or termination, respectively, of some individual strips. The surface is frequently overlain by additional (carbohydrate ?) material, which in some cases is exquisitely elaborate (e.g., see Bourrelly *et al.,* 1976; Rosowski, 1977). In *E. gracilis* the normally naked surface may be coated with a prominent wall during encystment induced by nitrogen starvation (Triemer, 1980), but the composition of the wall itself is not known.

As in all cells, the euglenoid surface is limited by a plasma membrane, but unlike most other eukaryotic cells, this membrane retains a high degree of visible organization. Freeze–fracture studies (Miller and Miller, 1978; Lefort-Tran *et al.,* 1980) clearly have demonstrated in *E. gracilis* that the membrane overlaying the surface ridges consists in part of arrays of well-ordered intramembrane striations (Fig. 1). These are arranged in parallel rows (major striation) oriented at about 34–37° to the direction of the strip, and individual rows are separated by

---

*See also this volume, Chapter 1.

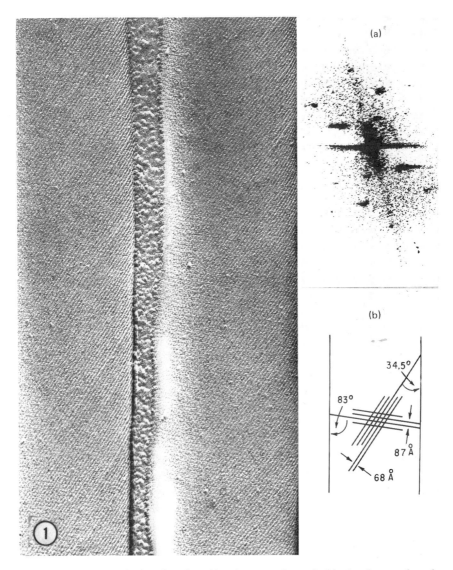

Fig. 1. Internal organization of *Euglena* ridge plasma membrane. In this view the outer face of the membrane is exposed, revealing the paracrystalline arrays of intramembrane striations. In the optical diffraction pattern (a) and its interpretation (b), a major and minor orientation of striations is apparent. ×123,000. (Reproduced with permission from Miller and Miller, 1978.)

about 63–68 Å. A less prominent orientation (minor striation) of these same particles has been detected at 57–83° to strip direction, with rows being 80–87 Å apart. As pointed out by Lefort-Tran *et al.*, the orientation of both major and minor striations must vary somewhat, since the strip is itself curved. Significantly, the plasma membrane region (i.e., groove) between surface strips does not contain particles or striations with detectable organization. However, during surface replication striations arranged in the minor orientation predominate in the area of the newly replicating cell surface (see following paragraphs and Lefort-Tran *et al.*, 1980). The biochemical nature of these dramatic arrays of intramembrane striations is, unfortunately, unknown. The striations seem to associate with the outer face (EF) of the membrane bilayer (Fig. 2), and under appropriate conditions, their complementary depression can be detected in the inner protoplasmic face (PF) of the membrane bilayer (Miller and Miller, 1978). From their position in the membrane, they are presumably hydrophobic, and from their apparent size, they would appear almost to span the membrane. Some perturbation in striation orientation can be induced by biotin starvation (Bre and Lefort-Tran, 1978), but in general visible intramembrane organization seems to be yet another indication of the unusual stability of the *Euglena* plasma membrane. Whether other euglenoids possess similar plasma membranes remains to be seen.

Internal to the membrane is a closely adhering layer of electron-scattering material. The thickness of this layer seems relatively constant for one species but varies from one form to another. A partial but not absolute correlation has been noted between layer thickness and whole organism rigidity (Mignot, 1966). The layer is not plastered amorphously to the plasma membrane but is deposited in parallel strands or ropes that run perpendicular to the direction of the surface strips (Lefort-Tran *et al.*, 1980). These strands are particularly evident in the groove region and may fuse and lose their identity in the raised strip area, or be further modified into "teeth" extending across the strip. The submembrane layer often appears to be interrupted in the notch (i.e., the lateral depression of the strip adjacent to the groove), apparently, in some cases, to accommodate the exit site of muciferous bodies or other possible extruded materials. This notched region has been described as a region of articulation or joining of interlocking strips in which muciferous bodies may provide lubrication for pellicle movements (cf. Leedale, 1967). Such imaginative ideas cannot be entirely rejected, but the apparent absence of secreted lubrication in some genera having active euglenoid movements, taken together with current concepts and clear examples of membrane fluidity in other organisms without secretions (e.g., Tamm, 1979), would seem to make the lubrication notion less compelling.

The role of microtubules in euglenoid surface organization is not fully resolved, although their consistent orientation relative to ridges and grooves is well-documented. In general, two sets of microtubules are associated with the ridge region, and a third set is related to the groove (Figs. 3–5). One set,

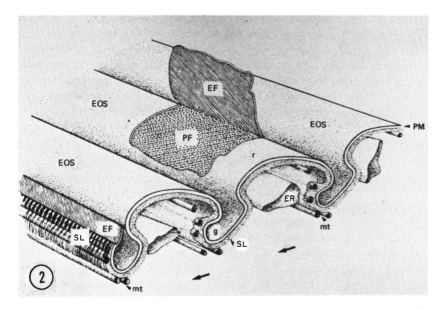

Fig. 2. Diagrammatic representation of the *Euglena* surface as revealed in thin section and by freeze-fracturing. The plasma membrane (PM) is folded into ridges (r) and grooves (g) and is underlain by a submembranous layer (SL), which appears to be rope-like in the groove region. Striations are present within the plasma membrane, especially in the outer, exterior face (EF) of the membrane bilayer over the ridges. No striations are visible on the outer cell surface (EOS). The protoplasmic face (PF) can be shown to have depressions complementary to the particles. Microtubules (mt) and endoplasmic reticulum (ER) are characteristically arranged on the posterior margins of the ridges. Arrows indicate the anterior end of the cell. (From Lefort-Tran *et al.*, 1980.)

generally [indicated as set A in Mignot (1966) and numbered as 2 and 3 in Fig. 3] consisting of two microtubules, is closely appressed to the ridge notch. Of these microtubules, 3 is conspicuously adorned with lateral appendages, which may be so dense that the microtubule itself is obscured. Portions of the submembrane layer appear to impinge on this microtubule as well. A second set of microtubules [set B in Mignot (1966) and number 1 of Fig. 3] is formed in association with the elevated portion of the ridge and generally occupies a position near the posterior margin of the ridge. The number of these microtubules varies from genus to genus, with a single microtubule generally found in the relatively narrow ridge of *E. gracilis,* and up to seventy microtubules found characteristic of those genera with broader ridges (Mignot, 1966). The third microtubule set [set C in Mignot (1966) and numbers 4–7 of Fig. 3] is localized internally and anteriorly to the groove and may consist of from one to five members, with frequent variation in the strips of a single individual. A role for all three microtubules sets either in cell movements or in shape maintenance might be expected by analogy with findings

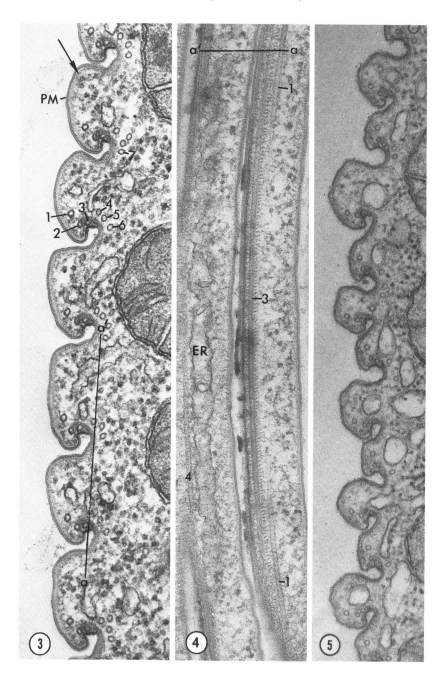

in other organisms. However, no direct evidence for either function has yet emerged, and the insensitivity of euglenoid microtubules to colchicine and to many other microtubule-affecting agents makes such data difficult to obtain. Similarly, no role for the extensive and consistent appearance of endoplasmic reticulum (ER) associated with the strips is as yet obvious. It is perhaps worth noting, however, that such ER cisternae are ideally positioned to mediate cation flux in a manner similar to that of the sarcoplasmic reticulum of striated muscle or, as has been recently demonstrated, for the ER intermingled with plant spindle microtubules (Hepler, 1979).

2. Biochemistry of Isolates

Isolation of *Euglena* surfaces has permitted not only the beginning of a biochemical characterization, but also has made possible a somewhat clearer picture of the composition which contributes directly to the rigid surface itself. Cavitation (sonication) of deflagellated cells, followed by equilibrium density centrifugation in sucrose, can yield pellicle fragments that are essentially free of cellular debris (Hofmann and Bouck, 1976). These fragments retain their ridge and groove organization (Fig. 8) and hence are representative of the native undisturbed structure. Interestingly, the microtubules, ER, and underlying cell cytoplasm are all absent in these isolates, and therefore surface form (ridges and grooves) must be due to one or both of the two structural components still present, i.e., the plasma membrane and the submembrane layer (Figs. 7–9). Treatment of isolated surfaces with neutral detergents and organic solvents has little effect on isolate appearance, but digestion with pepsin and other proteolytic enzymes destroys all semblance of ridges and grooves. These results are consistent with earlier observations on the sensitivity of whole euglenoids to similar enzymes (Kirk, 1964; Price and Bourke, 1966) and complement the observations of the high protein-lipid ratios characteristic of such isolates (Barras and Stone, 1965; Hofmann and Bouck, 1976).

Acrylamide gels of surface isolates, separated in the presence of SDS (sodium dodecyl sulfate), reveal a large number of peptides, but one dominant species with an apparent molecular weight of about 90,000 (Hofmann and Bouck, 1976). Three of these bands also stain with the periodic acid Schiff procedure (PAS),

---

Figs. 3–5. Sections through *Euglena* surfaces in transverse and longitudinal view. Line a–a' defines region viewed in Fig.4, and vice versa. The submembranous layer (arrow) is closely associated with the plasma membrane (PM). Microtubules are oriented in three sets (1; 2 and 3; 4–7, see text). Microtubule 3 is laterally adorned with fine projections (Fig. 4). ER, endoplasmic reticulum. In dividing cells (Fig. 5) new daughter ridges appear as upwellings between parental ridges and always contain at least microtubules 1, 2, and 3. Figs. 3 and 5, ×72,250; Fig. 4, ×63,750. (Reproduced with permission from Hofmann and Bouck, 1976.)

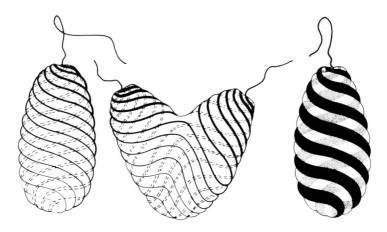

Fig. 6. Diagrammatic representation of a dividing *Euglena*. New strips (dark line) are intussus-cepted between parental ridges beginning at the pseudostome (mouth region) and progressively expand from anterior to posterior. Simultaneously, with strip development the cell undergoes fission, presumably along a surface groove. Separated daughter cells thus consist of equal amounts of old and new materials in alternate strips, which are ultimately indistinguishable from each other. (Modified from Hofmann and Bouck, 1976.)

indicating that they are glycoproteins. No enzymatic function has as yet been ascribed to any of the bands, nor is it as yet clear which band represents the submembrane layer and which the intramembrane particles, although the 90-K peptide, by reason of its relative abundance, must surely represent one or the other. The sugar composition of the membrane glycoproteins consists primarily of glucose (Bouck *et al.*, 1978).

A preliminary survey of the lipid fraction of these isolates has suggested the presence of abundant glycolipid and detectable amounts of phosphatidylcholine and phosphatidylethanolamine, which are the major phosphatides (S-J. Chen, personal communication, 1981). Some neutral lipids are also consistently extracted from isolates.

3. *Development and Physiology*

A fascinating problem in euglenoid surface behavior is the method of duplication during the binary fission of the cell. This process begins in metaphase with the separation of the single reservoir into two, with the concomitant appearance of two sets of flagella (Leedale, 1968). Both reservoirs are joined initially by a common canal, but eventually two canals appear and cell separation is initiated between the two canals. The mechanism of cell separation has not been well-resolved, although it is generally agreed that separation occurs along a pellicle strip, and that daughter cells rotate as they separate. This process has often been visualized as a tearing and resealing of dividing cells, but recently the line of

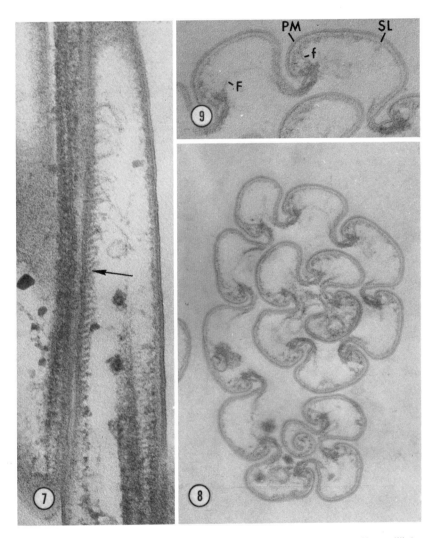

Figs. 7–9. Isolated surface strips obtained after sonication of whole cells followed by equilibrium centrifugation in sucrose gradients. Ridge and groove pattern is retained despite loss of microtubules, endoplasmic reticulum, and underlying cytoplasm. Submembranous layer (SL) and plasma membrane (PM) are still evident. Traversing fiber (F) is suggestive of the well-developed "teeth" described in other euglenoids. Arrow indicates rope-like appearance of submembranous layer in the groove region. (f) Additional fibril of unknown function. Fig. 7, ×81,000; Fig. 8, ×67,500; Fig. 9, ×108,000. (Reproduced with permission from Hofmann and Bouck, 1976.)

fission has been compared to the movement of a $-360°$ disclination, in which the "closed body can be cut in half without breaking the surface" (Harris, 1977). The obvious appeal of such a mechanism would seem to justify a more complete theoretical and experimental treatment of this model in future studies. In any case, the daughter cells eventually separate at their posterior regions and generally, although not always (Leedale, 1967), have identical helical (screw) symmetry.

Separation into two daughter cells does not require the synthesis of a completely new cell, but rather the systematic separation of two halves with apparently equivalent new and old surfaces. This is accomplished by the generation of surface strips within the groove of the parental strip (Fig. 5). Upwellings in the groove are accompanied by microtubules that will eventually constitute A and B of new strips, but which apparently were derived from set C of the original parental groove (Fig. 3). The process of intussusception is progressive, beginning at the anterior end of the cell, and it more or less accompanies cell fission. Eventually, the daughter upwellings expand and are indistinguishable from the original parental strips (Fig. 6).

Evidence that the strips are alternating old and new ones, and not the result of intermixing, is provided by immunological labeling studies (Hofmann and Bouck, 1976). Antibodies raised against purified surface strips and then tagged with fluorescein, ferritin, or latex spheres were found to uniformly coat the surface of *Euglena*. However, labeling followed by one or two cell duplications produces an antibody distribution in which only alternate strips or every fourth strip retained antibody. These results suggest that new materials (strips) are intussuscepted between existing strips and reaffirmed the older notion that each daughter cell receives equivalent amounts of parental surface. Additional conclusions suggested by these experiments are that the surface antigens marked with antibody are immobile and do not diffuse laterally, and that new surface strip antigens are formed *de novo,* and not by membrane flow, from adjoining strips. The experiments do not rule out the possibility that other unlabeled moieties are more mobile.

Freeze–fracture studies identify additional differences between old and newly forming daughter strips. Initially, the groove region is devoid of the regular paracrystalline array of intramembrane striations (Miller and Miller, 1978), but as the upwellings progress, regular particle orientation becomes evident (Lefort-Tran *et al.,* 1980). However, particle orientation follows the *minor* pattern of the parental ridges and presumably, only when mature, achieves the characteristic major pattern. These results taken together suggest that immobilization of surface antigens, final intramembrane particle orientation, and perhaps the deposition of the submembrane layer are temporally related events that ultimately determine the final mechanical properties (rigidity) of the surface complex.

Cells starved of vitamin B_{12} that are blocked in the G_2 or the end of the S phase of the cell cycle undergo no cell division and no strip replication, if the endogenous vitamin supply has been exhausted. Nonetheless, these cells survive and increase in the number of chloroplasts and in diameter without strip replication (Carell, 1969; Goetz et al., 1974). Intramembrane particle distribution in such cells shows some pattern discontinuities (Bre and Lefort-Tran, 1978), and such cells may be especially useful in future studies for examining the timing and control of strip replication, since surface duplication is reinitiated within a few hours of returning the cells to a B_{12}-rich medium.

B. Canal and Reservoir*

The canal and reservoir are invaginated regions of the cell that are significantly different in surface organization and function from the other portions of the cell. The surface ridges and grooves lose their distinctive organization as the plasma membrane descends (or, more properly, acquires ridges as it ascends the canal) into the canal (Fig. 10). This altered region has been termed the "transition" region (Miller and Miller, 1978). In freeze-fractured canal plasma membranes, the paracrystalline arrays of striations are no longer evident (Fig. 10), and particles appear predominately longitudinally, with no suggestion of the two-dimensional orientation that is characteristic of the free cell surface (Miller and Miller, 1978). The lower (proximal) portions of the canal are smooth. Viewing the canal only as a channel connecting the reservoir with the exterior of the organism is probably understating its real role. The circularly and longitudinally disposed microtubules (cf. Leedale and Hibberd, 1974), as well as the collection of microfilaments surrounding the canal suggest that the canal may be capable of contraction in order to block access to the reservoir or to control exit from it. The factors mediating the transition of one canal region to another are unknown and deserve continued study.

The reservoir appears to be the site of intense activity as judged by (1) the emptying of the contractile vacuole, (2) exocytosis of the Golgi system, (3) the occurence of mastigonemes along the walls, (4) the site within which the flagella orginate and exhibit their photoreceptors (paraflagellar bodies), and (5) the area of eyespot (stigma) localization. Some of these activities suggest that the plasma membrane is more fluid in the resevoir and, in fact, that the intramembrane particle distribution in freeze-fractured membranes (Miller and Miller, 1978) is random, that active exocytosis is evident, and that no submembrane layer is present. Wall materials in encysting cells are reportedly extruded in this region (Triemer, 1980) as well as the complex polysaccharide stalk of *Colacium* (Willey et al., 1977; see also Fig. 11).

*See also this volume, Chapter 1.

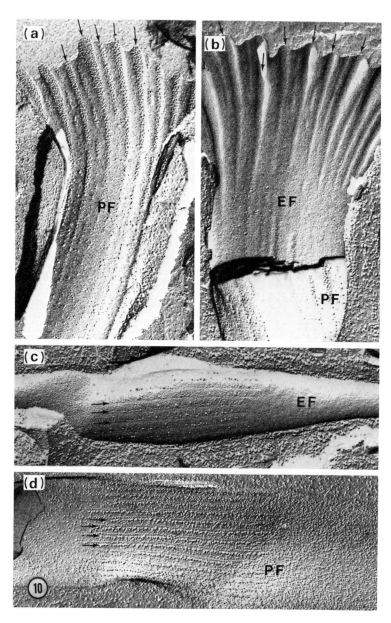

Fig. 10. (a) and (b): Freeze-fracture view of inner (PF) and outer (EF) intramembrane particle appearance in region of transition from ridged surface to smooth canal. Note ridges (arrows); patterns gradually flatten and lose their uniform particle distribution. (c) and (d): Micrographs illustrating EF and PF views of the canal with longitudinally arranged intramembrane particles in a smooth matrix. (a) and (b), ×41,225; (c) and (d), ×39,100. (Reproduced with permission from Miller and Miller, 1978.)

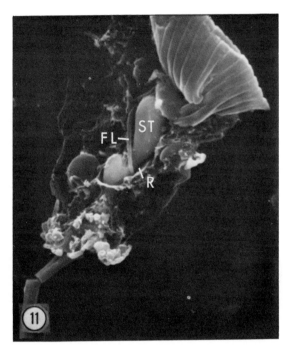

Fig. 11. Ruptured cell of *Colacium* in process of depositing stalk through the reservoir (R) and canal. Stalk material (ST) is initially found within the reservoir but ultimately is confined to the cell exterior. Flagellum (FL) can also be identified in this preparation. (Photograph and preparation courtesy of K. A. Ward, University of Illinois, Chicago. ×7500.)

III. Flagella

A. MASTIGONEMES

The euglenoid flagellar surface is extensively coated with fine filaments or mastigonemes. Earlier studies noted long mastigoneme filaments that are arranged unilaterally and a finer coating of "sheath"materials (cf. Leedale, 1967). More recently these observations have been extended, particularly with respect to the sheath, which is highly organized and maintains a distinct relationship to the flagellar membrane (Bouck *et al.,* 1978). The basic element of the sheath is about 240 to 300 nm in length and has been termed a mastigoneme "unit." The unit, which consists of loops, side arms, and filaments, lies flat [i.e., parallel to, but outside of, the flagellar membrane (Figs. 12–14)]. These units (about 7000/ flagellum) form a type of "picket fence" around the flagellum, and each layer or helix of pickets (units) overlaps the next layer or helix by one half of a unit

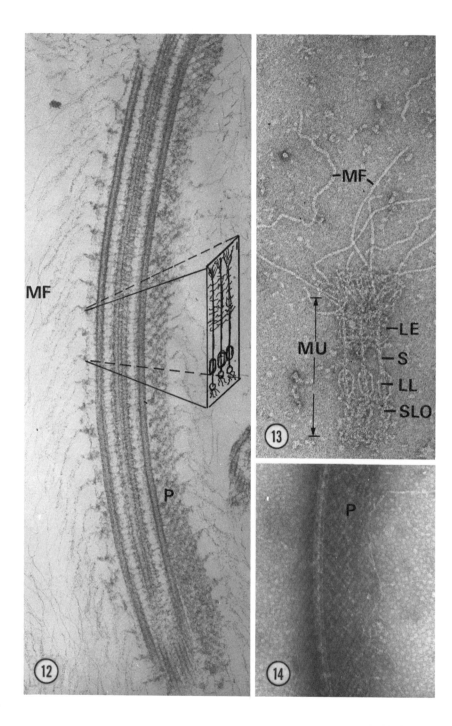

length. Thus, the units form tiers of one-half overlapping elements along the length of the flagellum. From each individual unit a group of filaments extends centrifugally, i.e., outward from the surface, and constitutes what was originally considered to be the sheath (Fig. 15). The sheath is now thought to consist of two longitudinally arranged half-sheaths, each half anchored to a different axonemal component (Bouck and Rogalski, 1981). The function and developmental origin of these mastigonemes is uncertain, although they obviously increase the effective surface area of the flagellum, and similar filaments line the inside of the reservoir. *Euglena* mastigonemes are complex biochemically and consist of a number of high-molecular-weight peptides and glycoproteins (Fig. 16G and H). The principal saccharide of the glycoproteins is the pentose sugar xylose, with trace amounts of glucose also detectable. As with other kinds of mastigonemes (Markey and Bouck, 1977), these are firmly attached to the internal axonemal components of the flagellum, as evidenced by their persistence after removal of the flagellar membrane with detergents (Fig. 12).

B. FLAGELLAR MEMBRANE

The flagellar membrane in *Euglena* is characterized by a fuzzy exterior that is distinct from the mastigonemes. This "fuzz" consists of fine 150-Å projections (Figs. 17 and 18) and is readily removed (solubilized) with low concentrations of neutral detergents, apparently without destroying membrane integrity. The solubilized fuzzy layer has an apparent molecular weight of around 90,000, as determined on SDS acrylamide gels and is strongly PAS-positive, suggesting it is a glycoprotein (Fig. 16C and D). Since the principal sugar moiety is xylose, the glycoprotein has been termed *xyloglycorien* (Rogalski and Bouck, 1980). Xyloglycorien is soluble in organic solvents as well as detergents, and it readily aggregates when frozen. Thus, it is believed to be at least partially hydrophobic and therefore partially immersed in the flagellar membrane when in its native state.

Since euglenoid flagellar membranes have not yet been purified of other flagel-

---

Figs. 12-14. Partially detergent-digested flagella from *E. gracilis* illustrating the arrangement of paraflagellar rod (P) and mastigoneme units (MU). As seen in Fig. 13, units consist of several distinct parts including lateral elements (LE), shaft (S), large loops (LL), and small loops (SLO). Mastigoneme filaments (MF) extend from one end of the unit. Units lie flat on the flagellar surface, but filaments extend outward as seen in Fig. 12. Note that the unit spacing seems to include two sets of mastigoneme filaments in Fig. 12, but this is due to the half-overlap of successive tiers of such units. Although the flagellar membrane has been removed with neutral detergents in Fig. 12, mastigoneme units still remain attached to the axoneme and paraflagellar rod. The paraflagellar rod is seen in Fig. 14 as firmly attached to an axonemal microtubule doublet, although other flagellar components have been dissociated from each other. Fig. 12, ×60,000; Fig. 13, ×113,000; Fig. 14, ×70,000. (Figs. 12 and 13 reproduced with permission from Bouck *et al.*, 1978.)

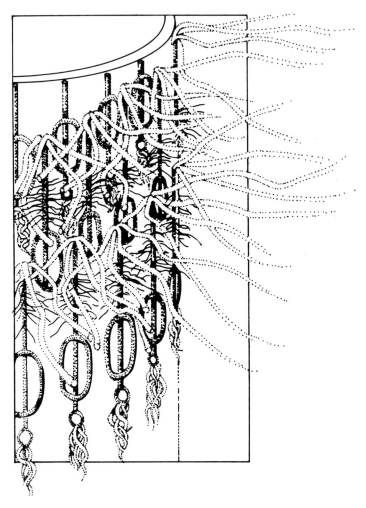

Fig. 15. Diagrammatic representation of a mastigoneme unit arrangement in half-overlapping staggered tiers on the *Euglena* flagellum. Mastigoneme filaments centrifugal to the surface compose the flagellar "sheath" of earlier workers. (Reproduced with permission from Bouck *et al.*, 1978.)

lar components, it is difficult at this juncture to further define membrane composition. However, chloroform–methanol extracts of whole purified flagella produce at least two major glycolipids on TLC plates, and the principal phosphatide seems to be phosphatidylethanolamine (Rogalski, 1980).

None of the unusual chlorosulfolipids reported in *Ochromonas* flagellar membrane (Chen *et al.*, 1976) have been identified in *Euglena*, thus circumventing the difficult interpretation of membrane organization, required to explain the *Ochromonas* results. From these preliminary *Euglena* analyses, euglenoid

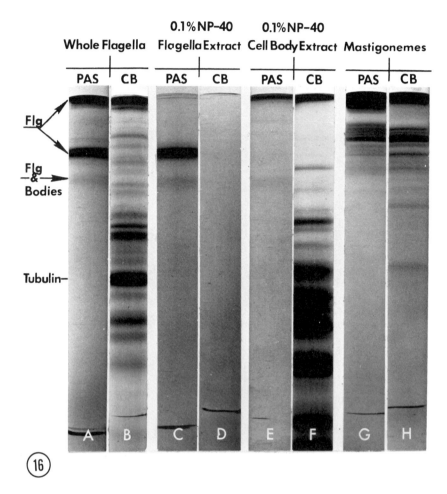

Fig. 16. SDS acrylamide gels stained for peptides (CB) or carbohydrates (PAS). The neutral detergent Nonidet P-40 (NP-40) extracts a glycoprotein (16-C) termed xyloglycorien, which seems to constitute a flagellar surface fuzzy layer. One PAS-positive band is common to both flagellar (Flg) and cell surfaces, but mastigonemes and xyloglycorien are specific to the flagellum only. (Reproduced with permission from Rogalski and Bouck, 1980.)

flagella would appear to have a conventional plasma membrane, rich in glycolipids and coated with glycoprotein.

C. AXONEME

In addition to the usual 9 + 2 microtubule arrangements of the axoneme, euglenoids are generally characterized by a prominent paraflagellar rod that lies parallel to and is attached to at least two microtubule doublets. The paraflagellar

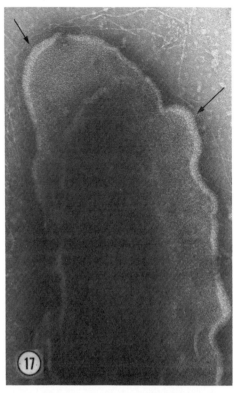

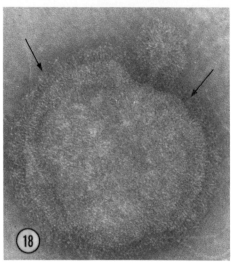

THIS IS NOT PART

rod in the locomotary flagellum (F1) is structurally different from the paraflagellar rod in the nonlocomotory flagellum (F2), when the latter contains such a structure. The F2 flagellum may contain a dramatic and approximately rectilinear crystalloid, which is particularly well-described in the publication of Mignot (1966). In the locomotory flagellum, by contrast, the paraflagellar rod always appears to consist of concentrically or helically arranged semicrystalline elements, which are still attached to the axomeme microtubules (Figs. 12 and 14). The function and composition of the paraflagellar rod are largely conjecture. ATPase activity as demonstrated cytochemically has been reported (Piccini *et al.*, 1975) in association with the paraflagellar rod of *Euglena*, but no obvious enhancement of high-molecular-weight dynein-like (flagellar ATPase) is seen in SDS gels of *Euglena* flagella, as might be expected from the large total volume occupied by the paraflagellar rod. In fact the only unidentified prominent polypeptide bands in *Euglena* flagella are one or two components (Fig. 16B), with somewhat higher molecular weights than those of tubulin. Attempts to solubilize the paraflagellar rod selectively from axonemal microtubules have not been successful, since both structures respond by resistence or solubilization to the same concentrations of tested detergents. The possibility that the paraaxial rod contains tubulin has not yet been ruled out.

One other important axonemal component, the parabasal body,* is generally present in the proximal portion of the flagellum. The role of this structure in light reception now seems well-established (e.g., see Checcucci *et al.*, 1976), but little is known as to how this light reception is subsequently transmitted into the organism's response. The paracrystalline nature of the parabasal body [which is separated from the axoneme by the parabasal rod (cf. Kivic and Vesk, 1972; Walne and Arnott, 1967)] has evoked a piezoelectric theory of impulse conduction (Bovee and Jahn, 1972), which seems particularly attractive.

Other flagellar components such as dynein arms, nexin, and radial links have not been critically examined in euglenoids, but can be recognized as described in other flagellates (Warner, 1972). *Euglena* microtubule protein has not yet been obtained in an assembly-competent form, although *Euglena* axonemes can clearly initiate assembly on non-*Euglena* tubulin (unpublished observation), similarly to that reported for *Chlamydomonas* axonemes and basal bodies (Snell *et al.*, 1974).

Figs. 17 and 18. Partially dissociated flagellar membranes obtained by repeated shearing of isolated flagella through a 25-guage hypodermic needle. Mastigonemes are removed during this process and the underlying membrane is thereby exposed. The membrane is clearly overlain by a fuzzy layer (xyloglycorien). Fig. 17, ×48,000; Fig. 18, ×275,000. (Reproduced with permission from Rogalski and Bouck, 1980.)

*See also this volume, Chapter 4.

D. KINETICS OF FLAGELLAR GROWTH

The important early paper of Rosenbaum and Child (1967) detailed regeneration kinetics in *Euglena* and several other flagellates following mechanical amputation of the flagellum. Regeneration could be measured shortly after amputation and subsequent growth followed deceleratory kinetics. However, in *Euglena* no flagellar growth zone could be demonstrated (in contrast to *Chlamydomonas*) in pulse-labeling experiments. Also, that the relatively short lag period before regeneration can be measured in *Euglena* is an interesting fact, in view of the long flagellar segment that extends from the basal bodies deep in the reservoir, through the canal, and finally outside the cell where it can be detected. Clearly, the flagellum has not regenerated from the basal body juncture during the period of time before it becomes visible outside the canal. Leedale *et al.* (1964) concludes that the flagellum in *E. spirogyra* tends to break at a region just above the parabasal body (i.e., near the reservoir–canal junction), but in *E. gracilis* the break seems to occur closer to the mouth region, a fact that is consistent with the rapid appearance of a visible flagellar stub after amputation. Thus, the labile flagellar junction normally associated with the flagellar base or transition zone of the flagellum in *Chlamydomonas* and other flagellates occurs elsewhere in *Euglena*. Furthermore, the break region in *Euglena* is not determined by the mechanical shearing of a canal-held flagellum because detachment occurs in the same region after cold shock. It may be of significance that flagellar waveform in the reservoir–canal region is essentially planar but in the free portion past the mouths describes the propagation of an interrupted helix (Holwill, 1966). These facts suggest that a functional transition zone may coincide with the general area of preferred flagellar breakage.

Also, unlike that of *Chlamydomonas,Ochromonas,* and many other flagellates, *Euglena* flagellar regeneration is relatively unaffected by the microtubule-assembly inhibitor colchicine. Concentrations up to $10^{-1}\ M$ (40 mg/ml) have very little effect other than a slight prolongation of the regeneration process (Silverman and Hikida, 1976). Whether this is due to detoxification by *Euglena*, difficulties in penetration, and/or *Euglena* tubulin insensitivity to colchicine is unknown, but the absence of effect mimics the response of the pellicle and mitotic microtubules. The protein synthesis inhibitor cycloheximide at 10 μg/ml inhibits regeneration after only a short flagellum has been regenerated. These results are unlike those obtained with *Chlamydomonas* and *Ochromonas* and suggest that the small pool of some peptides is limiting to flagellar growth.

IV. Surface Domains and Their Maintenance

Four reasonably distinct modifications of the euglenoid surface can be recognized from the data summarized in this chapter. These include the ridged external surface, the canal, the reservoir, and the flagella. To a surprising extent, the membrane associated with each region is different, although the surface membrane is continuous from one region to another. These differences are dramatically revealed in the freeze-fracture studies and are supported in part by biochemical and immunological observations. The canal has been described as a surface transition region (Miller and Miller, 1978) and is conveniently visualized as having a differentiated distal (mouth) region that generates the upwellings of the surface strips. These acquire a striated substructure and apparently have little surface mobility. The region between the surface upwellings perhaps remains fluid, and new membrane appears there during surface replication in the anterior to posterior direction as progressive upwellings. Initiation and termination of new strips may occur at various levels, perhaps depending on local conditions of membrane availability or interstrip fluidity, as evidenced by the reduction in strip numbers at anterior and posterior ends of the cell. The semirigid exposed surface ultimately formed in the mature cell is maintained either by the membrane or, more likely, the layer directly underlying the membrane. It is tempting to speculate that the submembrane layer may have a function similar to that of the red blood cell protein spectrin (Marchesi *et al.*, 1976) in providing rigidity and form to the cell surface. Whether the layer is involved in the movements of metaboly remains to be resolved, although if the spectrin analogy of the submembrane layer is valid, then spectrin–actin interactions described in the red blood cell (Tilney and Detmers, 1975) may be similar in *Euglena*. These are conjectures which await future studies, hopefully with reactivated isolated surface components. However, the more or less inverse relationship that exists between the thickness of the submembrane layer and the flexibility of the surface is difficult to explain if the submembrane layer is directly involved in movements and suggests that the motility mechanism may require components not yet identified or preserved in isolates.

Immunological studies also support the concept of a regionalization of *Euglena* surfaces. Antibodies directed against isolated cell surfaces (Hofmann and Bouck, 1976) do not bind to the flagella, reservoir, or canal; the saccharide content of isolated surfaces is predominately glucose, whereas that of the flagella is xylose. Furthermore, antibodies produced against two purified flagellar glycoproteins (xyloglycorien and mastigoneme) will not bind to the ridged cell surface (Figs. 19 and 20) but do interact under appropriate conditions with components of the reservoir (Rogalski, 1980). These immunological studies suggest that the *Euglena* surface consists of well-defined macrodomains in which there is little intermixing, at least between the ridge and grooved cell surface with the flagellar surface. Such surface differentiation is found also in some sperm cells and intestinal villi but generally has not been the case with unicellular organisms (cf. Rogalski and Bouck, 1980). How these surface domains are established and maintained are problems of major general theoretical interest, and euglenoids may well be excellent models for their resolution.

In replicating euglenoid cells, there is an obvious requirement for synthesis of

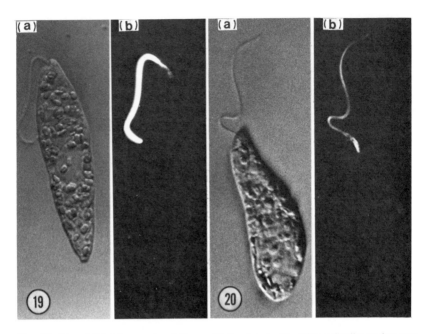

Fig. 19. (a) and (b): Micrographs of *E. gracilis* incubated with rabbit antibodies against mastigonemes followed by staining with flourescent goat–anti-rabbit antibody. Bright flagellar staining indicates mastigoneme binding of antibody. Only the flagellum retains mastigoneme antibody. ×2000. (Courtesy of A. Rogalski, University of Illinois, Chicago.)

Fig. 20. (a) and (b): Preparations similar to Fig. 19, but primary antibody is against xyloglycorien. Pattern of localization suggest xyloglycorien is also confined to the flagellum. Apparent thinner line of staining is probably due to membrane binding rather than to the broad sheath of mastigonemes. ×2000. (Courtesy of A. Rogalski, University of Illinois, Chicago.)

new specialized surfaces. The focal point for this activity would seem to be the reservoir. Immunological labeling of flagellar antigens suggests that both mastigonemes and xyloglycorien are derived from reservoir membrane pools that find their way to the highly ordered flagellar surface. Since the reservoir membrane is directly continuous with the flagellar membrane, this poses no conceptual problem, and, in fact, is consistent with the postulated origin of mastigonemes in other cells (e.g., Bouck, 1972). The source of membrane for replicating surface ridges is more difficult to assess but could conceivably also be reservoir-derived (as is the wall in encysting cells). Pulse-labeling experiments may help to resolve this problem. A final intriguing fact, which also remains to be explained, is how the identical glycoprotein apparently found in both flagella and ridges as well as groove surfaces is distributed (Fig. 16A and E). The existence of such overlap in itself suggests a common source of membrane origin requiring different methods of shuttling appropriate components to their final disposition.

References

Barras, D. R., and Stone, B. A. (1965). *Biochem. J.* **97,** 14–15.
Bouck, G. B. (1972). *Adv. Cell. Mol. Biol.* **2,** 237–271.
Bouck, G.B., and Rogalski, A. A. (1981). *Symp. Soc. Exp. Biol.* (in press).
Bouck, G. B., Rogalski, A., and Valaitis, A. (1978). *J. Cell Biol.* **77,** 805–826.
Bourrelly, P., Coute, A., and Rino, J. A. (1976). *Protistologica* **12,** 623–628.
Bovee, E. C., and Jahn, T. L. (1972). *J. Theor. Biol* **35,** 259–276.
Bre, M. H., and LeFort-Tran, M. (1978). *J. Ultrastruct. Res.* **64,** 362–376.
Buetow, D. E., ed. (1968). "The Biology of *Euglena,*" Vol. I, Academic Press, New York.
Carell, E. F. (1969). *J. Cell Biol.* **41,** 431–440.
Checcucci, A., Colombetti, G., Ferrara, R., and Lenci, F. (1976). *Photochem. Photobiol.* **23,** 51–54.
Chen, L. L., Pousada, M., and Haines, T. H. (1976). *J. Biol. Chem.* **251,** 1835–1842.
Goetz, G. H., Johnston, P. L., Dobrosielski-Vergona, K., and Carell, E. F. (1974). *J. Cell Biol.* **62,** 672–678.
Harris, W. F. (1977). *Sci. Am.* **237,** 130–145.
Hepler, P. K. (1979). *J. Cell Biol.* **83,** 372a.
Hofmann, C., and Bouck, G. B. (1976). *J. Cell Biol.* **69,** 693–715.
Holwill, M. E. J. (1966). *J. Exp. Biol.* **44,** 579–588.
Kirk, J. T. O. (1964). *J. Protozool.* **11,** 435–437.
Kivik, P. A., and Vesk, M. (1972). *Planta,* **105,** 1–14.
Leedale, G. F. (1964). *Br. Phycol. Bull.* **25,** 291–306.
Leedale, G. F. (1967). "Euglenoid Flagellates." Prentice-Hall, Englewood Cliffs, New Jersey.
Leedale, G. F. (1968). *In* "The Biology of *Euglena*" (D. E. Buetow, ed.) Vol. I, pp. 185–242. Academic Press, New York.
Leedale, G. F., and Hibberd, D. J. (1974). *Arch. Protistenk.* **116,** 319–345.
Leedale, G. F., Meeuse, B. J. B., and Pringsheim, E. G. (1965). *Arch. Mikrobiol.* **50,** 68–102.
Lefort-Tran, M., Bre, M. H., Ranck, J. L., and Pouphile, M. (1980). *J. Cell Sci.* **41,** 245–261.
Marchesi, V. T., Furthmayr, H., Tomita, M. (1976). *Ann. Rev. Biochem.* **45,** 667–697.
Markey, D. R., and Bouck, G. B. (1977). *J. Ultrastruct. Res.* **59,** 173–177.
Mignot, J. P. (1966). *Protistologica* **2,** 51–117.
Miller, K. R., and Miller G. J. (1978). *Protoplasma* **95,** 11–24.
Piccini, E., Albergoni, V., and Coppellotti, O. (1975). *J. Protozool.* **22,** 331–335.
Price, C. A., and Bourke, M. E. (1966). *J. Protozool.* **13,** 474–477.
Rogalski, A. (1980). Flagellar surface complex of *Euglena gracilis.* Ph.D. Thesis, University of Illinois, Chicago.
Rogalski, A., and Bouck, G. B. (1980). *J. Cell Biol.* **86,** 424–435.
Rosenbaum, J. L., and Child, F. M. (1967). *J. Cell Biol.* **34,** 345–364.
Rosowski, J. R. (1977). *J. Phycol.* **13,** 323–328.
Silverman, H., and Hikida, R. S. (1976). *Protoplasma* **87,** 237–252.
Snell, W. J., Dentler, W. L., Haimo, L. T., Binder, L. I., and Rosenbaum, J. L. (1974). *Science* **185,** 357–360.
Tamm, S. L. (1979). *J. Cell Biol.* **80,** 141–149.
Tilney, L. G., and Detmers, P. (1975). *J. Cell Biol.* **66,** 508–520.
Triemer, R. E. (1980). *J. Phycol.* **16,** 46–52.
Walne, P. L., and Arnott, H. J. (1967). *Planta* **77,** 325–353.
Warner, R. (1972). *Adv. Cell Mol. Biol.* **2,** 193–236.
Willey, R. L., Ward, K., Russin, W., and Wibel, R. (1977). *J. Phycol.* **13,** 349–353.

CHAPTER 3

CIRCADIAN AND INFRADIAN RHYTHMS

Leland N. Edmunds, Jr.

I. Introduction

Within the set of living systems, an organism is said to be adapted if it has thrived by means of differential reproductive success and if its adaptation is reflected in its total organization. Biological problems, therefore, pivot upon the complexities of biological organization. It is almost self-evident that the spatial organization and functioning of organisms are inextricably intertwined; of equal importance, however, is the temporal dimension. Indeed, the environment is highly periodic in many of its variations so that it is often advantageous, if not essential, for living systems to adapt to these periodicities. That organisms can and do measure *astronomical* time (as opposed to the purely private timekeeping

THE BIOLOGY OF *EUGLENA*, VOL. 3
Copyright © 1982 by Academic Press, Inc.
ISBN 0-12-139903-6

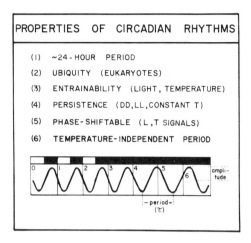

Fig. 1. Properties of circadian (*sensu stricto*) rhythms. A generalized sinusoidal oscillation under entraining (LD) and free-running (DD) environmental conditions is shown at the bottom over a time span of 7 days; white areas denote light spans and black bars, darkness. LL, constant light; DD constant darkness, L, light; T, temperature.

reflected in such variable-period physiological rhythms as heart beat or alpha brain waves) is demonstrated by four categories comprising quite diverse phenomena: (1) persistent rhythms having daily (circadian), tidal, lunar monthly, and yearly (circannual) periods; (2) the *Zeitgedächtnis,* or time sense of bees; (3) seasonal photoperiodism, during which many plants and animals perform a certain function at a quite specific time of the year by what may well be essentially a daily measurement of the length of the day (or night); and (4) celestial orientation and navigation, in which the sun, moon, or stars are used as direction givers, thereby implicating a timing system to compensate for the continuously, but predictably, shifting position of these celestial bodies. Inasmuch as the latter three types of functional biochronometry may represent evolutionarily more recent and sophisticated variations on a more basic adaptation to the 24-hour day, an understanding of the nature of circadian "clocks" may be crucial to the elucidation of these higher-level phenomena (Palmer *et al.,* 1976).

The generally accepted properties of circadian rhythms, having a period of about one day, are summarized in Fig. 1. Unless otherwise noted, the term *circadian* will refer only to those biological periodicities that have been demonstrated to persist for at least several cycles following removal of the synchronizing, or entraining, regimen of light, temperature, or other *Zeitgeber.* Our central thesis is that there is a selective premium for temporal adaptation, especially to solar periodicities, and that these adaptive features have been attained by organisms through some sort of timing mechanism, in particular, an *endogenous, autonomously oscillatory* biological clock which is responsive to and which can be reset and otherwise modulated by those environmental periodicities that the organism has encountered throughout its evolutionary history (Palmer *et al.*, 1976). The following books and symposia provide major treatment of biological rhythms: Frisch (1960); Goodwin (1963); Aschoff (1965); Menaker (1971); Sweeney (1969a; 1972); Bünning (1973); Pavlidis (1973); Scheving *et al.* (1974); Palmer *et al.* (1976); Goodwin (1976); Hastings and Schweiger (1976); Saunders (1976); Krieger (1979); Wever (1979); Scheving and Halberg (1980); and the *Proceedings of the XIII, XIV, and XV International Conferences of the International Society for Chronobiology* (in press).

A. TEMPORAL ORGANIZATION OF *Euglena*

Circadian organization is not restricted to multicellular organisms. Overt persisting circadian rhythms have been documented in a number of unicellular forms, including the green algae *Acetabularia, Chlamydomonas,* and *Chlorella,* the marine bioluminescent dinoflagellate *Gonyaulax,* the algal flagellate *Euglena,* and the ciliates *Paramecium* and *Tetrahymena.* The variables exhibiting rhythms include cell division, mating type reversal, luminescence, pattern formation, and enzymatic activity (reviewed by Bruce and Pittendrigh, 1957; Hastings, 1959; Bruce, 1965; Ehret and Wille, 1970; Vanden Driessche, 1970; Ehret, 1974; Edmunds, 1975; Schweiger and Schweiger, 1977; Wille, 1979). In each of these different microorganisms, furthermore, several different circadian rhythms have been observed concurrently (in some cases even in individual cells as well as in synchronous populations), with the attendant implication that many or all of the overt rhythms in a unicell may represent "hands" of a single underlying master pacemaker mechanism (see particularly McMurry and Hastings, 1972). These organisms, therefore, constitute attractive systems for the experimental investigation of the fine structure of circadian rhythms and the mechanism(s) whereby they are generated.

In particular, the temporal organization of *Euglena gracilis* Klebs (strain Z) has been intensively studied in a number of laboratories over the past 20 years (reviewed by Edmunds, 1975, 1978; Wille, 1979; Edmunds and Halberg, 1981), and this organism and *Gonyaulax* and *Tetrahymena* form a trio that have provided evidence for the so-called "G-E-T effect" described by Ehret and Wille (1970). This organism can be grown on a variety of different, completely defined media (Wolken, 1967), either photoautotrophically in the presence of CO_2, or organotrophically in the light or dark on carbon sources ranging from acetate, ethanol, and lactic and glycolic acids to glutamic and malic acids over a wide pH range (alkaline to acidic pH < 3.0). This versatility in growth mode, in conjunction with the fact that cell division can be easily synchronized by appropriate 24-hour lighting schedules (Cook and James, 1960; Edmunds, 1965a) and tem-

perature cycles (Terry and Edmunds, 1969, 1970a), has made *Euglena* a key
experimental organism for a variety of physiological and biochemical investigations (Buetow, 1968). Another important consideration is that populations of
Euglena can be maintained in the "stationary" phase of growth—the so-called
infradian growth mode (Ehret and Wille, 1970)—for days, weeks, and even
longer time spans with little or no net increase in cell concentration; circadian
outputs can thus be monitored while divorced from the driving force of the cell
division cycle, which itself can be modulated or gated by a circadian oscillator
(Edmunds, 1975,1978). Finally, a number of photosynthetic mutants (or even
completely bleached strains devoid of their chloroplast genome) have been isolated. These strains also exhibit light-entrainable circadian rhythms, but now
with the problem of the dual use of imposed light spans and signals—as an
energy source for growth, on the one hand, and as a cue for the timing
mechanism, on the other—effectively eliminated (Kirchstein, 1969; Jarrett and
Edmunds, 1970; Mitchell, 1971; Edmunds *et al.*, 1974,1976).

These important and useful characteristics, therefore, prompt a tabulation in
this chapter of a number of the circadian rhythms reported for *Euglena* throughout the literature and an indication of their phase relationships to each other,
followed by a survey of some of the major classes of rhythms. Only rhythms
having free-running periods of about 24 hours (circadian) or longer (infradian)
will be considered; purely driven (passive) oscillations, or those having higher
frequencies (i.e., ultradian rhythms), are beyond the scope of this review. A
broader, comprehensive review of circadian rhythms in protozoa has been published recently by Wille (1979).

B. TABULATION OF CIRCADIAN RHYTHMS IN *Euglena*

Many circadian rhythms that have been documented in *Euglena* are tabulated
in Table I. As is evident, a number of variables have been studied by repeated
sampling (usually every 1 or 2 hours). For the sake of convenience, they have
been grouped arbitrarily (and somewhat artificially) into physiological and
biochemical rhythms. In the majority of studies, the Z strain of *Euglena gracilis*
Klebs was utilized, though in several cases (such as the investigations of cell
division and mobility rhythms) various photosynthetic mutants or bleached
strains—obligate organotrophs—were used.

In each instance, the variable has been measured in synchronously dividing
cultures held under conventional, full-photoperiod, 24-hour cycles of light and
darkness (for example, *LD*: *8*,16, *LD*: *10*,14, *LD*: *12*,12, and *LD*: *14*,10) or of
temperature (for example, *18°/25°C*: *12*,12), and in certain cases, as for the
rhythm of cell division, under more exotic illumination regimens (for example,
LD: *1*,3, *LD*: *12*,36, and so-called two-point symmetric skeleton photoperiods,

Table I
CIRCADIAN RHYTHMS IN *Euglena gracilis* KLEBS[a]

Rhythm[b,c]	Strain[d]	Phase marker	ϕ^e	Reference[f]
A. Physiological				
Cell division	Z	Onset	CT 12–13 (180–195°)	Edmunds (1964, 1965a, 1975, 1978); Edmunds and Funch (1969a,b)
	P_4ZUL	Onset	CT 10–12 (150–180°)	Edmunds (1971, 1978); Edmunds *et al.* (1971, 1974, 1976); Jarrett and Edmunds (1970)
	P_7ZNgL	Onset	CT 10–12 (150–180°)	Mitchell (1971)
	W_6ZHL	Onset	CT 10–12 (150–180°)	Edmunds (1978); Edmunds *et al.* (1971)
	W_nZUL	Onset	CT 10–12 (150–180°)	Mitchell (1971)
	Y_9ZNalL	Onset	CT 10–11 (150–165°)	Edmunds *et al.* (1976)
Cell volume	Z	Maximum	CT 18–21 (270–315°)	K. Brinkmann and U. Kipry (unpublished)
Flagellated cells (%)	Z 1224-5/9, Göttingen	Maximum	CT 03 (45°)	Brinkmann (1966)
Motility, random (dark) *Dunkelbeweg-lichkeit*	Z 1224-5/9, Göttingen	Minimum	CT 18–21 (270–315°)	Schnabel (1968); Brinkmann (1966, 1971); Kreuels and Brinkmann (1979)
	W_nZHL 1224-5/25, Göttingen	Minimum	CT 12 (180°)	Kirchstein (1969)
pH (external medium)	Z			Brinkmann (personal communication, 1980)
Photokinesis (photoactivation of random motility)	Z	Maximum	CT 18–21 (270–315°)	Brinkmann (1976b)
Photosynthetic capacity $^{14}CO_2$ uptake	Z, ZR	Maximum	CT 06–12 (90–180°)	Walther and Edmunds (1973); Laval-Martin *et al.* (1979); Edmunds and Halberg (1981); Edmunds and Laval-Martin (1981)
O_2 evolution	Z	Maximum	CT 04–06 (60–90°)	Walther and Edmunds (1973); Lonergan and Sargent (1978, 1979)

(continued)

Table I (Continued)

Rhythm[b,c]	Strain[d]	Phase marker	ϕ^e	Reference[f]
Phototaxis (capacity)	Z	Maximum	CT 04–08 (60–120°)	Pohl (1948); Bruce and Pittendrigh (1956, 1958); Brinkmann (1966); Feldman (1967); Feldman and Bruce (1972)
Settling	Z	Maximum	CT 21–09[g] (315–135°)	Terry and Edmunds (1970b)
			CT 15 (225°)	Kiefner et al. (1974)
Shape	Z 1224-5/9, Göttingen	Maximum elongation	CT 03–09 (45–135°)	Brinkmann (1976b)

B. Biochemical

Amino acid incorporation (DL-[3-¹⁴C] phenylalanine)	Z	Maximum	CT 10–12 (150–180°)	Feldman (1968)
Gross metabolic variables[h]				
carotenoids	Z	Onset	ZT 0 (0°)	Cook (1961b); Edmunds (1965b)
chlorophyll *a*	Z	Onset	ZT 0 (0°)	Cook (1961b); Edmunds (1965b)
dry weight	Z	Onset	ZT 0 (0°)	Cook (1961b); Edmunds (1965b)
total protein	Z	Onset	ZT 0 (0°)	Cook (1961b); Edmunds (1965b)
total cellular RNA	Z	Onset	ZT 0 (0°)	Cook (1961b) Edmunds (1965b)
total cellular DNA	Z	Onset	ZT 08–09 (120–135°)	Cook (1961b); Edmunds (1964, 1965b)
Enzymatic activity[i]				
acid phosphatase	Z	Peak	ZT 06–08 (90–120°)	Sulzman and Edmunds (1972); Edmunds (1975)
alanine dehydrogenase	Z	Peak	CT 06–08 (90–120°)	Sulzman and Edmunds (1972, 1973); Edmunds et al. (1974); Edmunds (1975)
carbonic anhydrase	Z	Peak	ZT 06–08 (90–120°)	Lonergan and Sargent (1978)
glucose-6-phosphate-dehydrogenase	Z	Peak	ZT 06 (90°)	Sulzman and Edmunds (1972); Edmunds (1975)
glutamic dehydrogenase	Z	Peak	ZT 06–08 (90–120°)	Sulzman and Edmunds (1972); Edmunds (1975)

Rhythm[b,c]	Strain[d]	Phase marker	ϕ^e	Reference[f]
glyceraldehyde-3-phosphate dehydrogenase (NADP and NADPH-dependent)	Z	Peak	ZT 05–08 (75–120°)	Walther and Edmunds (1973); Lonergan and Sargent (1978)
lactic dehydrogenase	Z	Peak	ZT 06 (90°)	Sulzman and Edmunds (1972); Edmunds (1975)
L-serine deaminase	Z	Peak	ZT 06 (90°)	Sulzman and Edmunds (1972); Edmunds (1975)
L-threonine deaminase	Z	Peak	ZT 08 (120°)	Sulzman and Edmunds (1972); Edmunds (1975)
Susceptibility ethanol (pulses)	Z (1224-5/9, Göttingen)	Maximum	CT 03–06 (45–90°)	Brinkmann (1976a)
trichloro-acetic acid	Z (1224-5/9, Göttingen)	Maximum	CT 03–06 (45–90°)	Brinkmann (1976b)

[a] Modified from Edmunds and Halberg (1981).

[b] All rhythms listed have been shown to persist with a circadian period in DD (or LL) and constant temperature following prior synchronization or initiation unless otherwise indicated.

[c] All cultures were essentially nondividing (stationary, or long infradian), except for those in which the cell division rhythm itself was monitored. Consult reference for precise culture conditions.

[d] The wild-type strain is designated as Z strain. Unless otherwise noted, it was originally obtained from the American Type Culture Collection (#12716) and maintained at the State University of New York at Stony Brook since 1965 or at Princeton University. Also available from the Algal Collection at Indiana University (No. 753) and the Algensammlung Pringsheim at Göttingen (No. 1224-5/9). All other strains are photosynthetic mutants or completely bleached strains incapable of photosynthesis.

[e] Phase given in circadian time, CT (hours or degrees after subjective dawn, when the onset of light would have occurred had the synchronizing light cycle been continued). Synchronizing light cycles were either LD: 12,12 or LD: 10,14 [except in the case of the settling rhythm in which the cultures were synchronized by a 12:12 temperature cycle (18°/25°, LL) before release into LL, and CT 0 equivalent to onset of lower temperature]. In those instances where a free run was not monitored, phase is given in Zeitgeber (synchronizer) time (ZT 0, onset of light).

[f] Only key references are given; no attempt is made to survey all relevant or derivative publications.

[g] Rhythm synchronized by a 12:12 temperature cycle (18°/25°) in either dividing or nondividing cultures maintained in LL before release into constant conditions (25°, LL). CT 0, onset of lower temperature.

[h] These variables have been monitored only in light-synchronized (LD: 14, 10), synchronously dividing cultures. The phase marker, onset, here refers to the stage in the rhythm when values start to increase.

[i] Although all of the enzymes indicated undergo oscillations in activity in nondividing cultures in LD: 10, 14, only alanine dehydrogenase has been investigated in sufficient detail to demonstrate conclusively that its activity persists for long time spans in DD and constant temperature.

as *LD*: *3,6,3*:12).* Although some of the key gross metabolic factors (for example, total protein, RNA, and DNA) are given for the sake of reference, no attempt has been made to include all such reported rhythms since there are innumerable reports of biochemical variables that change across the cell cycle of *Euglena*. Rather, emphasis is given to those 24-hour rhythms which not only occur in synchronously dividing cultures, but which also (a) appear in nondividing (stationary, or long infradian) populations as well and (b) persist for extended time spans under conditions held constant with respect to illumination (or darkness), temperature, and other culture variables such as pH and aeration.

Under these circumstances, then, we have tabulated circadian rhythms (*sensu stricto*) that free-run with a period that usually only approximates 24 hours following removal of the synchronizer (also referred to as an entraining agent, or *Zeitgeber*). The exceptions given include the persisting rhythm of cell division itself which, of course, must be monitored in a dividing culture, and certain biochemical variables [for example, DNA, RNA, enzymatic activities (see Table 1, footnotes *h* and *i*)] whose long-term assays are quite laborious. It is important to note, however, that in these cases it is quite likely that the oscillations would have been found to persist had the variables in question actually been measured, if for no other reason than that they have all been shown to exhibit 24-hour fluctuations across the cell cycle in *LD*: *10*,14, *LD*: *12*,12, and *LD*: *14*,10, and that cell division itself manifests a persisting circadian rhythmicity in LL or DD.* Furthermore, all enzyme activities listed oscillate in nondividing cultures subjected to appropriate LD cycles. Finally, the common enzyme alanine dehydrogenase, which had been studied in detail (Sulzman and Edmunds, 1972,1973), was shown to undergo rhythmic changes in activity of relatively large amplitude for at least 14 days in DD (but not in LL) in nondividing cultures of *Euglena* grown organotrophically and initially synchronized by an imposed *LD*: *10*,14 cycle.

It is evident even from the qualitatively determined time relations of the rhythms [correlated in Table I to the onset of the light, i.e., *Zeitgeber* time zero (*ZT* 0, also referred to as *ST* 0) under synchronizing regimes, or circadian time zero (*CT* 0, i.e., "subjective dawn") under constant illumination or darkness during a free-run] that they scan the entire timespan of 24 hours (equivalent to 360°). It is important to note, however, that in this tabulation the phase marker on a rhythm (a feature chosen to indicate phase relations, sometimes referred to as phase reference point) is arbitrarily and often visually selected by the experimenter (usually, but not always, as the peak or maximum value of the oscillation, or its onset). The timing of such markers is not necessarily equal to the quantitatively defined and determined acrophase (ϕ) described in the following section.

---

*LD denotes a repetitive cycle of light and darkness; thus, *LD*: *10*,14 represents a diurnal light cycle having a period (*T*) of 24 hours and comprising 10 hours of light followed by 14 hours of darkness. LL denotes continuous illumination; DD, constant darkness; and D–L (or L–D) a single transition from dark to light (or from light to dark).

C. Phase Relations Among Circadian Rhythms in *Euglena*

As we have seen in Table I, a number of variables in *Euglena* have been studied by repeated sampling of cell populations under synchronizing conditions with conventional 24-hour light and temperature cycles; data are available also for many of these variables under conditions of continuous light (or darkness) and constant temperature. It is thus possible to attempt to quantify the time relations among these diverse circadian rhythms in order to determine which processes are roughly coincident or in phase and which are out of phase during synchronization. Such quantitative examination of the circadian time structure of *Euglena* is particularly important for determining whether any of these phase angles are altered with statistical significance under conditions of presumed constancy with regard to environmental inputs (i.e., "free-running" conditions). Indeed, internal desynchronization of a number of free-running physiological circadian rhythms, with attendant implication of a spectrum of different frequencies, has been demonstrated under certain experimental conditions in man and several other mammals, but not in a unicell (McMurry and Hastings, 1972). The demonstration of circadian dyschronism (or dysphasia) at the cellular level of organization would be most provocative (Laval-Martin *et al.*, 1979), having important implications as to the multiple pacemaker problem and the nature of the time-keeping mechanism(s) (see Section III, C).

One widely used method for quantitative and objective assessment of time relationships inherent in time series data is the so-called *cosinor analysis* developed by Halberg and his colleagues (Halberg *et al.*, 1967,1972; Nelson *et al.*, 1979), the aim of which is to estimate statistically the "microscopic" parameters of a biological rhythm by the least squares fit of a cosine curve. This statistical method enables one to compute values for the circadian mesor (M, rhythm adjusted average midway between the highest and lowest values of the fitted curve); amplitude (A, half the peak-to-trough difference of the fitted curve, i.e., the difference between the maximum and the mesor); and acrophase [ϕ, the lag of the peak in the fitted curve from an arbitrarily chosen zero time, which may be local midnight or some other time-point, expressed in hours or degrees, with 360° equivalent to the period of fitted curve, for example, 24 hours. In the cases to be discussed here, this zero time will be designated as the onset of the light in an imposed LD cycle (usually LD: *10*, 14) on the basis of both computational and biological considerations].* Although cosinor analysis, as well as every other, even more sophisticated method, has certain shortcomings (for example, estimates of parameters may be biased if the circadian function as represented by the raw data in actuality is not a cosine function), in most instances the cosine

---

*The values for M, A, and ϕ are usually expressed together with their 95% confidence limits (CL) or standard errors (SE), together with the probability (p), so that the rhythm is statistically significant (so regarded if an F test of zero amplitude for the fitted curve yields a p value $<$ .05) and the percent rhythm, PR (percentage of variability accounted for by the fitted curve).

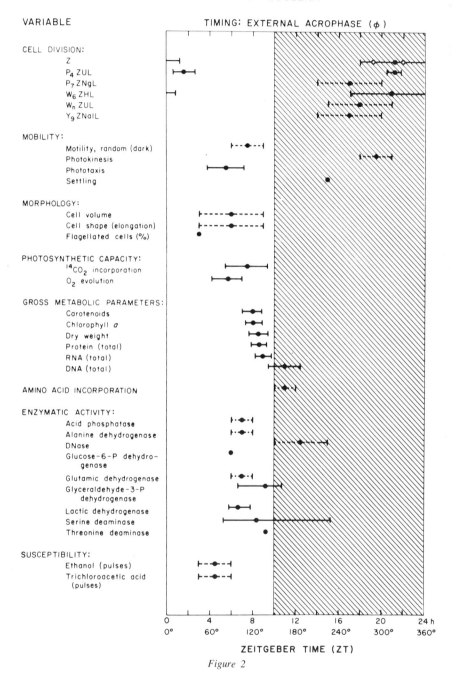

Figure 2

approximation has proved a useful and successful first step in the statistical analysis of biological rhythms. Certainly, it is more desirable than simply relying on an intuitive, cursory visual analysis of "macroscopic" raw data.

Accordingly, a number of the circadian rhythms tabulated for *Euglena* in Table I have been analyzed by the cosinor method and the values for the various parameters computed (Edmunds and Halberg, 1981). A convenient way to illustrate the time relationships of the periodicities thus analyzed to the LD cycle imposed during synchronization (usually by *LD: 10*,14) is by means of an acrophase (ϕ) chart, as illustrated in Fig. 2. Such a chart also allows one to visualize the time relations among the various rhythms during both entrainment and free-running conditions.

In this presentation, actual acrophases* (computed by cosinor analysis or subjectively estimated from published chronograms) are utilized rather than the variable phase reference points (given in Table I) whose timing may not be the same. For example, the phase marker for the cell division and the phototaxis rhythms are taken usually as the onset (minimum) and the maximum values, respectively; in contrast, computed acrophase values can be directly compared because they always indicate the timing of peak values in the rhythm as approximated by the cosinor model.† In addition, similar types of rhythms are grouped. Thus, cell division rhythms in the wild-type and the several mutants all show ϕ's in the latter part of the dark span (or during late subjective light in LL or DD), whereas most synthetic activities (gross metabolic variables and enzymatic activity) show high values toward the end of the light span (or in the late subjective day in the case of alanine dehydrogenase, which was also monitored under free-running conditions of constant illumination and temperature).

Fig. 2. Acrophase (ϕ) chart for *Euglena* showing the time relations among the circadian rhythms listed in Table I. The bars extending to the right and left of the acrophase points give the 95% confidence interval, when calculated by cosinor analysis. In all cases acrophases are used (in contrast to the variable and arbitrarily chosen phase markers for timing itemized in Table I). Single points or points with dashed lines and brackets indicate subjective estimates of the acrophase from published data without benefit of statistical analysis. The data for gross metabolic variables, although originally obtained in *LD: 14*,10, have been adjusted to correspond to a synchronizing *LD: 10*,14 cycle used for most of the other rhythms. (From Edmunds and Halberg, 1981.)

*Note that ϕ represents the so-called "external acrophase" and in this review relates the peaks of the fitted curve to the onset of light in the synchronizing LD cycle (Halberg *et al.*, 1973) and to "subjective dawn" in DD and LL when the light would have come on had the synchronizer been continued. This notation corresponds to ψ (phase angle difference between the entraining *Zeitgeber* and the biological oscillation) in the "circadian vocabulary" introduced by Aschoff *et al.* (1965).

†Further standardization has been attempted by reference to an *LD: 10*,14 cycle throughout, although in several cases measurements were made in an *LD: 12*,12 or *LD: 14*,10 regimen. In these instances ϕ could differ possibly by 1 hour or so (especially if the rhythm were not keyed to the onset of light but rather to the light–dark transition) from the values that would be obtained in *LD: 10*,14.

Clearly, then, the *Euglena* system provides an excellent case for *temporal differentiation:* A large number of diverse behavioral, physiological, and biochemical activities are partitioned in time (24-hour time) thus providing dimensions for both environmental adaptation and, what is at least equally important, for functional integration in time (Edmunds, 1975; Edmunds and Halberg, 1981).

II. Survey of Selected Classes of Circadian Rhythms in *Euglena*

In Section I a number of circadian rhythms were tabulated for the *Euglena* system and their phase relationships were given in an acrophase chart. In this section some of the major classes of rhythms will be examined in greater detail, although space precludes a comprehensive treatment. Sufficient key references have been given (Table I), however, to lead an interested reader to the archival sources and critical reviews.

A. RHYTHMS OF CELL DIVISION

One of the most intensively studied persisting circadian rhythms in *Euglena* is that of cell division (reviewed by Edmunds, 1975, 1978; Edmunds and Adams, 1981)—an event that occurs only once per life cycle in an individual cell, but one which occurs repetitively in a population. Because of mounting interest in the relation of cell cycles to circadian (and other) oscillators, with the possibility that they share at least certain elements in common (Edmunds, 1978; Wille, 1979; Edmunds and Adams, 1981), a major portion of this survey will be devoted to this topic.

1. *Cell Cycle Clocks: Deterministic and Probabilistic Models*

The cell division cycle (cdc) of a typical microorganism such as *Euglena* comprises a series of relatively discrete morphological and biochemical events, although the specific elements may vary among different systems. Ordinarily the cdc is loosely considered to begin with the completion of one cell division and end with the completion of the next; the time taken for one such cycle is termed the generation time (g). Progress through the cdc is usually monitored by observing the overt processes of DNA replication [synthetic (S) period] and cell division [(D), or mitotic (M), period]. Thus, the cdc has been divided classically into four consecutive intervals: G_1-S-G_2-M, where G_1 and G_2, respectively, designate the gaps between the completion of division and the onset of DNA replication, and between the end of replication and the onset of mitosis. (A phase-duration map for these cdc states in a synchronized culture of *Euglena* is given in Fig.20.)

We now know that both G_1 and G_2 (as well as S and M) can be subdivided into many smaller steps, such as the expression of a particular enzyme or the replication of mitochondrial DNA [see the book by Mitchison (1971) for a general treatment of the subject]. In a sense then, the cdc is a time-measuring device, or clock, of sorts.*

Attempts to describe the cell division cycle (cdc) fall into two broad categories: deterministic and indeterminate or probabilistic. Within the former approach, which is historically the older, there are two possible types of mechanisms for ordering a fixed sequence of cell cycle events relative to each other (Mitchison, 1974). On the one hand, there may be a direct causal connection between one event and the next, so that it would be necessary for the earlier event in the cdc to be completed before the following could occur. In contrast to this sequential type of approach, there is the possibility that no direct causal connection exists between any two events, but that they are ordered by some master timing mechanism that operates on one or more key events (control points) of the cdc, such as the initiation of DNA synthesis or mitosis.

Such deterministic models, however, have difficulty in accounting for the large variances commonly observed in generation time (as great as 20% of the mean in mammalian cell systems), rendering timekeeping relatively imprecise and leading to the rapid decay of synchrony in phased cultures (Engelberg, 1964). In an effort to explain this variability, other attempts to characterize the cell cycle traverse have considered a portion of the cdc to be indeterminate or have turned to probabilistic descriptions. Thus, Smith and Martin (1973), observing that the fraction of cells that have divided as time progresses decreases exponentially, have suggested that the S and G_2 portions of the cdc (which they term the B phase), are deterministic and invariant, whereas other parts are probabilistic (the A state, or "waiting" phase, of G_1). According to this transition probability model, newly divided cells enter the A phase in which their activity is no longer actively directed to proliferation; they then have a certain degree of probability of again entering the B phase, wherein DNA synthesis and mitosis occur once more. Gilbert (1978) has taken a similar theoretical approach, giving particular attention to various perturbations that may trigger the transition of the cell from the "quiescent" state into the more highly dynamic one of replication and the possible relation to differentiation and cancer.

Finally, Klevecz (1976) has proposed a "quantal" subcycle, G_q, for mammalian cells, whose traverse time is about 3 to 4 hours. This cycle would be appended to the deterministic $S + G_2 + M$ pathway at a point i. The exit of a cell from G_q would be probabilistic in the sense that there could be an indefinite number of G_q cycles depending on environmental conditions (e.g., cell density,

---

*The word "clock" is used as a generic term denoting a timekeeping device, without any bias as to the underlying mechanism. This definition was adopted at the Dahlem Conference on "The Molecular Basis of Circadian Rhythms" in 1975 (J. W. Hastings, H.-G. Schweiger, eds. 1976).

nutritional variables, and mitosis-stimulating factors). This model has the virtue of explaining the heterogeneity of G_1 (and hence, of the generation time of individual cells) as arising from the "gated entry of cells into the S phase" and gains credence from two sets of observations: (1) the distribution of possible generation times in populations of mitotically selected cells taken from synchronous cultures, or from randomly dividing cultures, as observed by time-lapse video tape microscopy, did not appear to be continuous but rather was "quantized" in multiples of 3–4 hours; and (2) the activity of a number of enzymes that have no obligatory relation with other periodic events such as DNA synthesis oscillated with periods also of 3–4 hours, even if DNA and RNA synthesis were inhibited (Klevecz, 1969, 1976). In a sense, then, the G_q subcycle would constitute a *cellular clock* with a basic period of about 3–4 hours (at least in cells whose cycle time is less than 24 hours), and cdc times would be multiples of this fundamental period (increasing, for example, at higher cell densities or lower temperatures).

In the preceding discussion of deterministic models of the cdc, we noted that some sort of master timing mechanism might exist. Inasmuch as mitosis (M) is a periodic event of short duration relative to the total length of the cdc, it is not surprising that various types of oscillatory systems, or biological clocks, have been proposed to underlie the cdc. Indeed, even on the hypothesis that the cdc is merely a linear array of discrete metabolic states, each causing the next, and that it is controlled simply by the sequential transcription and subsequent translation of genes linearly ordered on the chromosomes, an oscillator could be formulated by invoking a recycling component that would initiate another time-metering, transcriptional cycle (Ehret and Trucco, 1967).

Likewise, several biochemical oscillators have been hypothesized to control the cdc, such as those thought to underlie the high degree of natural mitotic synchrony observed to occur every 10 to 12 hours in the nuclei (as many as 10^8) within the syncytial plasmodia of the myxomycete *Physarum polycephalum*. Plasmodia of different stages, or phases, of the cdc can be fused, their nuclei and cytoplasm intermingle, and the fused pair then undergoes mitosis at some intermediate phase (Rusch *et al.*, 1966; Sachsenmaier *et al.*, 1972; Sachsenmaier, 1976). These facts, on the one hand, have led Sachsenmaier and co-workers to characterize the timing mechanism of mitosis as an "hourglass" or *discontinuous, extreme relaxation oscillator* in which mitotic initiator molecules ("mitogen") are formed more or less continuously and proportionately to the increase in plasmodial mass during G_2. These molecules would be counted by combining stoichiometrically with a given number of nuclear receptor sites. Mitosis would be initiated at a critical initiator–nuclei ratio, and the clock would be reset by each nuclear division (or by a related, obligatory event), at which time the number of nuclear sites would also double in a stepwise manner (Fig. 3A). In this model, the event M is an essential feature of the oscillatory system.

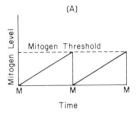

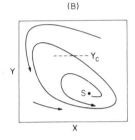

Fig. 3. (A) Diagrammatic representation of a simple, discontinuous "hourglass" relaxation oscillator underlying mitosis (M). Mitogen accumulates linearly during the cell division cycle and triggers mitosis at a critical threshold level, whereupon it is destroyed, the "clock" is reset, and mitogen reaccumulates once again during the next cycle. (From Tyson and Kauffman, 1975; redrawn by permission.) (B) Diagrammatic representation of a continuous biochemical oscaillator exhibiting limit cycle behavior. Two interacting substances (X and Y) autonomously fluctuate; mitosis occurs when one of them (Y) attains a threshold level (Y_c). S denotes the point of singularity, which is unstable with respect to small pertubations. Trajectories spiral out to the closed curve, the limit cycle. (From Tyson and Sachsenmaier, 1978; modified by permission.)

Given the same set of facts, Kauffman and Wille (1975,1976) have put foward an alternative model: The timing mechanism for mitosis is a *continuous limit cycle oscillator,* analogous to that proposed for circadian rhythms, which would gate mitosis and DNA replication. On this hypothesis, two (or more) interacting components (X, Y) fluctuate autonomously, and mitosis would be triggered if one of them (an initiator) reached a threshold level (Fig. 3B). The critical distinctions between this model and Sachsenmaier's hourglass model are that in the limit cycle (a) mitosis does not function as an essential component of the oscillator system—it is downstream from the clock itself—and thus, under certain conditions, the system may continue to oscillate at a subthreshold level even if mitosis is blocked and does not occur; and (b) although the amplitude and period do not depend on initial conditions (i.e., the limit cycle is stable, resisting and recovering from most perturbations), a perturbation given at a unique, "singularity" point results in a phaseless, timeless (motionless) state. Although the experimental evidence obtained thus far does not rigorously exclude either hypothesis in *Physarum,* a recent critical study (Tyson and Sachsenmaier, 1978) favors the discontinuous relaxation oscillator.

Finally, there is the possibility that Klevecz's (1976) G_q quantal subcycle might itself constitute a limit cycle oscillator (Klevecz, 1978) having a 3-to-4 hour period [whereas that of Kauffman and Wille (1975) would have a period equal to that of the cdc]. The recent finding that Chinese hamster cells with an 8.5-hour cdc yield a biphasic phase response curve for high serum pulses, heat shock, or ionizing radiation, with both advances and delays in the timing of cell division, is consistent with this notion (Klevecz *et al.,* 1978, 1980).

Regardless of the underlying controlling mechanisms and generative "subcycles," however, the overall cdc is a clock in the sense that specific events correspond to the numerals on the dial, i.e., phase reference points (ϕ_r's), and the generation time (g) of the cell reflects the period of (τ) of the timing process(es), or oscillation(s). As is well known, the duration of the cell division cycle itself is quite variable, for it is markedly affected by alterations in temperature, illumination, and nutrients in the medium; consequently, the underlying oscillator would appear to be imprecise. For example, under the very best of all possible worlds, the generation time of a culture of the unicellular algal flagellate *Euglena* may be as short as 8 hours; more typically, g ranges from 12–30 hours in the laboratory, and the upper limit approaches infinity as the stationary phase of population increase is entered. It may be that g is nothing more than a summation of all the individual reaction times of the constituent processes, and that the cell cycle is simply a statistical averaging machine. Yet, we shall soon see that this is an oversimplification.

2. *Synchronously Dividing Populations*

In exponentially increasing cultures of microorganisms, the phase points of the individual cell cycles of the cell population are distributed randomly;* they are *developmentally asynchronous* (Fig. 4A). A striking contrast is afforded by *developmentally synchronous* populations: Cell division (as well as some of the other events comprising the cell cycle) in cultures of numerous protists, algal unicells, and cells dissociated from plant and animal tissues can be synchronized by a variety of inductive treatments so that there is a one-to-one mapping of similar phase points of the cell cycles throughout the culture with respect to time (Fig. 4B), as is illustrated for *Euglena* in Fig. 5 (reviewed by Mitchison, 1974; Edmunds, 1975,1978). These techniques include the addition of inhibitors and selective lethal agents to asynchronous cultures. The operational principle here is that a particular stage of the cell cycle is differentially sensitive to the agent; the cells accumulate at this stage, therefore, and after release from the inhibitor

---

*Actually, they are skewed toward G_1; a majority of the cells are younger because as soon as they attain "old age" (M) they divide to form two (or more) young daughter cells.

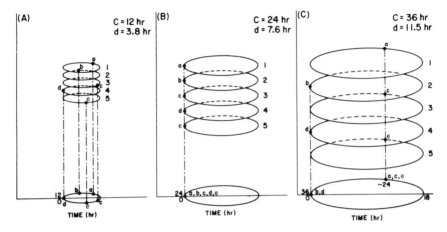

Fig. 4. Schematic representation of developmentally asynchronous and developmentally synchronous cultures in various modes of growth. The stacked "doughnuts" depict individual cell division cycles (cdc); solid points (a–e) indicate the landmark of cell division that terminates the cdc. The circumference (C) of the cycles (taken as circles but projected as ellipses) is equal to the generation time (g), and is calculated as πd where d is the diameter of the circle. (A) Exponentially growing culture in ultradian growth mode ($g < 24$ hours); the population is developmentally asynchronous, with divisions occurring continuously. (B) Synchronized (entrained) culture in circadian growth mode ($g \simeq 24$ hours); the population is developmentally synchronous, with divisions being confined to relatively narrow intervals ("windows" or "gates") 24 hours apart. (C) Culture in infradian growth mode ($g > 24$ hours); the population is developmentally asynchronous but is synchronized, nevertheless, with regard to the circadian oscillations in each cell since divisions, when they do occur, are clustered at intervals of approximately 24 hours. The lengths of the individual cdc's here only average 36 hours (and actually may be discontinuously distributed), and the ellipses should vary in size to reflect the variance (see text). (Adapted from Edmunds, 1978.)

(usually by washing it out), they divide synchronously for several rounds until the synchrony decays due to extrinsic, microenvironmental fluctuations and to intrinsic, karyotypic heterogeneity (Engelberg, 1964). This concept of cell cycle blockage points can be easily extended to embrace those synchronization methods in which the culture is starved or allowed to enter the stationary phase and then rejuvenated by the addition of fresh medium or even to the use of single or repetitive heat (or cold) shock, as has been documented, for example, in *Euglena* (Terry and Edmunds, 1970a).

Indeed, a general model for such synchronization by shifts in environmental conditions has been proposed by Campbell (1957, 1964). It was assumed that under a given set of extrinsic conditions the cell progresses through a sequence of stages, and that under a different set of conditions the cell cycle consists of the same sequence, but that the relative time spent between such stages is different. If a series of shifts is performed between these two sets of conditions of such

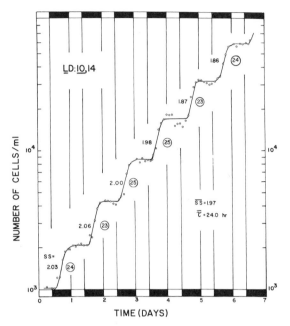

Fig. 5. Entrainment of the cell division rhythm in a population of *Euglena* grown photoauto-trophically at 25°C in *LD: 10,*14. Ordinate: cell concentration (cells/ml); abscissa: elapsed time (days). Stepsizes (*ss,* ratio of number of cells per milliliter following a division burst to that just before the onset of divisions) are indicated for successive steps; the period of the oscillations is also given in hours (encircled just to the right of each division step). The average period ($\bar{\tau}$) of the rhythm in the culture is essentially identical to that of the synchronizing LD cycle, and a doubling of cell number ($\overline{ss} \cong 2.00$) usually occurs every 24 hours. [After Edmunds and Funch (1969a) *Science* **165,** 500–503. Copyright 1969 by the American Association for the Advancement of Science; Edmunds and Funch (1969b).]

duration that one period of the regime corresponds to a doubling of cell number, the model predicts a gradual attainment of complete synchronization. In the well-synchronized culture, therefore, a majority of the cells pass through the same developmental stage at the same time, and that which is determined for the entire culture can be assumed to obtain as a first approximation for the individual cells.

In addition to the techniques just described, it is well-known that appropriately chosen light cycles (LD) can synchronize cultures of microorganisms; indeed, it is the method of choice for most photosynthetic unicellular algae (Edmunds, 1974b, 1975) including *Euglena* (Cook and James, 1960; Edmunds, 1965a). Photoautotrophically grown cultures of *Euglena* can be synchronized by empiri-cally chosen repetitive, 24-hour LD cycles so that cell division is confined almost entirely to the dark intervals; typical synchrony obtained in *LD: 10,*14 is shown in Fig. 5. In this case, the population doubles (factorial increase, or $\overline{ss} \simeq 2.0$) at each step, the period ($\bar{\tau}$) of the rhythm in the population is exactly 24 hours [matching that (T) of the imposed LD cycle], and, by inference (see Fig. 4b), the length of the individual cdc's also must average 24 hours (cell death is insignificant). Long-term synchronous cultures can be obtained (Fig. 6) using continuous or semi-continuous culture techniques (Terry and Edmunds, 1969) in which the cell titer is maintained at a constant level without reaching limiting conditions [as occurs in batch cultures (see Fig. 5)]. If one reduces the total duration of the light interval (for example, $LD: 8,16$) within a 24-hour framework, the amplitude of the rhythm of cell division in the population is proportionately reduced ($\overline{ss}$ < 2.0), but the culture continues to be synchronized (Fig. 7) in the sense of event simultaneity (Edmunds, 1968; Edmunds and Halberg, 1981). The average individual cdc is now lengthened, however, to approximately 36 hours ($\overline{ss}$ = 1.68), and the culture is no longer developmentally synchronous to the extent of a one-to-one correspondence between the cdc stages of all the constituent cells (see Fig. 4C). Nevertheless, cell divisions, when they do occur, do so during the dark only, at intervals of 24 hours, and the rhythmicity observed in $LD: 8,16$ (and $LD: 10,14$) stands in sharp contrast to the asynchronous, exponential growth curve obtained in LL (see Fig. 4A) having a minimum doubling time of 12–14 hours (Edmunds, 1965a).

Although it would seem plausible to ascribe this light-induced division synchrony to the same mechanism responsible for that obtained by inhibitor blocks and heat shock, involving repetitive shifts between two sets of environmental conditions (i.e., light and darkness) to which specific stages of the cell cycle are

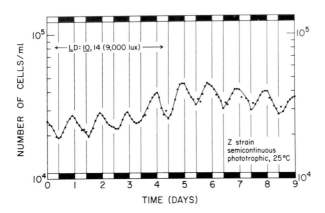

Fig. 6. Entrained rhythm of cell division in *Euglena* (Z strain) grown photoautotrophically at 25°C in $LD: 10,14$ in semicontinuous culture. Compare with Fig. 5. Note that the rate of dilution with fresh medium approximately balances the overall growth rate of the culture. (From Edmunds and Halberg, 1981.)

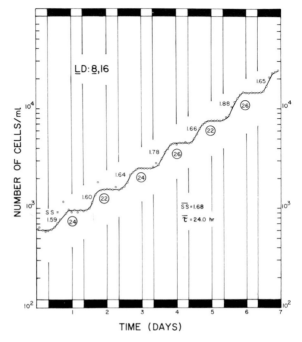

TIME (DAYS)

Fig. 7. Entrainment of the cell division rhythm in a population of *Euglena* grown photoauto-trophically in *LD*: *8, 16.* Labels as for Fig. 5. Although $\bar{\tau}$ of the rhythm is precisely that of the synchronizing LD cycle, $\overline{ss}$ of the successive fission bursts is substantially less than 2.0, indicating that not all of the cells divide during any one cycle. [After Edmunds and Funch (1969a) *Science* **165,** 500–503. Copyright 1969 by the American Association for the Advancement of Science; Edmunds and Funch (1969b).]

differentially sensitive, it will be seen shortly that this type of model does not adequately explain the observed facts and, indeed, may not even be relevant at all under certain experimental conditions. Rather, we have assumed as a working hypothesis that an endogenous, light-entrainable, circadian clock—having all those properties (Fig. 1) that characterize the oscillatory mechanism(s) under-lying the many persisting 24-hour rhythms documented for *Euglena* (Table I)—may underlie division rhythmicity in light-synchronized cultures, which "gates" cell division to restricted intervals of time during successive 24-hour time spans.

3. Clocked Cell Cycles

The synchronization of the rhythm of cell division in *Euglena* by diurnal, full-photoperiod LD cycles (Figs. 5–7) is consistent with the notion that a puta-tive circadian clock is entrained by the imposed light regime and, in turn, phases cell division to the dark intervals occurring every 24 hours (perhaps by acting on one or more key control points of the cdc); nevertheless, it does not demand it. Light (or darkness) could be acting by directly inhibiting (or promoting) division, and periodic shifts between light and dark would synchronize the culture. A number of other observations in a variety of experimental organisms, however, render this seemingly straightforward hypothesis unlikely.

These lines of evidence (Edmunds, 1974b,1975,1978) include the following: (a) entrainment by LD cycles having $T \neq 24$ hours (for example, LD: $10,10$) may also occur within certain limits (Edmunds and Funch, 1969b; Ledoigt and Calvayrac, 1979); (b) appropriate temperature cycles (for example, $18°/25°C$: $12,12$ or $28°/35°C$: $12,12$) will entrain the rhythm in cultures maintained in LL (Terry and Edmunds, 1970a); (c) "skeleton" photoperiods comprising the framework of a normal, full-photoperiod cycle (for example, LD: $3,6,3:12$) will also synchronize the rhythm to a precise 24-hour period (Edmunds and Funch, 1969b), as shown in Fig. 8; (d) high-frequency (for example, LD: $1,3$) LD cycles (Fig. 9A) and even random illumination regimes (Fig. 9B) induce circadian division periodicities (Edmunds and Funch, 1969a,b); (e) short light signals (3-hour, 7700 lux), given at different circadian times to the free-running rhythm

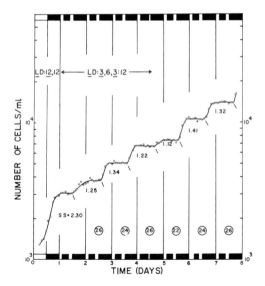

Fig. 8. Entrainment of the cell division rhythm in a population of Euglena grown photoautotrophically at 25°C by an LD: 3,6,3:12 "skeleton" photoperiod following exposure to an LD: 12,12 full-photoperiod synchronizing regime. Other labels as for Fig. 5, except that the successive period lengths in hours are encircled at the bottom of this figure. (From Edmunds and Funch, 1969b.)

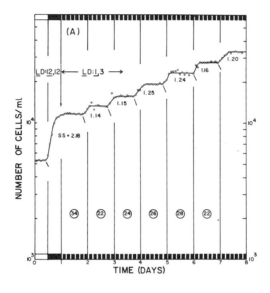

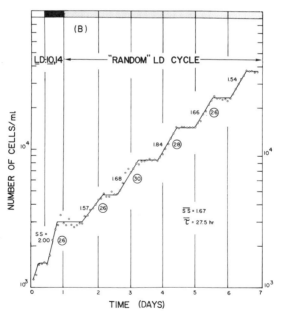

Figure 9

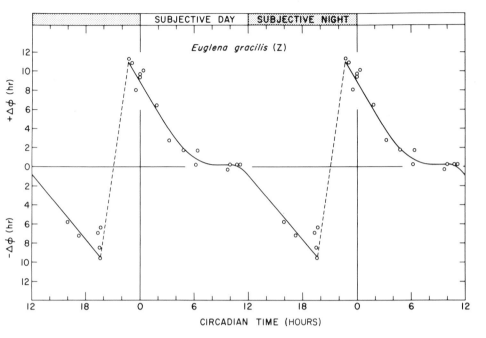

Fig. 10. Phase-response curve for the action of 3-hour light signals (7500 lux) on the "free-running" rhythm of cell division in photoautotrophic cultures of *Euglena* maintained in *LD: 3,3* at 25°C. The steady-state phase shift (ordinate) is double-plotted as a function of the circadian time (normalized to 24 hours) at at which the signal was given. (From L. N. Edmunds, Jr., D. E. Tay, and D. L. Laval-Martin, unpublished data; Edmunds, 1981; Edmunds *et al.*, 1982.)

of cell division in autotrophic cultures maintained in *LD: 3,3* (cf. Fig. 39), generate steady-state phase shifts whose sign and magnitude are dependent on the subjective time at which the pulse is applied to the system (Fig. 10); (f) rhythmic cell division will persist (Fig. 11) for a number of days ($\tau \simeq$ 24 hours) in the autotrophically grown Z strain, batch cultured under dim LL (Edmunds, 1966). This last series of experiments was perhaps the most definitive but was restricted

---

Fig. 9. (A) Effects of a high-frequency *LD: 1,3* cycle on the cell division rhythm in photoautotrophic cultures of *Euglena* at 25°C following initial synchronization by an *LD: 12,12* regime. Other labels as for Fig. 5 (Edmunds and Funch, 1969b). (B) Persisting, "free-running," circadian rhythmicity in the cell division rhythm of cultures exposed to a random light–dark cycle after synchronization by *LD: 10,14*. The average free-running period of the population rhythm is 27.5 hours. The average stepsize ($\overline{ss}$ = 1.67) is almost identical to that found in *LD: 8,16* (Fig. 7), which affords the same total duration of illumination in a given 24-hour time span. Similar results are obtained upon exposure of an exponentially growing culture (no prior synchronization) to the random regime. [From Edmunds and Funch (1969) *Science* **165,** 500–503: Copyright 1969 by the American Association for the Advancement of Science.]

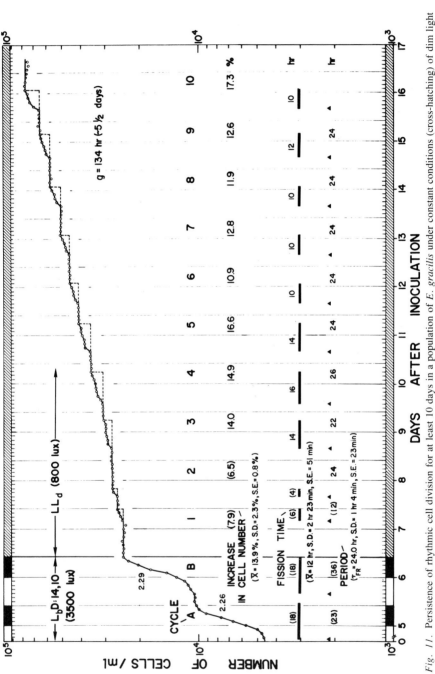

Fig. 11. Persistence of rhythmic cell division for at least 10 days in a population of *E. gracilis* under constant conditions (cross-hatching) of dim light (LL$_d$, 800 lux) and temperature (25°C), following 6 days in *LD*: *14*, *10* (L$_b$D, 3500 lux). For each division burst, the increase in cell number (%), fission time (hours), and period, τ_{FR} (hours), are given; *x*, *S.D.*, and *S.E.* denote, respectively, the mean, standard deviation, and standard error. Step size is shown for cycles A and B. The generation time (*g*) of the population in dim light is indicated. (From Edmunds, 1966.)

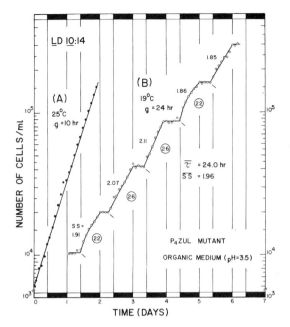

TIME (DAYS)

Fig. 12. Population growth of a photosynthetic mutant (P_4ZUL) of *Euglena* grown photoor-ganotrophically on a defined medium containing glutamate and malate in LD: *10*,14. (A) Exponen-tial increase in cell number [generation time (g) of 10 hours] at 25°C. (B) Phasing of the cell division rhythm at 19°C (g of 24 hours). Other labels as for Fig. 5. [After Jarrett and Edmunds (1970) *Science* **167**, 1730–1733: Copyright 1970 by the American Association for the Advancement of Science.]

by the low light intensities (800 lux) that had to be utilized.* Nevertheless, although the division bursts were relatively small (on the average only 17% of the population underwent fission in a given step and $g \simeq 5$ days), those cells that did divide did so during their "subjective night" at the times that they would have experienced darkness had the entraining LD cycle been continued.

Even more conclusive evidence for the implication of a basic circadian oscil-lator in the control of the cdc has been obtained by utilizing photosynthetic mutants of *Euglena* (obligate heterotrophs), thereby circumventing the problem of the dual use of light as an energy source for photosynthesis, on the one hand, and as a timing cue for the circadian clock, on the other (Edmunds, 1975,1978). Representative results for the uv light-induced P_4ZUL mutant (Jarrett and Ed-munds, 1970) grown on a medium containing glutamic and malic acids as carbon sources under LD: *10*,14 are shown in Fig. 12. Entrainment to a 24-hour period

*Because of the requirement of light for photosynthesis, one cannot use DD; nor can one impose LL of those moderate intensities (3500–7000 lux) normally used in the entraining LD cycle, for the population reverts to asynchronous, exponential growth.

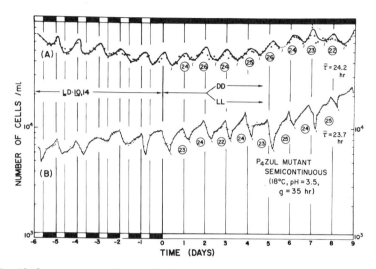

Fig. 13. Long-term, persisting circadian rhythm of cell division in two different semicontinuous cultures of the P_4ZUL photosynthetic mutant of *Euglena* grown on low pH glutamate–malate medium at 18°C. The cultures were first synchronized by LD: *10*,14 (6 monitored days shown) and then placed either in DD (curve A) or in LL (curve B); the first nine cycles under constant conditions are indicated. The overall generation time (g) of both cultures was calculated from the known dilution rate to be about 35 hours. Successive period lengths are encircled just below each "free-running" cycle and the average period ($\bar{\tau}$) is given to the right for each curve. (From Edmunds *et al.*, 1974.)

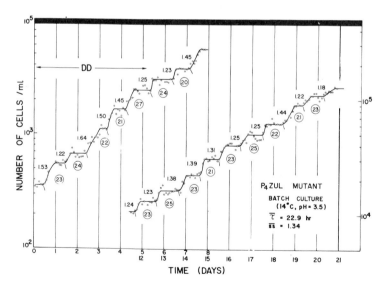

Fig. 14. Persisting circadian rhythm of cell division in a batch culture of the P_4ZUL photosynthetic mutant of *Euglena* grown at 14°C in DD for 21 days following entrainment by an LD cycle. Compare with cultures grown at 18° or 19°C (Figs. 12 and 13). (From Edmunds *et al.*, 1971.)

was not possible with cultures in the fast-growing ultradian ($g < 24$ hours) growth mode (Ehret *et al.*, 1977): At 25°C, the doubling time of the exponential growth curve obtained was about 10 hours. If the growth temperature was reduced to 19°C [yielding an exponential curve with $g \simeq 24$–26 hours in DD or LL (not shown)], however, the culture was synchronized. In this infradian ($g > 24$ hours) growth mode (Ehret *et al.*, 1977), divisions were set back or delayed for 8–10 hours at 24-hour intervals. Furthermore, the rhythmicity persisted with a circadian period (22–23 hours) in DD (not shown) for as long as 5 days in batch culture (Jarrett and Edmunds, 1970) and for at least 10–14 days in continuous culture (Fig. 13) in DD or in LL (Edmunds *et al.*, 1971, 1974). The period of the oscillation is temperature-compensated over a range of at least 7°C within physiologically permissive conditions (allowing measurable cell division to occur but not encroaching upon the ultradian domain); thus, batch cultures of the P_4ZUL mutant synchronized by LD at 14°C displayed a persisting rhythm having $\bar{\tau}$ of 22.9 hours (Fig. 14), although the amplitude was reduced ($\overline{ss} = 1.34$) as expected (Edmunds *et al.*, 1971). Even a single D-L transition (Jarrett and

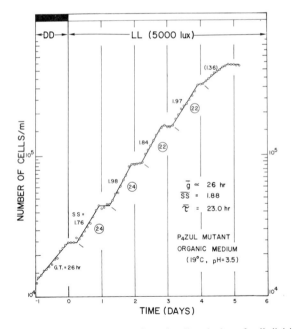

Fig. 15. Initiation of a persisting, free-running, circadian rhythm of cell division by a single transition from darkness to continuous light in cultures of the P_4ZUL photosynthetic mutant of *Euglena* grown at 19°C on glutamate–malate medium. The culture had been increasing exponentially with a generation time ($\bar{g}$) of about 26 hours during the preceding interval of darkness (DD). Other labels as for Fig. 5. [After Jarrett and Edmunds (1970) *Science* **167,** 1730–1733. Copyright 1970 by the American Association for the Advancement of Science.]

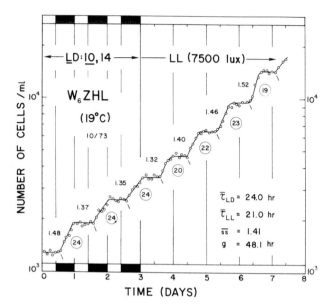

Fig. 16. Synchronization of the cell division rhythm and its persistence for at least 5 days under constant illumination in the aplastidic, heat-bleached W_6ZHL mutant of *E. gracilis* batch-cultured photoorganotropically at 19°C under an *LD: 10,14* cycle and subsequently released into LL (3500 lux). Labels and notation as for Fig. 5. Note that although the period of the population rhythm is about 24 hours, the period of the individual cell cycle (equivalent to the generation time, *g*) is approximately 48 hours. (After Edmunds, 1974b; Edmunds *et al.*, 1976.)

Edmunds, 1970) was sufficient to induce rhythmicity, which then persisted under LL with $\bar{\tau}$ = 23.0 hours (Fig. 15). These results have been extended to the naladixic acid-induced Y_9ZNalL photosynthetic mutant of *Euglena* as well as the white, heat-bleached W_6ZHL strain (Fig. 16) that totally lacks chloroplasts (Edmunds, 1975; Edmunds *et al.*, 1976), and are consistent with the data of Mitchell (1971) for the pale green nitrosoguanidine-mutagenized P_7ZNgL mutant and the white, uv-bleached W_nZUL strain. In the studies of all five different photosynthetic mutants of *Euglena*, cell division is confined primarily to the dark spans of LD cycles (or to subjective night in DD or LL). This timing agrees with findings of other workers on the algae *Chlamydomonas*, *Chlorella*, *Gonyaulax*, and *Gymnodinium* and the ciliates *Paramecium* and *Tetrahymena* (see reviews by Edmunds, 1975;1978).

To summarize the foregoing data conveniently, a cosinor display for these long-term persisting cell division rhythms in the Z strain and the P_4ZUL and W_6ZHL photosynthetic mutants is shown in Fig. 17 (Edmunds and Halberg, 1981). The *p* values for the rhythms of cell division monitored in continuous culture are all < .001. It should be noted that these long trains of data enable one

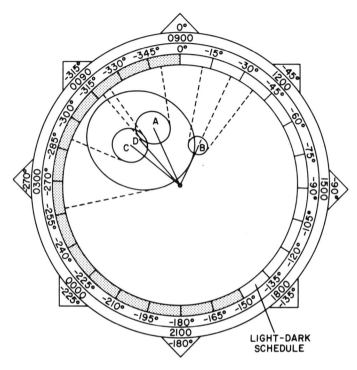

Fig. 17. Single cosinor displays for synchronized rhythms (A–E) of cell division in *Euglena:* (A) Z strain (Fig. 6); (B) P_4ZUL mutant (Fig. 13, curve A); (C) P_4ZUL mutant (Fig. 13, curve B); and (D) W_6ZHL mutant (Fig. 16). The length and direction of the lines (vectors) directed outward from the center of the figure respectively represent a given rhythm's amplitude (*A*) and its acrophase (ϕ) in relation to the synchronizing light cycle, as read in degrees (middle circular scale; $0° =$ onset of L). The circular or elliptical regions at the tip of the vectors indicate the joint 95% confidence regions for ϕ and *A*. Mesors and amplitudes expressed as (*A*) $\log_{10}$ (cells/ml), mantissa only. (From Edmunds and Halberg, 1981.)

to perform a more thorough frequency analysis by the least squares method (Edmunds and Halberg, 1981). One can conduct a chronobiological ''serial section'' by fitting a fixed-period cosine curve to consecutive data intervals displaced by increments throughout the time series, so as to ascertain continuously (by moving point estimates) the mesor, amplitude, and acrophase of a rhythm (Halberg *et al.,* 1967, 1972; Nelson *et al.,* 1979).

Finally, we have observed (Edmunds *et al.,* 1976; Edmunds, 1977) that cultures of W_6ZHL and P_4ZUL may gradually lose both their capacity to exhibit division synchrony in LL or DD and even to be entrained by imposed LD cycles (Fig. 18), reverting to random exponential growth ($g > 24$ hours). These properties could be restored, however, by the addition of certain sulfur-containing

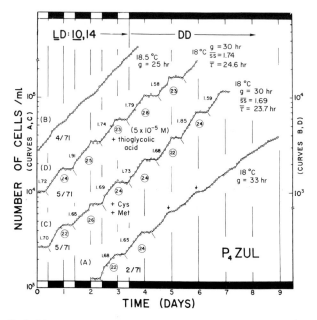

Fig. 18. Gradual loss and subsequent restoration of circadian division rhythmicity in organotrophically batch-cultured populations of the ultraviolet-light-induced P_4ZUL photosynthetic mutant
of *Euglena*. (A) Initial loss of persistence of the rhythm in DD. (B) Loss of capacity for entrainment
of the rhythm by an *LD: 10,*14 cycle a few months later. (C) Restoration of the capacity for both
entrainment and a DD free run by the addition of cysteine and methionine (1×10^{-5} *M*) to the
medium. (D) Restoration of synchrony by the addition of thioglycolic acid (5×10^{-5} *M*) to the
medium. The generation time (*g*), the mean stepsize ($\overline{ss}$), or fractional population increase for the
successive division bursts, and the mean free-running period ($\overline{\tau}$) in DD for the growth curves are
indicated (using the onsets of division bursts as the phase reference points). (From Edmunds *et al.,*
1976; reproduced by permission.)

compounds (e.g., cysteine and methionine) to the medium at the onset of the
experiment. If these substances were added at various times to an arrhythmic
P_4ZUL culture in LL (following prior exposure to *LD: 10,*14), periodic division
was likewise induced (Fig. 19), whose phase was precisely that predicted on the
assumption that the underlying clock had been running undisturbed (but unex-
pressed) throughout the experiment, merely having been uncoupled from division itself until the sulfur compounds were added (Edmunds *et al.,* 1976).

These results, then, taken together with those of numerous other systems
(Edmunds, 1975, 1978), implicate a master timer—in this case a circadian
oscillator—that, while entrainable by appropriate LD (or temperature) cycles,
can itself modulate the cdc and phase or "gate" cell division (and probably other
marker events) to intervals of approximately 24 hours. We shall consider further
this complex relationship between circadian and mitotic clocks in Part III.

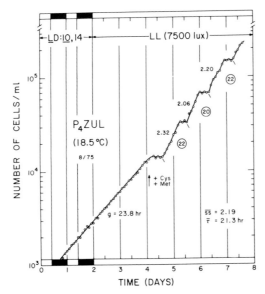

Fig. 19. Induction of circadian division rhythmicity in an asynchronously dividing (exponentially increasing) population of the P_4ZUL mutant of *Euglena* organotrophically batch-cultured in LD: *10*,14 and subsequently released into LL by the addition of cysteine (1×10^{-5} *M*) and methionine (5×10^{-5} *M*) at the time indicated by the arrow (during the third day in LL). The phase of the elicited rhythms is comparable to that usually found (see Fig. 18) in LD with divisions occurring in the subjective "night". Other notation as for Fig. 18. (From Edmunds *et al.*, 1976; reproduced by permission.)

4. *Cell Cycle Variability*

Thus far we have considered the cdc as a time-measuring process and its modulation by a circadian clock that appears to lie outside of the cdc itself. Implicit in such a discussion is the notion of some degree of precision and accuracy on the part of the basic oscillator(s) or timing sequence(s), and yet we are faced paradoxically with a disturbing amount of variability in the length of the cdc of most cells. This variability in traverse time of the complete cell division cycle has been observed with bacteria, yeasts, algae, and cultured animal cells and has been extended to other cdc stages, such as the entry of mammalian G_1 cells into S [reviewed by Edmunds and Adams (1981)]. How, then, can this dilemma be resolved?

The synchronous cell division observed in populations of *Euglena* (Figs. 5 and 6) entrained by LD cycles, as well as of other microorganisms, is far from perfect (Edmunds, 1965a), despite the fact that a doubling of cell number occurs every 24 hours in LD: *10*,14; divisions occur over a 8–12 hour timespan in the culture with some cells dividing almost at the beginning of darkness and others several hours later. A similar situation was found for the reduced steps in LD: $8, 16$ (Fig. 7) as well as for the P_4ZUL photosynthetic mutant in LD: $10, 14$ (Fig. 12), in which divisions seemed to be delayed 8–10 hours every day and then took place over a relatively long 12- to 14-hour interval. This spread might be anticipated given the fact that individual cells, grown photoautotrophically on minimal medium with CO_2 as the sole carbon source, show a variation in g from 8–24 hours, and from about 10.5–22 hours on proteose peptone-supplemented medium, although pairs of daughter cells appear to have closely related values of g (Cook and Cook, 1962). The decay in synchrony upon removal of the synchronizing regime, at least in bacterial and mammalian cell cultures, has been attributed to this variance in cdc lengths (Engelberg, 1964), presumably arising from cellular heterogeneity or stochastic processes. Indeed, Smith and Martin (1973) conclude that perfect synchrony can never be attained because of the probabilistic component (the A state) that they hypothesize to be a part of G_1.

That this variability in the cdc of *Euglena* is not merely a proportional, extension of all the constituent cdc stages is clear from the results of phase distribution studies (K. Adams and L. Edmunds, unpublished); a typical population phase-duration map is shown in Fig. 20. Thus, the duration of mitosis is independent of the rest of the cdc and averages approximately 1 hour. Over a wide range of cell concentrations, the duration of cytokinesis is also independent of the lengths of the earlier stages of the cdc, averaging 30 minutes to 1 hour.* Furthermore, studies on the accumulation of total cellular DNA in the population across the cdc (Edmunds, 1964, 1965b; K. Adams and L. Edmunds, unpublished) indicate that neither the duration of S nor the intervening G_2 phase can account for the observed variation in generation times (Fig. 20). As concluded for other cell types, it would seem, therefore, that the differences in the lengths of the cdc among individual cells of a population must result primarily from variation in the classically defined G_1 phase or some substage such as A, G_q, or even G_0 (sometimes called "proliferative rest").

The preceding discussion deals only with low level variability in the length of the cdc. We are confronted with a higher level of variability, however, in slowly growing, synchronously dividing cell populations in which not every cell divides at each step (i.e., where $\overline{ss} < 2.0$), but in which the steps themselves occur at periodic (e.g., circadian) intervals with virtually no division taking place between the steps. An individual cell may take several days (during which it will experience several LD cycles) to complete the cdc, and yet it will somehow be programmed to complete the terminal acts of mitosis and cytokinesis at night. Since in the fast-growing (ultradian) mode it would normally be expected to complete its cdc within the range from approximately 8–24 hours, some sort of

---

*Stirring rate, however, as well as other culture conditions, may have a significant effect on the time required for the separation of pairs of daughter cells.

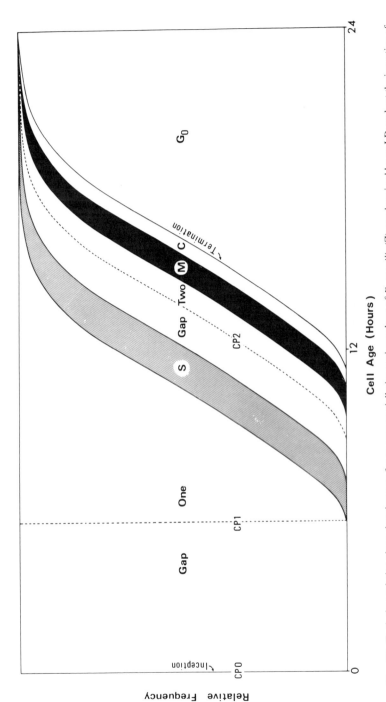

Fig. 20. Typical population phase-duration map for an exponentially increasing culture of *E. gracilis* (Z) synchronized by an LD cycle at the inception of the cdc. Note that the mean durations of S (1–2 hours), Gap Two (2–3 hours), M (60 minutes), and C (40 minutes) for a given cell age are independent of the duration of Gap One. Gap One is shown subdivided into a common phase (left) and a variable phase (right) by the hypothetical control point locus CP1. [From Edmunds and Adams (1981) *Science* **211**, 1002–1013. Copyright 1981 by the American Association for the Advancement of Science.]

mechanism, coupled in some manner to the LD cycle, must "stretch" the cdc successively day by day to account for these very long infradian cell cycles. Just such a situation was observed in photoautotrophic cultures of *Euglena* entrained by *LD: 8,16* (Fig. 7), in which the doubling time for the population was about 36 hours ($ss \simeq 1.68$), as well as for the longer interdivision intervals in high fre- quency cycles (Edmunds and Funch, 1969b; see Fig. 9A), random illumination regimes (Edmunds and Funch, 1969a,b; see Fig. 9B), dim LL (Fig. 11), and other conditions in which $g >> 24$ hours, but bursts of cell division in the population occur predictably* with circadian periods. Finally, in *LD: 12,36*, in which a doubling occurs during the first 8–10 hours of darkness every 48 hours (effectively yielding $g = 48$ hours), long plateaus of some 36-hours duration occur during which no cell number increase is observed (Edmunds, 1971).

On the other hand, the phased division in *LD: 10,14* for the P_4ZUL mutant (and others) cultured at 19°C (Fig. 12) yielded rather different growth curves. In this case, although divisions were set back to 8–10 hours (plateaus), the increases in cell number, when they did occur, took place at a rate ($g \simeq 14$–16 hours) greater than that found for asynchronous, exponential growth (not shown) in LL or DD ($g \simeq 24$–26 hours), as reflected in the increased slopes of the growth steps, which were markedly loglinear as compared with the sigmoidal steps obtained with photoautotrophic cultures of the Z strain (Figs. 5 and 7). The net result of this compensatory process was that a doubling of cell number ($ss \simeq 2.0$) still occurred every 24 hours, and cell concentration attained approximately the same value as it would have by uninterrupted growth in DD or LL (without prior entrainment), at the same time conserving the developmental asynchrony† nor- mally characterizing such exponential growth. This conservation of developmen- tal asynchrony rules out any simple model employing a localized block or transi- tion point in a determinate sequence (Mitchison, 1971, p. 220) behind which cells would be expected to accumulate in synchrony, and the paradox can be resolved only if it is assumed that the simultaneous prolongation of the individual cdc's by 8–10 hours, once every circadian cycle, is accompanied by the reduc- tion or complete suppression of any G_1 variability. A similar analysis can be made for the persisting rhythm observed in LL (or DD) following a D–L transition (Fig. 15), as well as for curves found for low-temperature pulses (13°C) of various durations imposed on exponentially increasing, photoautotrophic cul- tures of *Euglena* maintained in LL at 25°C (K. Adams, unpublished).

---

*The timing of the bursts can be calculated from the expression $t_d = k + n\bar{\tau}$, where t_d is the midpoint of the observed fission burst, k is a constant (empirically determined to be about 12 hours after the onset of light in the prior entraining LD cycle), n is an integer denoting the nth free-running (or entrained) circadian cycle in LL or DD (or LD), and $\bar{\tau}$ is the free-running (or entrained) period of the cell division rhythm as exhibited by the population (see Edmunds, 1966,1978).

†In this case, the phased division observed in DD would not be truly "synchronous" in the customary sense, despite the discontinuous doubling patterns.

All of these observations on synchronized cultures of *Euglena*, then, appear to demand that individual cdc's be either delayed (prolonged) or advanced (shortened), or both, in order to cluster cell divisions (and probably other events of the cdc) at periodic intervals separated by longer timespans during which no cell division takes place. Since this gating is observed for many days under conditions held constant with respect to illumination (LL or DD), temperature, and other variables (see Section III,C), the master timer or oscillator hypothesized to underlie the cdc (at least in the infradian growth mode) must perform this function, even on probabilistic models for cdc control.

B. RHYTHMS OF CELL MOTILITY

Pohl (1948) first demonstrated a circadian rhythm of phototactic response in nondividing (stationary-, or infradian-phase) populations of autotrophically grown *E. gracilis* Klebs. The rhythm showed a maximal response during the day and a minimal response at night. Since then, this and associated motility rhythms in *Euglena* have been intensively investigated.

1. Rhythm of Phototaxis*

Bruce and Pittendrigh (1956,1958) introduced a high degree of automation into the assay system that allowed simultaneous recording from a number of different cultures in small, transparent Falcon flasks, which were placed in cabinets furnished with timers capable of imposing LD and temperature cycles on the culture. Each cabinet contained a small incandescent "day lamp" used to establish LD cycles and to elicit phase shifts, and a microscope "test lamp" located beneath the culture and carefully focussed, which was used to assay the phototactic response.* (See also Chapter 5 of this volume). The latter provided a narrow, vertical beam of light intercepting only about 2% of the culture volume as it passed through the algal suspension after passing through some heat-absorbing glass filters. This light, typically turned on about 30 minutes every 2 hours, was proportionately diminished by the motile cells and then fell on a photocell placed on the opposite side of the culture; the output current from the

---

*This system is complicated by the fact that the rhythm can be assayed only by repeated short exposures to light (i.e., the "test light") so that the usual constant DD conditions do not exist. Indeed, unless the test lamp was turned on at least 10 minutes every 2 hours, the rhythm rapidly damped out (Bruce and Pittendrigh, 1956). It appears that sufficient energy is required for photosynthesis and normal maintenance operations and turnover. The fact that the rhythm persisted in DD with a period not exactly equal to 24 hours or to an integral multiple of the period tends to rule out the possibility of entrainment by the test light by frequency demultiplication and to support the endogenous nature of the rhythm. This finding that the test lamp apparently did not affect either the phase or the period of the rhythm was further supported by the results of Feldman (1967) and Brinkmann (1966).

latter was automatically recorded on a chart that moved only when the test light
was on. The test light, therefore, served the following two functions: (a) to act as
the attracting stimulus that elicited the phototactic response, and (b) to measure,
by virtue of the transmitted light, the number of cells attracted and the speed of
their attraction. The recorded pattern was a measure of the aggregation of
Euglena cells in the beam and was termed the *phototactic response*, which is
actually a composite of both general motile and light-oriented behaviors.

The pattern of phototactic response exhibited by a culture of *Euglena* during
synchronous, autotrophic growth (Edmunds unpublished; see Edmunds and Halberg, 1981) in *LD: 14*,10 is illustrated in Fig. 21. It is evident that the response
to the test light was minimal during the dark period; furthermore, the response
began to decrease well before the actual transition from light to dark, as if the
cells anticipated its coming. An alternative explanation, however, would seem
possible for the observed pattern of rhythmic response. Since the cells all divide
in the dark period during this regime (Edmunds, 1965a; see Fig. 5), during which
time the flagellum itself is replicated, one might argue that the cells are unable to
swim actively toward the light and settle toward the bottom, thereby generating
the apparent minimal response. This hypothesis was effectively ruled out by the
observations of Bruce and Pittendrigh (1956, 1958) and Feldman (1967), which
indicated that (a) the rhythm continues to be entrained to a 24-hour period by *LD:
12*,12, although the population had entered the stationary phase where little, if
any, net increase in cell number occurred; and that (b) the rhythm free-ran with a
circadian periodicity (ranging from 23.6–24.3 hours) for as long as 14 days in

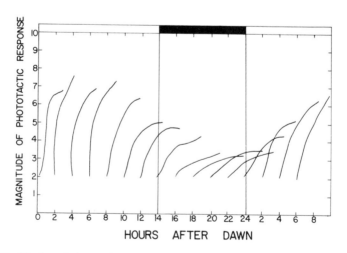

Fig. 21. Rhythm of phototactic response of *Euglena* during synchronized photoautotrophic
growth in *LD: 10*,14. The rhythm persists with a circadian period in DD (L. N. Edmunds, unpublished data; Edmunds and Halberg, 1981).

DD (except for the test light), under which conditions essentially no net increase in cell number was found (see Fig. 23a). These results demonstrate, then, that the postulated circadian oscillation underlying the response is conceptually distinct from the cell developmental cycle.

The free-running period of the rhythm of photoaxis in *Euglena* is seemingly independent of temperature, or nearly so. Bruce and Pittendrigh (1956) maintained cultures at 16.7°, 18.5°, 23.0°, 26.0°, and 33.0°C and found $\bar{\tau}$ to be 26.2, 25.5, 23.5, 23.8, and 23.2 hours, respectively. The Q_{10} over this 16°C range in temperature was calculated to be 1.01 to 1.10. This virtual temperature-independence in a unicellular clock, was considered to be achieved by virtue of a temperature-compensating mechanism within the organism. Component parts of the clock, however, might well be temperature-dependent, as suggested by the demonstration (Bruce, 1960) that an approximately sinusoidal temperature cycle (18°–31°C) can modify the phase relation between the rhythm and a simultaneously imposed *LD: 12,12* cycle in a manner dependent upon the phase angle difference between the two *Zeitgeber*.

Bruce (1960) investigated the entrainment of the *Euglena* phototactic rhythm by 24-hour LD cycles having various photofractions (i.e., ratio of light duration to entire cycle length). He found that in *LD: 2,22; LD: 4,20; LD: 6,18; LD: 8,16; LD: 10,14; LD: 12,12; LD: 14,10; LD: 16,8; LD: 18,6;* and in *LD: 20, 4* (with photofractions = 1/12, 1/6, 1/4, 1/3, 5/12, 1/2, 7/12, 2/3, 3/4, and 5/6, respectively), the rhythm was entrained to a precise 24-hour period. Furthermore, the phase of the rhythm depended chiefly on the light–dark transition (where the phase reference point was taken as the onset of the typical daytime phototactic response). For photofractions greater than 5/6 (i.e., cycles having more than 20 hours of light) the rhythm damped out.

Bruce and Pittendrigh (1956, 1957) also found that the rhythm would entrain to *LD: 3, 3; LD: 8,8;* and *LD: 24,24,* where $T = 6, 16,$ and 48 hours, respectively. It therefore appears that the rhythm has unusually wide limits of entrainment. On the other hand, the system exhibited frequency demultiplication to a 24-hour period when exposed to *LD: 2,10; LD: 12,36;* and *LD: 15,33* regimes, all of which consist of relatively short light periods followed by long dark periods. Indeed, as Bruce (1960) points out, there is always the possibility that for rhythms in which a population of individuals is assayed, what might appear to be entrainment to a short period (for example, 6 hours as was found for *LD: 3,3*) might in fact be frequency demultiplication of *separate subpopulations* to 24-hour periods but with differing phase relationships (for example, four subgroups, each 90° out of phase with the next). In most cases, however, this did not seem to occur, for when the rhythm was allowed to persist in DD after entrainment to the high frequency cycle, the period was usually circadian.

Bruce and Pittendrigh (1958) and Feldman (1967) have demonstrated that the free-running *Euglena* phototactic rhythm in DD can be reset by light pulses of

12, 8, 4, and 2 hours duration. As has been found for higher organisms, the magnitude of the phase shift ($\Delta\phi$) depends on the phase of the internal oscillation at which the pulse is given and on the duration of the pulse. Inasmuch as 8- and 12-hour signals were objectionably long, whereas the amplitude of the curve for 2-hour pulses was quite small, 4-hour perturbations were used (although the general shape of the curves was similar); a phase response curve for such 4-hour signals given on the second day of DD free run (Feldman, 1967) is shown in Fig. 22. There was essentially no phase shift if the pulse was given during the early subjective day (the first half of the normalized circadian cycle); greater and greater delays ($-\Delta\phi$) resulted with progression from the late subjective night; and there was a short time during the late subjective night when slight advances ($+\Delta\phi$) occurred. This daily fluctuation in the sensitivity to light could be interpreted as a defining property of the postulated underlying endogenous oscillation

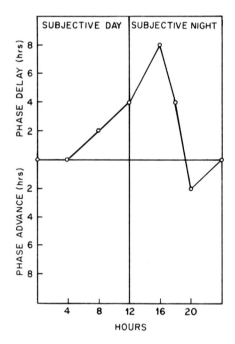

Fig. 22. Phase-response curve for the rhythm of phototaxis in *Euglena* for 4-hour light pulses given on the second day of DD. Points indicate the time of the beginning of the light pulse. Each point is based on the results of at least four independent experiments (except hour 18, for which there is only one experiment). Abscissa, circadian time. Circadian time 0 (subjective dawn) is the time when the lights would have come on the first day of DD if the previous light–dark cycle had been continued. The first half of the circadian cycle is called the subjective day and the second half is called the subjective night. (From Feldman, 1967; reproduced by permission.)

and could therefore be used to assay the phase of the clock itself (as contrasted with simply observing the overt rhythm of phototaxis).

The phototactic rhythm in *Euglena* has been subjected to treatment by a variety of chemical substances, such as the respiratory inhibitor KCN, the mitotic inhibitor phenylurethane, the adenine growth factor analogue 2,6-diaminopurine sulfate, the pyrimidine and nucleic acid analogue 2-amino-4-methylpyrimidine, adenine, guanine, thymine, cytosine, and uracil, and the growth factors gibberelic acid and kinetin, without consistently affecting the phase or period (Bruce and Pittendrigh, 1960).

Deuterium oxide (D_2O), or "heavy water," which has been shown to reversibly lengthen τ in a variety of organisms, ranging from unicellular algae to sandy beach isopods, fruit flies, pigeons, mice, and higher plants, and which also slows down several ultradian (high frequency) biological rhythms having periods in the millisecond range (see review by Enright, 1971), similarly alters both the period and the phase of the *Euglena* phototactic rhythm (Bruce and Pittendrigh, 1960). Cultures adapted to 20% D_2O showed no alteration in period, but the period of those maintained in 25% D_2O for 3 weeks and then in 45% for 1 month increased to 27.0 hours; one culture living in 95% D_2O increased its period to 26.6 hours, whereas a replicate increased to 27.7 hours. Obviously there was no correspondence between concentration of D_2O and the period lengthening. When cultures were readapted back to H_2O, the period of the persistent phototactic rhythm returned to about 24 hours again. As Enright (1971) has speculated, heavy water may retard biological rhythms by influencing the ionic balance across the cellular membrane: alteration of nerve activity of higher animals and of related membrane processes of simpler organisms. Inasmuch as high frequency rhythms originate in pacemakers dependent on diffusion processes involving ion exchange, it is quite possible that biological clocks having longer periods (e.g., circadian rhythms) are also based on diffusion-dependent pacemakers.

Distressing in all such studies, however, is the fact that the action of D_2O is quite general and nonspecific. Although its period lengthening effects are consistent with notions of membrane oscillators, they could favor equally well kinetic models, in which rate constants of elementary reactions directly determine the circadian period, or even transcriptional, tapereading models. [For example, time-consuming diffusion circuits between nucleus and cytoplasm are essential elements of Ehret and Trucco's (1967) chronon model for circadian timekeeping in two-compartment, two-envelope eukaryotic systems.] Indeed, Kreuels and Brinkmann (1979) have found in a comparative study of the effects of D_2O on the cell-bound circadian oscillator underlying the motility rhythm in *Euglena*, on the cell-free glycolytic oscillator of yeast, and on the Belousov–Zhabotinsky chemical reaction having a known network structure (Kreuels *et al.*, 1979), that although the circadian and glycolytic oscillations were slowed down to an extent depending on the D_2O concentration, the period of the chemical reaction was either lengthened or shortened, respectively, at high or low catalyst concentrations. They conclude, therefore, that the generalized period-lengthening effect of D_2O does not demand a distinct membrane model (or a simple kinetic model, for that matter) but rather some sort of more complex network approach.

More specific (but still not well understood) biochemical perturbations of the

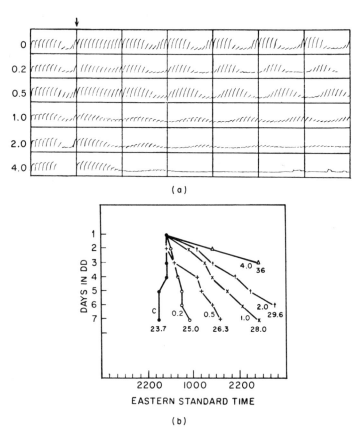

(a)

(b)

Fig. 23. Period lengthening of the free-running rhythm of phototaxis in *Euglena* by cycloheximide. (a) Photograph of original records of phototactic response of cultures to which cycloheximide was added at subjective dawn of the second day of DD (indicated by the arrow). The numbers at the left are the concentrations (μg/ml) of drug added to each culture; control (C), no drug added. Vertical lines, 24 hours apart, indicate 1000 EST. (b) The clock hours at which successive minima occurred are plotted for successive days for each of the cultures shown in (a). The numbers next to the data lines indicate the concentrations (μg/ml) of drug added; C, control. The numbers below the data lines indicate the period length (hours). There is some slight variability (up to about 1 hour) in the exact amount of lengthening of the period induced by the highest concentrations of the drug (2 and 4 μg/ml). (From Feldman, 1967; reproduced by permission.)

circadian phototactic rhythm in *Euglena* have been obtained by altering the nutritional conditions. Thus, Feldman and Bruce (1972) have reported that acetate (100 m*M*) added to autotrophic *E. gracilis* cultures caused an increase in the period length of the circadian rhythm of phototaxis to 27 hours. In addition, acetate "pulses" (10 m*M*) administered at different phases of the circadian rhythm of phototaxis induced phase shifts. In fact, a 36-hour exposure resulted in a nearly arrhythmic culture. Other carbon sources added as a pulse (succinate, lactate, and pyruvate) caused first a temporary cessation of the rhythm and then a resumption with variable phase shifts. Feldman and Bruce (1972) argue that, since each carbon source is effective by itself, the changes are probably caused

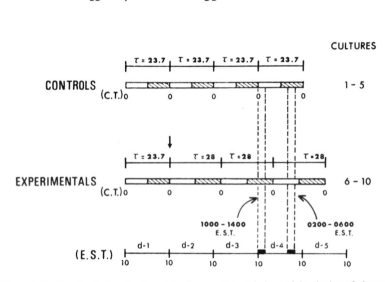

Fig. 24. Testing for the lengthening of the free-running period (τ) of the rhythm of phototactic response in *Euglena* after the addition of cycloheximide in constant darkness. The phase of the underlying clock was assayed by measuring the phase-resetting effects of 4-hour light pulses in reference to those predicted by the phase-response curve determined previously for this rhythm. The bottom line indicates Eastern Standard Time (i.e., absolute elapsed time); in the prior light cycle, lights were on from 1000 to 2000 EST; d-1, d-2, etc, indicate successive days in constant darkness. The bar labeled "controls" indicates circadian times (*C.T.*) of subjective day (open bars) and subjective night (hatched bars) of the control cultures; similar notation for the "experimentals" in which cycloheximide was added at subjective dawn of the second day of DD (arrow). Solid blocks on the EST scale at day 4 of DD give times at which each of the two 4-hour light pulses were administered to separate cultures. Note that the first pulse strikes the control cultures in their early subjective day, whereas the second pulse hits during the subjective night; but the same pulses strike the experimentals with the reverse-phase relationship, indicating that the phase had been changed due to the lengthening of the period of each cycle by the inhibitor. (From Feldman, 1967; reproduced by permission.)

by a general metabolic switch (i.e., from autotrophy to mixotrophy) rather than by any specific effect of the carbon source. These results are also compatible with the suggestion by Brinkmann (1966) that a phase shift may occur when cells are induced to make a transition from one metabolic state to another, particularly if the component reactions of each ensemble (e.g., photochemical events or dark reactions) are characterized by slightly different Q_{10} values in response to changes in steady-state temperature. Feldman and Bruce (1972) caution, however, that since the Q_{10} of a given dark reaction (such as protein synthesis) can differ in the two states, temperature experiments alone do not suffice for deciding whether the pathways include photochemical reactions, dark reactions, or both.

Finally, and perhaps most specifically, the period of rhythm of phototactic response in *Euglena* has been shown to increase (reversibly) by the addition of the strong inhibitor of protein synthesis, cycloheximide (actidione), as illustrated in Fig. 23 (Feldman, 1967). Furthermore, the effects of this drug appear to be on the clock itself rather than on some parameter controlled by the clock or on some "uncoupling" mechanism between the clock and the parameter (Feldman, 1967). This was confirmed (Fig. 24) by experiments that assayed the position of the light-sensitive oscillation by 4-hour resetting light signals after cycloheximide addition in DD, as predicted from the phase responsive curve (Fig. 22) already determined for this unicell. Yet even these relatively "clean" results tending to implicate protein synthesis in circadian clock function must be interpreted cautiously (Feldman, 1967): cycloheximide not only may have other primary effects besides the inhibition of protein synthesis, but may also produce secondary metabolic effects such as alteration of pools of intermediates, energy levels, or membrane composition that would modify only indirectly the circadian oscillator (Sargent, 1976).

2. Dark Motility Rhythm (*Dunkelbeweglichkeit*)

In stationary infradian cultures of *E. gracilis* (strain 1224-5/9), a diffusion gradient in the vertical distribution of cells as indicated by differences in the sedimentation equilibrium has been observed (Brinkmann, 1966, 1971; Schnabel, 1968). This gradient is apparently a function of random cell mobility, which itself fluctuates with a circadian periodicity that will persist for as long as 3 months in DD (see Fig. 25). The dark motility rhythm (*Dunkelbeweglichkeit*) damps out in LL but may be induced again by a single transition from LL to DD. The assaying test-light cycle (*LD: 20′,100′*) did not entrain the circadian rhythm either directly or by frequency demultiplication (Schnabel, 1968).

In some respects, this dark motility rhythm is more appropriate than the phototactic response previously described for determining the effects of light on the hypothesized underlying circadian oscillator itself. In the first place, the random motility rhythm undoubtedly contributed to rhythm of phototactic re-

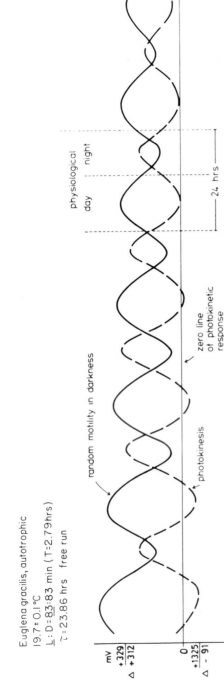

Fig. 25. An original trace of the random motility rhythm in DD ($\bar{\tau}$ = 23.9 hours) and the photokinetic response in *Euglena* cultured autotrophically at 19.7°C and assayed with an *LD: 83′, 83′* cycle (*T* = 2.8 hours). The photokinetic response smoothly passes the border of the positive and negative portions of the random motility curve several times, with an increasing and then decreasing level of reactivity. This may suggest the operation of a threshold system where the threshold itself fluctuates with the circadian cycle, as opposed to different reaction mechanisms for positive and negative photokinesis. (Unpublished data, courtesy of K. Brinkmann, personal communication, 1980).

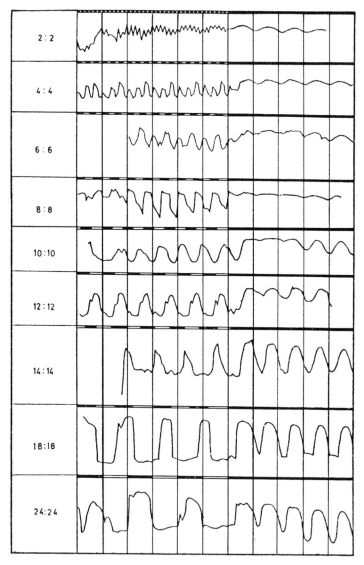

Fig. 26. Effects of light cycles of different period lengths on the circadian rhythm of random motility in autotrophic cultures of *E. gracilis* in the "stationary" (infradian) growth phase. (From Schnabel, 1968; reproduced by permission.)

sponse measured by Bruce and Pittendrigh (1956, 1958) during stationary conditions in DD (Fig. 25). Also, with the latter it is difficult to distinguish between the influence of light on the oscillator on the one hand, and on the response itself, on the other. In contrast, the assay for the dark motility rhythm is independent of phototaxis and photokinesis.

Schnabel (1968) examined the effects of a wide variety of LD cycles upon the motility rhythm in mixotrophic and in autotrophic cultures. She discovered that LD: *24,24*; LD: *18,18*; LD: *14,14*; LD: *12,12*; LD: *10,10*; LD: *8,8* and perhaps LD: *6,6* cycles entrained the rhythm to periods of 48, 36, 28, 24, 20, 16, and 12 hours, respectively (Fig. 26). Upon release of the cultures from these *Zeitgeber* into constant conditions (DD), the rhythm reverted to a circadian periodicity. Thus, the range of entrainment was limited to a *Zeitgeber* period of about 16 hours for LD ratios of 1:1. With shorter *Zeitgeber* periods (for example, LD: *2,2*

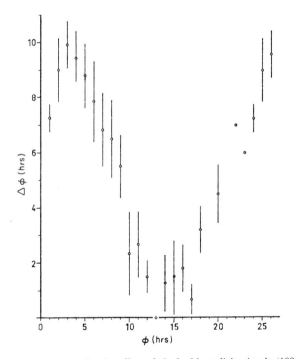

Fig. 27. Phase-response curve for the effect of single 6-hour light signals (1000 lux) on the circadian rhythm of random motility in autotrophic cultures of *E. gracilis* in the infradian growth phase. Abscissa: hours after the minimum point of the free-running rhythm at which the light pulse was given. Ordinate: phase shift, in hours, after reestablishment of the previous frequency. The final phase difference between the shifted rhythm and the control is plotted, neglecting the sign of the phase shift. (Schnabel, 1968; reproduced by permission.)

and *LD: 2,4*), the circadian component also was exhibited; she found no evidence of frequency demultiplication. A close examination of her figures, however, reveals that the circadian envelope of the rhythm in these high frequency LD cycles is itself apparently modulated by short, sharp peaks which approximately correspond with the dark periods. A colorless, obligatorily heterotrophic mutant (*E. gracilis,* strain 1224–5/25) also exhibits a light-entrainable circadian rhythm of random motility in DD just as the green mixotrophic and autotrophic cultures did (Kirschstein, 1969). Finally, Schnabel (1968) has determined the phase response curve for this rhythm using 6-hour light pulses (1000 lux) in free-running conditions (DD): Allowing for different period lengths, the response curves are nearly identical in both autotrophic (Fig. 27) and mixotrophic green cultures. Few, if any, transients occur.

The range of entrainment of the motility rhythm by sinusoidal temperature cycles having driving periods varying from 4.8 to 55.7 hours has been explored

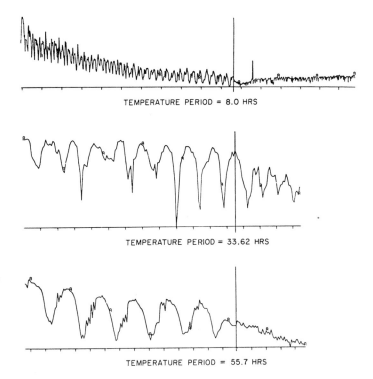

Fig. 28. Effects of sinusoidal temperature cycles of different lengths on the circadian rhythm of random motility in autotrophic cultures of *E. gracilis.* Horizontal: days. Vertical line: time at which cultures were released from the temperature regime into constant temperature. (Unpublished data, courtesy of T. Kreuels and K. Brinkmann, personal communication, 1980.)

also (unpublished data, T. Kreuels and K. Brinkmann, personal communication, 1980). There were several interesting findings: (1) All temperature cycles used synchronized the overt biological rhythm. (2) The synchronized rhythm reverted to its free-running circadian period upon removal of the temperature cycle only when the period of the latter had been in the range of approximately 19–36 hours. Although passive enforcement did occur with temperature cycles having periods of 4.8–19 hours, or 36–55.7 hours, no rhythmicity was observed in the subsequent free run (Fig. 28). This is in contrast (Fig. 26) to the circadian period always seen in the free run following the imposition of very short or long LD cycles (Schnabel, 1968). (3) The phase difference between the temperature cycle and the synchronized rhythm itself changed within the range of entrainment (19–36 hours) but was more or less constant in temperature cycles having periods greater than 36 hours.

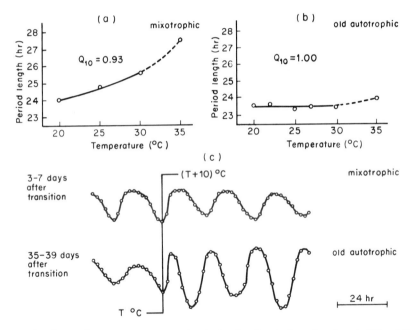

Fig. 29. Types of temperature compensation in the circadian rhythm of random motility of *E. gracilis.* In the two upper graphs (a) and (b) the free-running frequencies in young mixotrophic and old autotrophic cultures are compared as a function of different constant temperatures. The lower graphs (c) illustrate the phase response of the two cultures after a single temperature jump of + 10°C given at a minimum of the cycle. The mixotrophic culture represents the frequency-sensitive type of temperature compensation, and the old autotrophic culture represents the phase-sensitive type of temperature compensation. "Transition" means transition from LD: *12,*12 to the test-light program of LD: *20′,*100′. (From Brinkmann, 1971; reproduced from "Biochronometry," 1971, with the permission of the National Academy of Sciences, Washington, D.C.)

It is interesting to compare the effects of different constant temperatures on the free-running period of the motility rhythm of *Euglena* cultures grown under different modes of nutrition. Brinkmann (1966, 1971) has reported that under autotrophic conditions (minimal medium) the free-running period in DD was relatively independent of temperature (23.5 ± 0.3 hours) in the range between $15°$ and $35°C$, whereas a sudden temperature increase of $5°C$ or more caused a transitory increase (Fig. 29). In contrast, in mixotrophic culture (medium containing peptone, glucose, and citric and malonic acids), τ increased with increasing constant temperature, whereas the phase was not affected by sudden temperature changes. The phase response curve for a single $10°C$ temperature step-up given at various circadian times to a free-running, old autotrophic culture has been determined (Brinkmann, 1966, 1971); it was remarkably similar to that found for 6-hour light pulses (Schnabel, 1968; see Fig. 27). Brinkmann (1966, 1971) has suggested that the different behavior of the two types of cultures in response to different steady-state temperatures results from participation in the autotrophic oscillations of two types of reactions, those with a Q_{10} of 1 (photochemical) and those with a Q_{10} of 2 (dark reactions), whereas in mixotrophic cells the dark reactions predominate. It now appears (K. Brinkmann, personal communication, 1980) that the degree of dependence of energy conservation on temperature in turn determines the type of response of the circadian oscillation to

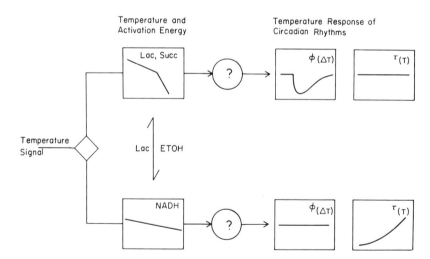

Fig. 30. Metabolic switching between the two types of response to temperature step-ups [$\phi_{(\Delta T)}$] and changes in steady-state constant temperature [$\tau_{(T)}$] as correlated with the temperature-dependence of energy conservation in cultures of *E. gracilis* by the addition of lactate (Lac) or ethanol (ETOH). Arrhenius plots for the oxidation of lactate and succinate (Succ) or NADH (see left insets) were obtained with isolated mitochondria; activation energies were 15–25 kcal/mole and 10 kcal/mole, respectively. (Unpublished data, courtesy of K. Brinkmann, personal communication, 1980.)

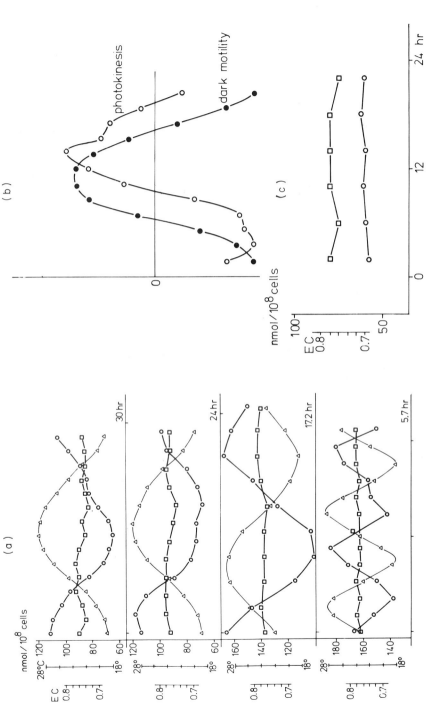

Fig. 31. Energy charge, EC(□) and glucose-6-phosphate (○) concentration during (a) various sinusoidal temperature cycles (△) and during a free run at 27.5°C (c) when circadian rhythms (τ = 23 hours) of photokinesis (○) and dark motility (●) are exhibited (b) in autotrophic cultures of *Euglena*. (Courtesy of K. Brinkmann, personal communication, 1980.)

different steady-state temperatures. Thus, the Arrhenius plots for the oxidation of lactate and succinate, or of NADH, have been obtained with isolated mitochondria; the activation energies for the former were 15 to 25 kcal/mole, and 10 kcal/mole for the latter. The type of response to temperature steps-up or to different steady-state constant temperatures can be switched from one to the other by the addition of either lactate or ethanol to the medium (Fig. 30).

Finally, several other circadian rhythms that may have some bearing on the

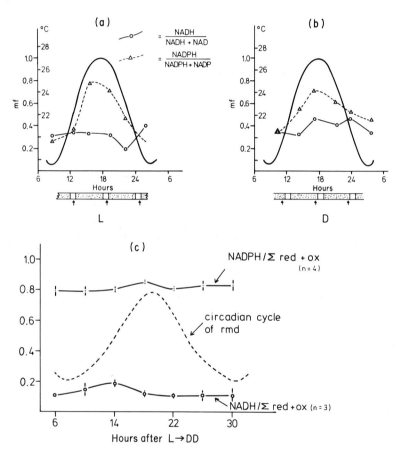

Fig. 32. Enforced oscillations in the redox state of NADP (a and b) but not of NAD, by an 18-hour sinusoidal temperature cycle in late-stationary, mixotrophic (Glu–malate) cultures of *E. gracilis* grown at 21°C in LD: *20',100'*. The amplitude of the curve is greater when the cultures were assayed at the end of the light pulses (a) than at the end of the dark intervals (b). Solid curves: circadian cycle of random dark motility. (c) No significant free-running rhythm in the redox state of pyrimidines at constant temperature, despite the presence of a circadian cycle of random dark motility (dashed curve). (Courtesy of K. Brinkmann, personal communication, 1980.)

random motility rhythm have been investigated recently. Thus, Brinkmann and co-workers have measured the energy charge (EC) and the concentration of glucose-6-phosphate (G6P) under various sinusoidal temperature cycles within and beyond the limits of entrainment for the motility rhythm: in contrast to G6P, EC never synchronized (Fig. 31a). On the other hand, neither EC nor G6P showed rhythmicity in autotrophic cultures during a free run at a constant 27.5°C (Fig. 31c), indicating that these variables cannot be directly responsible for generating the overt circadian rhythm. Similarly, although the redox state of NADP (but not NAD) are synchronized by a temperature cycle (Fig. 32a), there is no significant oscillation in the redox state of pyrimidine nucleotides during a free run at constant temperature. The temperature-enforced NADPH cycle shows a peak coinciding with low G6P, indicating flux regulation of the pentose-phosphate cycle in response to temperature (K. Brinkmann, personal communication, 1980). Recently, Kreuels and Brinkmann (personal communication, 1980) have found that the circadian period of the motility rhythm can be lengthened significantly by the addition of urea to the culture (Fig. 33), the effect

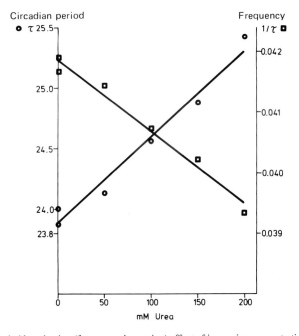

Fig. 33. Period lengthening (frequency-decreasing) effect of increasing concentrations of urea on the free-running circadian rhythm of random dark motility in *Euglena*. Single data records (10–14 days each). Periods (frequencies) calculated via periodograms. (T. Kreuels and K. Brinkmann, personal communication, 1980.)

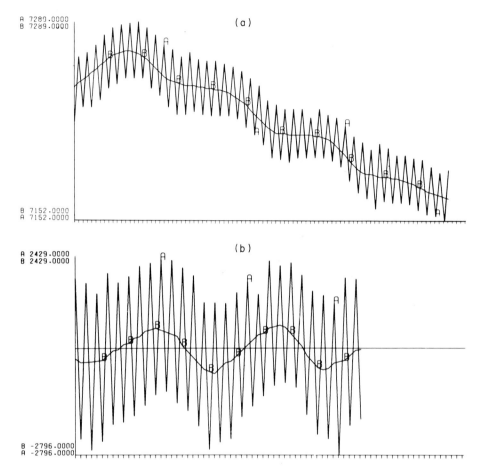

Fig. 34. Circadian rhythm of external pH in weakly buffered (30 μM phosphate) suspensions of *E. gracilis* autotrophically cultured at 20°C in *LD: 1,1* following an *LD: 12,12* cycle (the traces begin at the end of the last main photoperiod). (a) The original computer record showing direct responses (A) to the short light intervals (pH span 7.152–7.289) was put through a low-pass filter (solid line B) to remove the direct light response. (b) High-pass filtered version to remove trend remaining after low-pass filtering, to show the circadian component of the low-amplitude pH changes. (Courtesy of K. Brinkmann, personal communication, 1980.)

being linear with concentration both in the period and frequency domain. Saturation was not obtained even with urea concentrations as high as 200 m*M;* above this concentration, no regular oscillations occurred, even though *Euglena* could grow in medium containing 400 m*M* urea. Urea was shown to penetrate the cells but not to be metabolized at all, suggesting that the period-lengthening effect perhaps was due to structural changes (possibly by affecting the hydrogen bonds of macromolecules or water structures, or by weakening hydrophobic interactions). Finally, a low-amplitude circadian oscillation of external pH in weakly buffered suspensions of *E. gracilis* has been reported (Fig. 34). Preliminary evidence (K. Brinkmann, personal communication, 1980) suggests that this rhythm is due to the activity of a plasmalemma proton pump and is not generated by the inverse-phase, circadian rhythm of photosynthesis (O_2 evolution). A similar conclusion was drawn by Hoffmans and Brinkmann (1979) for *Chlamydomonas reinhardii*, whose photosynthetic activity inhibited by DCMU [3-(3,4-dichlorophenyl)-1,1-dimethyl urea] still evidenced a pH rhythm with an unchanged period.

3. *Rhythm of Cell "Settling"*

In addition to the light-oriented phototactic response and the dark random-motility rhythm discussed in the previous two subsections, we have discovered (Terry and Edmunds 1970a,b) a 24-hour rhythm of cell "settling" in *Euglena*, which may occur either concurrently with, or in the absence of cell division. This rhythm was reflected by an apparent change in cell concentration during the growth phase of populations of *Euglena* autotrophically batch-cultured in LL and synchronized by 24-hour temperature cycles alternating either 18°/25°C or 28°/35°C (Terry and Edmunds, 1970a), although the heat-synchronized division bursts in the population tended to mask the response. This was indicated by small dips that appear in the plateau portion of the synchronous growth curve. In such cases, the motility rhythm maintained the same phase angle difference in relation to either entraining temperature cycle (i.e., motility maxima occurred during the warmer period) regardless of whether or not this was the period of maximum cell division, whereas the cell division rhythm had a 180° phase difference between the 18°/25°C and 28°/35°C regimes.

In temperature-cycled, stationary cultures, the maxima of the rhythm still always occurred during the warmer phase of the temperature cycle. Stationary cultures in constant temperature sometimes did not exhibit the settling rhythm but did maintain a constant cell population for many days, indicating that cell death (necessarily synchronous) was not responsible for the rhythmic dips observed. Rather, the cells actually tended to settle out of the liquid phase and adhered to the vessel walls and then subsequently detached themselves and reentered the medium. The cultures were magnetically stirred so that the distribution of the cells in the medium was homogeneous; this was further substantiated by automatically monitoring cell number at two different levels in the vessel by a dual sampling system (Terry and Edmunds, 1970b). The cultures were held in LL (7500 lux) throughout the experiments; this intensity of constant illumination would normally be expected to cause the rhythms of phototactic response and dark mobility to damp out (Bruce and Pittendrigh, 1956; Brinkmann, 1966).

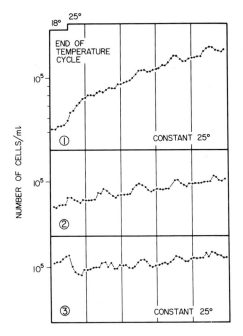

Fig. 35. Persisting circadian rhythm in cell "settling" in autotrophic cultures of *E. gracilis* during the infradian growth phase. Successive sections, reading from top to bottom, indicate the apparent cell concentration each day following transfer of the culture from an *18°/25°C: 12,12* temperature cycle to a constant 25°C in LL. (The step in the top line of the figure at the left side gives the time of the transition from the final 18°C interval of the temperature cycle to constant 25°C.) (From Terry and Edmunds, 1970b.)

Attachment could be prevented by vigorous agitation on a rotary shaker (Terry and Edmunds, 1970b).

The settling rhythm persisted (Fig. 35) for as long as 9 days in infradian cultures released from the 18°/25°C cycle and held at a constant 25°C (always in LL), strongly suggesting that the rhythm is endogenous (Terry and Edmunds, 1970b). Finally, we have evidence that the settling rhythm may also occur during both the growth and stationary phases of wild-type, P_4ZUL, and W_6ZHL cultures maintained in DD at 25°C following entrainment by a *LD: 10,14* cycle under organotrophic conditions (L. N. Edmunds, unpublished).

C. RHYTHMS OF PHOTOSYNTHETIC CAPACITY

Circadian rhythms in photosynthesis have been observed in a number of higher plants as well as in algae. Several studies have reported a rhythm of photosynthetic capacity during the cell cycle of *Euglena,* although the results were often

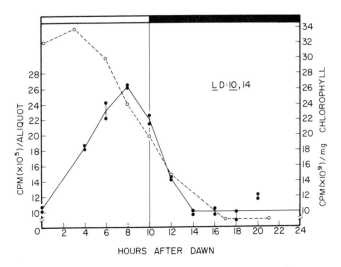

Fig. 36. Diurnal changes in photosynthetic capacity during the cell developmental cycle in auto-trophically batch-cultured *E. gracilis* synchronized by a *LD: 10*,14 at 25°C. Aliquots (10 ml) of the master culture were exposed to one μCi of NaH^{14}CO$_3$ for 10 minutes in saturating light and were then fixed. Photosynthetic capacity: per aliquot, solid line; per mg chlorophyll, dashed line. Double points indicate duplicate determinations. (From Walther and Edmunds, 1973.)

conflicting (Lövlie and Farfaglio, 1965; Cook, 1966; Codd and Merrett, 1971). Walther and Edmunds (1973) demonstrated a clear 24-hour rhythm in the capacity of *Euglena* to fix CO$_2$ in cultures synchronized by appropriately chosen light (12,000 lux) cycles (*LD: 10*,14) at 25° on minimal medium. Samples were then taken at various times and their ability to incorporate NaH^{14}CO$_3$ at saturating light was measured. Photosynthetic capacity (CO$_2$ fixation) was found to vary in a cyclic manner during the cell division cycle, reaching a peak 2 hours before the onset of darkness and then sharply decreasing to a basal level (Fig. 36). A similar change in O$_2$ evolution was seen (Walther and Edmunds, 1973). This diurnal rhythm also occurred in nondividing cultures in *LD: 10*,14 (Fig. 37) and could thus be divorced from the driving force of the cell division cycle. At the time, however, the rhythm was found to be only weakly persistent under continuous dim LL (750 lux) or in high frequency LD cycles [*LD: 2*,4 (12,000 lux)].

This lack of long-term persistence of the photosynthetic capacity rhythm and hence the inability to conclude that an endogenous circadian clock underlies the overt periodicity could have been due to the nature of the light conditions chosen. The 750-lux LL utilized completely suppressed cell division and, by implication, perturbed the overall physiology of the cell. This was reflected in the greatly decreased rate of CO$_2$ fixation observed, making the assay of the rhythm virtually impossible. Higher intensities of illumination could not be used, since circadian

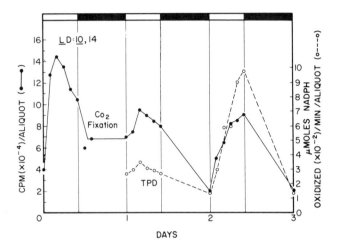

DAYS

Fig. 37. Oscillations (24-hour) in photosynthetic capacity (NaH^{14}CO$_3$ incorporation) and in the activity of glyceraldehyde-3-phosphate dehydrogenase activity (TPD) in nondividing stationary cultures of *Euglena* in *LD*: *10*, 14 at 25°C on minimal salt medium. (From Walther and Edmunds, 1973.)

rhythms often damp out under such conditions, and DD could hardly be employed in a phototrophic system. The attempt to circumvent this quandary by imposing a 6-hour light–dark cycle (*LD*: *2*, 4) was open to a similar shortcoming since the cells received only 8 hours of light (12,000 lux) in a 24-hour time span (in contrast to the 10 hours afforded by the synchronizing *LD*: *10*, 14 regime). In some instances, cultures exhibited direct entrainment to this shorter period (Walther and Edmunds, 1973).

More recently, we have managed to solve this problem by monitoring photosynthetic capacity (CO$_2$ fixation, PC) and chlorophyll content (Chl) in three strains of *E. gracilis* synchronized by an *LD*: *12*, 12 (7000 lux) cycle during both exponential growth and stationary phases and then released into an *LD*: *1/3*, 1/3 (7000 lux) regimen (Laval-Martin *et al.*, 1979). This particular high frequency, 40-minute cycle was selected so as to afford an amount of light during a 24-hour timespan identical to that received in the *LD*: *12*, 12 cycle. Direct entrainment of the cultures to a 40-minute period was not to be expected; the cell division rhythm in dividing cultures of *Euglena* is known to persist with a circadian period under these conditions (Edmunds and Funch, 1969a,b). For all intents and purposes, the high-frequency cycle is perceived by the cells as continuous illumination, at least with respect to their rhythmic output. Typical results are shown in the large panel of nine graphs comprising Fig. 38 for a DCMU-resistant strain (Z$_R$) of *Euglena* in both the exponential and infradian (stationary) growth phases; similar results obtain for the Z strain (Laval-Martin *et al.*, 1979).

From these results the following generalizations can be made: (a) there is a clear-cut 24-hour rhythm in CO$_2$ uptake in dividing cultures synchronized by *LD*:

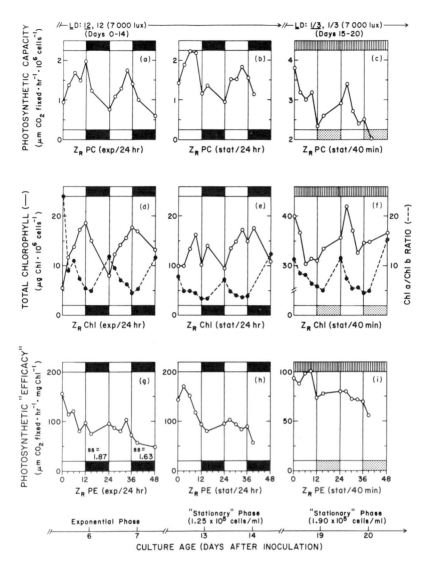

Fig. 38. Photosynthetic rhythms in photoautotrophically batch-cultured DCMU-resistant strain (Z_R) of *E. gracilis* (Z). Photosynthetic capacity (a–c), total chlorophyll (d–f), the Chl *a*/Chl *b* ratio (d–f), and photosynthetic efficacy (g–i) were monitored at various intervals during a 20-day time span after initial inoculation of the culture and are expressed in the units given on their respective ordinates. Each variable was measured under three sets of conditions: during the exponential phase of growth across one or more synchronous growth cycles in LD: *12,12* (7000 lux), exp/24hours: panels a, d, and g, left column (successive step sizes of 1.25 and 1.63); during the stationary (infradian, or slowly dividing) phase in LD: *12*, 12 (7000 lux), sta/24 hours: panels b, e, and h, middle column (cell concentration = 1.9×10^5 cells ml^{-1}); and finally during the stationary phase (cell concentration = 1.9×10^5 cells ml^{-1}) in a high frequency (40 min) LD: *1/3*,1/3 (7000 lux) cycle, stat/40 minutes: panels, c, f, and i, right column. The light regimen imposed is shown on an absolute time scale at the top of the figure; absolute culture age is given at the bottom. (From Laval-Martin *et al.*, 1979.)

12,12, confirming the earlier work of Walther and Edmunds (1973); (b) this rhythm is also clearly manifested by nondividing (infradian phase) cultures in *LD*: *12*,12; (c) a non-damped, high-amplitude, persisting circadian rhythm of photosynthetic capacity is apparent under the high-frequency cycle for at least some 5–6 days after the *LD*: *12*,12 cycle had been discontinued. The free-

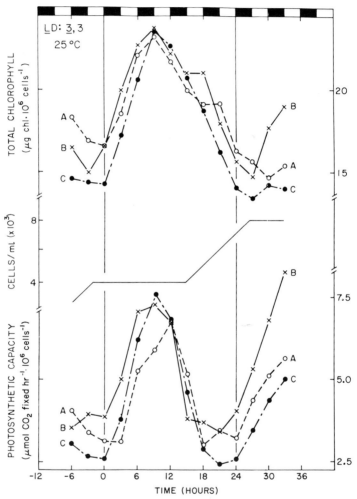

Fig. 39. "Free-running" circadian photosynthetic rhythms in *E. gracilis* (Z) maintained in a short-photoperiod *LD*: *3*,3 cycle during the exponential phase of growth. The data for each of the three replicate cultures (A,B,C) have been shifted in real time so that the peaks of the curves for total chlorophyll (Chl) are aligned in order to facilitate comparison. A generalized growth curve is also given in the center of the diagram. (Modified from Edmunds, 1980b; Edmunds and Laval-Martin, 1981.)

running period (τ) was estimated to be approximately 27.5 hours, corresponding quite closely to that of the cell division rhythm under rather similar conditions (Edmunds and Funch, 1969a,b).

That the LD: $20',20'$ high-frequency cycle is not unique, perhaps artifactually generating the observed circadian rhythm in PC, is demonstrated in Fig. 39. Measurements of PC, Chl, and cell number were made over a 40-hour timespan in three replicate, synchronously dividing cultures of *Euglena* cultured autotrophically in LD: $3,3$ (having a period of 6 hours) from the moment of inoculation for at least 1 week. All three variables evidenced a precise and reproducible circadian rhythm having a period of 27 to 29 hours. The rhythm in PC, as well as in total Chl, also occurs in nondividing or slowly dividing cultures in the infradian phase (Fig. 40), although there does not appear to be a precise corre-

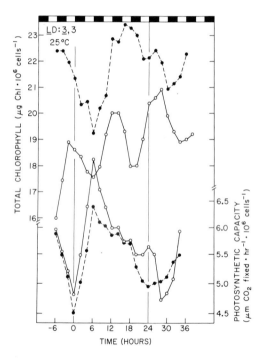

Fig. 40. Circadian rhythms of photosynthetic capacity and total chlorophyll content in "stationary"-(infradian) phase, autotrophic cultures of two different strains of *E. gracilis* [Z_A (●) and Z_B (○)] maintained in LD: $3,3$. The data for each of the strains have been shifted in real time so that the onsets of the increase in photosynthetic capacity are aligned (time 0) in order to facilitate comparison. All data are moving three-point averages (reducing the amplitude by about 30%). Note that the peaks in capacity and in total chlorophyll content do not correspond and that the free-running periods are not the same; indeed, strain Z_B appears to have either a bimodal circadian or an ultradian rhythm in chlorophyll content. (From Edmunds and Laval-Martin, 1981.)

spondence between the two rhythms (Edmunds and Laval-Martin, 1981). Thus, both LD: $20',20'$ and LD: $3,3$ elicit similar results, even if applied from the outset of the experiment, and the circadian rhythms observed can for all intents and purposes be described as free-running.

Lonergan and Sargent (1978) have also reported a circadian rhythm of O_2 evolution in *E. gracilis* (Z), which persisted for at least 5 days in dim LL and constant temperature (Fig. 41A) but damped out in bright LL. The phase of this rhythm could be shifted by a pulse of bright light, and the period length was unchanged over a 10°C span of growth temperature. Although the O_2 rhythm was found in both exponentially dividing and stationary phase cultures, CO_2 uptake was clearly rhythmic only in the latter. This may have been due to the sensitivity of the infrared CO_2 analyzer used for the assay.

Although it is beyond the scope of this paper to examine in-depth the biochemical and molecular bases of circadian clocks, it is perhaps useful to categorize at least four ways by which the rhythm in PC could be generated: (a) rhythmicity in the total number of cells in the culture or those carrying out photosynthesis; (b) rhythmicity in the total amount of chlorophyll species in the cell; (c) rhythmicity in the activity or concentration of enzymes regulating electron transport between the photosystems or dark reactions, or both; or (d) rhythmicity in the coupling between photochemical events within the photosynthetic units in the thylakoid membranes (Edmunds, 1980b).

The first, trivial explanation that variations in cell number might generate the rhythm in PC has been effectively eliminated by the demonstration (Walther and Edmunds, 1973; Lonergan and Sargent, 1978; Laval-Martin *et al.*, 1979) that a circadian oscillation is found in cultures of *Euglena* in which cell concentration is constant. Further, Sweeney (1960) has reported a rhythm in PC in single cells of *Gonyaulax*.

The second possibility—changing concentration of total chlorophyll—has more merit, although in most systems examined such variations have not been observed in the stationary growth phase. In *Euglena,* contrary to initial expectations (Walther and Edmunds, 1973), we have recently observed an endogenous circadian rhythm in total Chl in the absence of significant cell division in high frequency LD regimes (Laval-Martin *et al.*, 1979; Edmunds and Laval-Martin, 1981). These observations could reflect changes in the functional role of the pigment, but, in themselves, they are not sufficient to explain the rhythm in PC because (a) there was no quantitative relationship between the amplitudes of the rhythms of PC and Chl, and (b) in cultures maintained in LD: $1/3, 1/3$, the phase relationship between the rhythm of PC and that of Chl content varied, suggesting the possibility of desynchronization among different circadian rhythms driven by some master clock, or even the existence of a multioscillator system in a unicellular organism (Laval-Martin *et al.*, 1979; see Section III,C).

With regard to the third hypothesis (enzymatic variations), a number of the

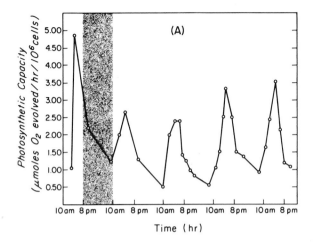

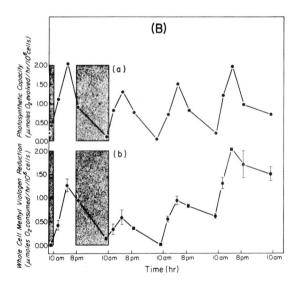

Fig. 41. (A) Persistence of photosynthetic capacity rhythm (O_2 evolution) in *E. gracilis* in constant conditions. The division-synchronized culture was grown in 10-hours of light and 14-hours of darkness (shade) and then exposed to constant dim light (2×10^3 ergs cm^{-2} sec^{-1}) at 25°C after the last dark period. (From Lonergan and Sargent, 1978.) (B) Photosynthetic capacity and whole chain photoelectron flow in intact cells: (a) photosynthetic capacity; (b) rate of whole chain electron flow as measured by a H_2O to methyl viologen assay in the presence of gramicidin D. Bars represent standard deviation for replicates. New cells were used for each replicate. (From Lonergan and Sargent, 1979; reproduced by permission of The American Society of Plant Physiologists.)

photosynthetic dark reactions comprising the Calvin scheme have been examined in several algal systems, both during the cell cycle and in nondividing cultures. To summarize, no enzyme has achieved a consensus as a clear candidate responsible for generating the PC rhythm. For example, ribulose-1,5-diphosphate carboxylase, although sometimes showing fluctuations in activity, does not show sufficient correspondence to satisfy the rates of CO_2 fixation at all stages investigated in either *Acetabularia* (Hellebust *et al.*, 1967), *Euglena* (Codd and Merrett, 1971; Walther and Edmunds, 1973), or *Gonyaulax* (Sweeney, 1969b, 1972). Similarly, although an earlier finding (Walther and Edmunds, 1973) of changes in the activity of glyceraldehyde-3-phosphate dehydrogenase suggested a possible mechanism for generating photosynthetic oscillations in *Euglena,* this does not appear to be the case (Lonergan and Sargent, 1978). Indeed, in *Acetabularia* none of the Calvin cycle enzymes were found to have rhythmic activities (Hellebust *et al.*, 1967). Finally, although carbonic anhydrase (associated with the transport of CO_2 through the cell) activity was found to be rhythmic in cultures of *Euglena* maintained in *LD*: *10,*14 with peak activity occurring at the time of the fastest rate of O_2 evolution, the rhythm in enzyme activity disappeared under constant conditions in which the photosynthetic rhythm persisted (Lonergan and Sargent, 1978).

Lastly, several lines of evidence have been brought to bear on the fourth type of mechanism involving photochemical events and variations in the light reactions that might generate the observed circadian rhythmicity in photosynthesis. Many earlier observations appear to rule out a causal role of the light reactions in *Acetabularia* (Hellebust *et al.*, 1967), *Euglena* (Walther and Edmunds, 1973), and *Gonyaulax* (Sweeney, 1960, 1965). There are, however, several indications that some parameters of the light reactions do oscillate in synchronous cultures of the green algae *Chlorella, Chlamydomonas, Skeletonema,* and *Scenedesmus.* Thus, Senger and co-workers (1975) observed that the rhythmic changes in PC during the cell cycle were closely paralleled by changes in quantum yield, whereas fluorescence yield followed an inverse pattern. They concluded that the reoxidation capacity of the plastoquinones is the regulating switch of the electron transport chain and thus of the rhythm in PC. Furthermore, recent work with *Acetabularia* (Vanden Driessche *et al.*, 1976) suggests circadian fluctuations in Hill reaction activity. Finally, although the individual activities of photosystem I and photosystem II in *Euglena* do not appear to change significantly with time of day (Walther and Edmunds, 1973; Lonergan and Sargent, 1979), the rate of light-induced electron flow through the entire electron chain (water to methyl viologen) was rhythmic both in whole cells and in isolated chloroplasts (Fig. 41B), with the highest rate of flow coinciding with the highest rate of O_2 evolution (Lonergan and Sargent, 1979). Evidence consistent with the notion that the coordination of the two photosystems may be the site of circadian control of photosynthetic rhythms in *Euglena* was obtained from studies of low-temperature fluorescence emission from systems I and II following preillumination, respectively, with light wavelengths of 710 or 650 nm, whereas there was no indication that changes in total Chl, the Chl a/b ratio, or the size of the photosynthetic units were responsible (Lonergan and Sargent, 1979).

Still further evidence for this fourth type of mechanism for generating circadian photosynthetic rhythms has come from an intensive examination of the *Gonyaulax* system and the circadian rhythm of O_2 evolution. Thus, Prézelin and Sweeney (1977) have found from their study of the photosynthesis–irradiance curves in this dinoflagellate that there is a temporal change in the relative quantum yield of photosynthesis in dim LL; Chl content, half-saturation constants, and the size and number of particles on the thylakoid freeze–fracture faces, however, were constant. They suggested that circadian ion fluxes across the thylakoid membrane generate reversible conformational changes that couple and uncouple entire photosynthetic units in the membrane and thus induce a circadian rhythmicity in PC of the cells. More recently, Sweeney *et al.* (1979) have reported that the intensity of Chl *a* fluorescence during the early part of fluorescence induction at O (initial fluorescence) and P (peak fluorescence) was higher during the day phase of the circadian cycle than during subjective night in continuous LL, and that it was positively correlated with the rate of O_2 evolution. Furthermore, this circadian rhythm in fluorescence in LL persisted in the presence of 10 μM DCMU, which blocks electron flow from photosystem II (PSII) during photosynthesis, indicating that the cause is not due to a change in the net electron transport between PSII and photosystem I (PSI). Although the observed temporal changes in fluorescence could arise from differences in the efficiency of spillover energy from the strongly fluorescent PSII to the weakly fluorescent PSI, such spillover should occur unimpaired at 77°K, but this was not the case: the rhythmicity in PC was abolished at this temperature (Sweeney *et al.,* 1979). The authors favor the possibility that changes in the efficiency of radiationless transitions generate the overt rhythm of photosynthesis, with a lower rate constant occurring during the subjective day. All of these results are consistent with the importance of structural changes in photosynthetic membranes and may well hold for the *Euglena* system also.

D. Oscillations in Gross Metabolic Parameters

The general patterns of biosynthesis of gross biochemical constituents have been mapped across the cell cycle in culture of *Euglena* synchronized by *LD*: *14*,10 under photoautotrophic conditions of growth (Cook, 1961; Edmunds, 1965b). Dry weight, total protein, chlorophyll *a,* carotenoids, soluble protein, RNA, DNA, and total phosphorus were determined (Edmunds, 1965b) at intervals of 2 hours during synchronous growth of *Euglena* at a cell concentration of 5–10 × 10⁴ cells/ml, maintained by periodic dilution. Detailed analyses of the

soluble proteins (DEAE-cellulose fractionation) and of the intracellular distribution of phosphorus were made also.

In general, a linear doubling of each of the major components occurred during
the photophase, as illustrated in Fig. 42. There was no net synthesis in the dark.
Since cell number doubled during the dark scotophase (see Fig. 5), a halving of
the amounts of the variables in each cell occurred. It is apparent that not all
patterns of biosynthesis proceed in the same manner. Different absolute rates of
synthesis were found for the different compounds, with a doubling of a given
substance sometimes being completed before the end of the photophase (e.g.,
carotenoids, chlorophyll a). In most work subsequent to these earlier experiments, we have reduced the photophase from LD: $14,10$ to LD: $10,14$. The most
notable exception to the linear increase observed for most variables was the
discontinuous synthesis of DNA (Fig. 42) during the photophase in LD: $14,10$.
DNA replicated only during the last 6 hours of the photophase, commencing
approximately at 8 hours after the onset of light. Although DNA replication is a
necessary condition for cellular division, it is not necessarily a sufficient one in
Euglena (Edmunds, 1964). Most of the foregoing results agree well with those of
Cook (1961).

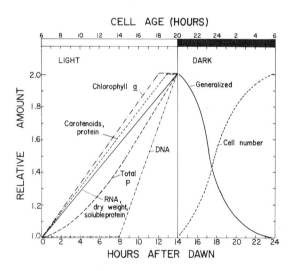

Fig. 42. Summary of the gross biosynthetic patterns in synchronized *E. gracilis* (Z) grown
photoautotrophically in LD: $14,10$ at 25°C. The relative amounts of the various cell components are
given on the ordinate as a function of hours after the onset of the light period (lower abscissa), and the
actual age of the cell, assuming that it completes fission on the average at the fourth hour of the dark
period and is therefore 6 hours old at the beginning of the following light period (upper abscissa). Cell
number is constant throughout the light period and doubles during darkness as shown. (From Edmunds, 1965b.)

Cosinor analysis of the data (Edmunds and Halberg, 1981) indicate that these diurnal rhythms are highly significant (with values for $p < .001$). All the variables (except for DNA) have their acrophases clustered at about $-180°$ to $-195°$, toward the end of the synthetic (light) span and just preceding the onset of cell division in the population. These results suggest that the G_1 phase of the cycle lasts about 8 hours, that the S phase is approximately 6–7 hours, and that the G_2 phase is relatively short (cf. Fig. 20). Individual cells, of course, show considerable variation in the timing of these processes.

Presumably, similar biosynthetic temporal patterns would obtain under free-running conditions inasmuch as the overt rhythm of cell division persists under constant dim LL (Edmunds, 1966; see Fig. 11). Their assay, however, would be complicated by the fact that in the population only a small number of cells divide during any given 24-hour span, with the result that the culture is developmentally asynchronous. Better results might be achieved with the photosynthetic mutants where stepsizes of 2.0 can be routinely attained (Edmunds *et al.*, 1976; Jarrett and Edmunds, 1970; see also Figs. 12 and 16).

In addition to the periodic oscillations observed in gross metabolic variables, several other interesting circadian rhythms in the biochemical state of *Euglena* have been documented (see Table I). For example, Brinkmann (1976b) has found that long-term exposure of *Euglena* to ethanol (10 to 100 mM for several weeks), which might act at the level of membrane(s), increased τ of the random motility rhythm of autotrophic cultures. Correlated metabolic studies demonstrated induction of the glyoxalate pathway for the breakdown of ethanol, an increase in the cellular ATP/ADP ratio, and an increase in respiratory activity for 2 days following addition of ethanol to the medium. Furthermore, only alcohols that could be metabolized by oxidative pathways were able to affect τ, suggesting that the activity and temperature activation of the mitochondrial electron transport chain perhaps are more important than membranes per se in the regulation of the circadian period. Indeed, Brinkmann (1976a) also found in a study of the susceptibility of *Euglena* to pulses of high concentrations (1.0–1.5 M) of various alcohols (Fig. 43) that the dynamic state of the underlying oscillator was not affected, and that one cannot conclude on the basis of his results that membranes are an essential part of the circadian clock (reviewed by Edmunds, 1980a); rather, he ascribed (Brinkmann, 1976a) the susceptibility rhythm to alcohols in *Euglena* to the fluctuating sensitivity of membranes controlled by another oscillatory process.

In a later study, Brinkmann (1976b) reported a circadian rhythm in the energy of activation required to produce acid denaturation of the outer cellular membrane in free-running cultures of *E. gracilis* previously entrained by LD: *12, 12* and then released into bright LL. The susceptibility of the membrane fluctuated synchronously with the motoric activity of the cells, but approximately 180° out-of-phase with the susceptibility of the circadian clock to phase-shifting light signals (Fig. 43). By examining the dependence of the acid-denaturation energy of activation on temperature, it was shown that the sensitive component of the membrane is probably a protein that undergoes circadian turnover in the membrane. These findings provide direct support for circadian regulation of the outer plasma membrane and perhaps identify the cell membrane as the primary target

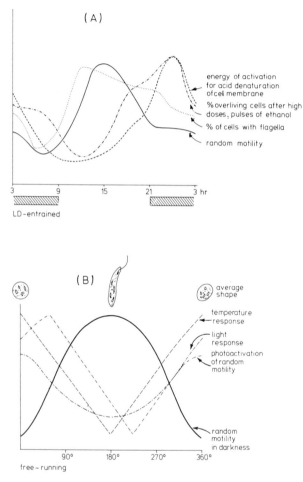

Fig. 43. Daily fluctuation of several cell parameters of nondividing autotrophic populations of *E. gracilis.* (A) Observed during entrainment by *LD: 12,*12 cycle. (B) Obtained under nondividing LD cycles of 20:100 minutes. By definition the phase scale runs from one minimum of random motility to the next one approximately 24 hours later. "Temperature response" and "light response" denote the phase shift of the circadian rhythm after a step-up in temperature or a step-down in light, applied at the indicated phase of the circadian cycle. (From Brinkmann, 1976b; reproduced by permission of Springer-Verlag.)

for mediating environmental signals to the circadian clock (see Wille, 1979). Finally, Feldman (1968) has observed a circadian rhythm of amino acid ([^{14}C]-phenylalanine) incorporation in autotrophic, nondividing cultures of *E. gracilis* entrained by *LD*: *12,12* and released into DD.

E. PERIODIC CHANGES IN ENZYME ACTIVITY

In addition to temporally mapping the patterns of various gross cellular constituents, the activities of a number of enzymes have been monitored during the cell cycle in synchronously dividing cultures of *E. gracilis* (Z), photoautotrophically batch-cultured in *LD*: *10,14*; certain of these enzymes might contribute to some of the concomitant overt physiological rhythms routinely observed (reviewed by Edmunds, 1974b, 1975). Furthermore, a circadian oscillation in enzyme activity can quite validly itself be considered as an index of an underlying biological pacemaker. Some of these periodic enzymes are listed in Table I.

Although the activity of many of these enzymes increased more or less linearly throughout the light interval and then leveled off (on a per aliquot basis) with the onset of darkness and ensuing cell division in the synchronized population, this was not always the case. Thus, deoxyribonuclease activity remained at nearly the same level for about 5–6 hours after the onset of light and then sharply increased until the onset of dark some 5 hours later, at which time it leveled off (Walther and Edmunds, 1970). This discontinuous increase paralleled that of total DNA (Edmunds 1964, 1965b; see Fig. 42), though occurring slightly before the latter; the two events are possibly associated. Likewise, both NADH- and NADPH-dependent glyceraldehyde-3-phosphate dehydrogenase (GPD or triose phosphate dehydrogenase) exhibited periodic increases during the cell cycle (Walther and Edmunds, 1973). In this case, the activity of the enzyme peaked between 6 and 8 hours after the onset of light in *LD*: *10,14* and then commenced to decrease well before the onset of dark and cell division back to its base level (on an aliquot basis). The close correlation of the rhythmic changes in GPD with the circadian rhythm of photosynthetic capacity was particularly striking [especially in stationary culture (see Fig. 37), although Lonergan and Sargent (1978) find no rhythmicity in either LD or LL and suggest that enzyme activity was not saturated in the earlier assay]. It seems likely that most, if not all, of the enzymes listed in Table I will continue to oscillate under LL (or DD) and constant temperature, if one can draw a parallel from the persistence under these conditions of the circadian rhythm in cell division.

The *Euglena* system has been simplified by divorcing autogenous enzyme oscillations from those directly generated by the driving force of the cell cycle itself (whereby replication of successive genes would lead to an ordered, temporally differentiated expression of enzyme activities). This was accomplished by using light-synchronized, photoorganotrophically batch-cultured wild-type

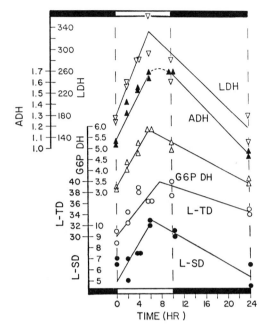

Fig. 44. Oscillations (24-hour) in the activities of several enzymes in nondividing, stationary cultures of *Euglena* in LD: *10,*14 at 19°C on low pH glutamate–malate medium. Enzyme activities are given on the ordinates. (●) L-serine deaminase (L-SD) activity (μmoles pyruvate produced $\cdot$ min^{-1}); (○) L-threonine deaminase (L-TD) activity (μmoles α-ketobutyrate produced $\cdot$ min^{-1}); (△) glucose-6-phosphate dehydrogenase (G6P DH) activity (μmoles $\times$ 10^2 NADH oxidized $\cdot$ min^{-1}); (▲) alanine dehydrogenase (ADH activity μmoles $\times$ 10^2 NADH oxidized $\cdot$ min^1); (▽) lactic dehydrogenase (LDH) activity (μmoles $\times$ 10^2 NADH oxidized $\cdot$ min^{-1}). (From F. M. Sulzman and L. N. Edmunds, unpublished; Edmunds, 1974b.)

Euglena that had reached stationary growth stage at which essentially no net change in cell number occurs (Edmunds, 1974b, 1975). These cultures had been previously grown and were then subsequently maintained in *LD: 10,*14 at 19°C. Relatively large-amplitude oscillations were found in the activities of alanine, lactic, and glucose-6-phosphate dehydrogenases, and L-serine and L-threonine deaminases (with maxima usually occurring during light) entrained to a 24-hour period by the imposed LD cycle (Fig. 44). Similar findings (Fig. 37) have been made (Walther and Edmunds, 1973) for glyceraldehyde-3-phosphate dehydrogenase in stationary cultures of *Euglena* photoautotrophically grown and maintained in *LD: 10,*14 at 25°C (but see Lonergan and Sargent, 1978). These rhythmic changes in enzyme activity, therefore, were effectively divorced from the cell cycle and periodic replication of the genome. Even more interesting, however, was the finding (Edmunds *et al.,* 1974; Sulzman and Edmunds, 1972,

1973) that the activity of alanine dehydrogenase (ADH) continues to oscillate in these nondividing (infradian) cultures for at least 14 days in DD (but not in LL), and thus constitutes an overt circadian rhythm in itself (Fig. 45).

Experiments have been initiated to determine the nature of the biochemical clock mechanism that generates the observed oscillations in alanine dehydrogenase activity (Sulzman and Edmunds, 1973). The possibility that fluctuations in pools of substrates or products could change the stability of the enzyme during its extraction and thus trivially generate the observed rhythm was ruled out. Results from mixing experiments likewise did not suggest the presence of fluctuating pools of effector molecules that could produce the rhythm by altering the activity of the enzyme, nor were there differences in pH optimum, K_m value, or electrophoretic mobility on polyacrylamide gel of enzyme extracted at different phases of the oscillation (Sulzman and Edmunds, 1973). The application of low doses of cycloheximide suppressed the oscillation in enzyme activity, but, following removal of the inhibitor after 12 hours, the rhythm resumed with no apparent change in phase. On the other hand, activity determinations of alanine

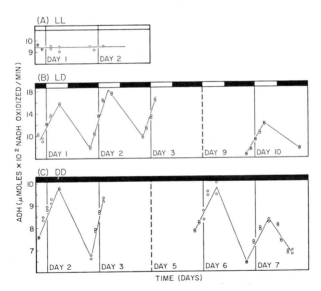

Fig. 45. Activity of alanine dehydrogenase (ADH) in nondividing, organotrophically cultured *Euglena* in various lighting regimes. (A) ADH activity in culture maintained under constant bright illumination (LL). (B) ADH activity in culture in LD: *10*,14 cycles. Data from day 1, day 2, day 3, and day 10 are shown. (C) ADH activity in constant darkness (DD) after many cycles of LD: *10*,14. Data from day 2, day 3, day 6, and day 7 following the transition from LD to DD are shown. Vertical lines are 24 hours apart. Double points are duplicate determinations. (From Sulzman and Edmunds, 1972.)

dehydrogenase extracted from the maximum and minimum points of the rhythm and partially purified by ammonium sulphate fractionation and polyacrylamide gel electrophoresis suggest that periodic *de novo* synthesis and degradation of alanine dehydrogenase may generate the observed variations in its activity. Such oscillations in enzyme activity, in turn, could result from the sequential transcription of a long, polycistronic "chronon" [hypothesised by Ehret and Trucco (1967) to be the basis for circadian timekeeping] or from end-product repression (see Donachie and Masters, 1969), or from membrane controlled cyclical processes (See Edmunds and Cirillo, 1974; Edmunds, 1980a), to name a few possible mechanisms.

III. General Considerations: Problems and Prospects

Now that we have completed the survey of selected classes of circadian rhythms in *Euglena,* it is perhaps instructive to note some of the more pressing unsolved problems concerning circadian clocks that might be approached using the *Euglena* system in the hopes of enticing others into entering the field and its dismal morass of ignorance. It is obvious that these topics, embracing entire disciplines in themselves, can be merely touched upon in a most superficial manner, and that others necessarily will be slighted.

A. ULTRADIAN, CIRCADIAN, AND INFRADIAN INTERFACES

1. The "Circadian-Infradian Rule"

Our review of the circadian temporal organization in *Euglena* amply documents the point that in cells not proceeding through their normal cell developmental cycle (cdc), as in cells of the plateau (infradian) stage, where little, if any, cell division is occurring (i.e., where the developmental sequence culminating in mitosis has been blocked or arrested), the circadian clock cycle continues to operate, as evidenced by the manifestation of numerous overt circadian rhythms (Table I). These overt rhythmicities can be abolished by changing the environmental conditions (see Fig. 12A) so that the overall g of the culture is less than 24 hours (ultradian growth mode) as, for example, by raising the temperature in organotrophic cultures, or by increasing the intensity or duration of illumination, or by introducing utilizable organic carbon sources into photoautotrophic cultures (Jarrett and Edmunds, 1970; Edmunds, 1978, 1981). Presumably, the clock mechanism is either operating at a higher frequency matching that of the fast-cycling cdc (with a lower limit equal to that of the minimum possible g for a species, about 8 hours for *Euglena* cultures). Alternatively, the oscillator may be operating and merely uncoupled from the cdc, or perhaps "stopped," or even absent (Ehret *et al.,* 1977; Edmunds, 1978, 1981). In any case, circadian rhythms would not be (and, indeed, have never been) observed in cultures of

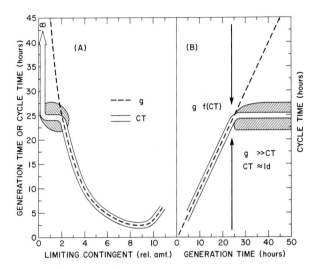

Fig. 46. The Circadian–Infradian Rule. (A) Diagrammatic representation of generation time *g* (broken line) and cell cycle time *CT* (open path) to an environmental factor (limiting contingent). Only in lethal environments does the *CT* exceed circadian values (approaching infinity, arrow). Hatched areas represent the range for *CT*. (B) *g* is a function of *CT* only during ultradian growth. As *g* approaches infinity, *CT* remains approximately constant (circadian). (From Ehret and Dobra, 1977.)

microorganisms in the ultradian growth mode [although Feldman (personal communication, 1981) has found that the conidiation rhythm in *Neurospora* may be an exception].*

Conversely, as *g* exceeds 24 hours (e.g., by lowering the temperature or by nutritional limitation), the periods of the basic oscillator and that of the cdc start to diverge in the other direction (Ehret *et al.*, 1977; Edmunds, 1978, 1981). In the limiting case, where *g* approaches infinity (very slowly dividing, stationary cultures), low-amplitude division bursts occur at circadian intervals in the population (but cdc > 24 hours), along with numerous other cyclic physiological and biochemical events that are not necessarily related to the cdc (reviewed by Edmunds and Halberg, 1981).

This notion of a distinction between *g* and the circadian cycle in slowly-dividing cell populations is diagrammatically represented in Fig. 46 (Ehret and Dobra, 1977; Ehret *et al.*, 1977). As *g* → ∞, the circadian cycle (here, their *CT*) approximates 24 hours and is nearly temperature-independent. Thus, although cell division cycles, like circadian rhythms, are endogenous, self-

---

*Ultradian cultures, however, can be phased (if not truly synchronized) by LD cycles whose period is also less than 24 hours. Using *LD: 6,6* (T = 12 hours), Ledoigt and Calvayrac (1979) have phased *Euglena* cultures growing (g = 12 hours) on lactate so that a doubling of cell number occurred every 12 hours during darkness.

sustaining oscillations, they differ in that the period (τ) of the circadian oscillation is not a function of the period (g) of the cdc, nor is it a function of the
temperature (it arrives at an apparently genetically determined limit value of
about 1 day, circadian); in contrast, the length of the cdc can take on all values
from 24 hours to infinity. [Only in lethal environments would CT exceed circadian values (surely a null case) as represented in Fig. 46A.]

But what is happening to the cellular circadian clock during its replication at
intervals less than 24 hours? Ehret et $al.$ (1977) conclude that during ultradian
growth, cell cycle time (CT) is a function of g (as well as temperature). That
is to say, $CT = g =$ the length of the interdivisional period (Fig. 46B). According to this scheme, then, the basic iteration period (CT) can take on all
values between the shortest g possible under optimal growth conditions for a
given cell and approximately 24 hours. More precise answers await further
elucidation of biological clock mechanisms.

2. Quantal Cell Cycles

In Section II (A,1 and A,4) the notion of a fundamental quantal cell cycle (G_q)
was discussed as formulated by Klevecz (1976).

Klevecz's model is particularly interesting when applied to the case (see Fig.
7) of a synchronized population of $Euglena$ (or other cell type) in which not all
cells divide in any given burst, i.e., where $\overline{ss} < 2.00$ (Edmunds and Funch,
1969a,b). In the "stacked doughnut" diagram presented earlier (Fig. 4), this
situation was described as developmentally asynchronous, since not all of the
marker points (cytokinesis) showed a one-to-one correspondence. Indeed, we
have no idea to what degree other events in the cdc show a similar registry or lack
thereof, much less what the phases of the cells that are not dividing in a given
burst are. By Klevecz's model, one could assume that individual cells have
different g's (a necessary inference from the $Euglena$ data) because the G_q
subcycle undergoes a varying number of revolutions for different cells.

If the notion of the G_q cellular clock is valid, then one might expect to find
repeated occurrences of some event in slowly dividing cells as G_q undergoes
successive revolutions. Indeed, Klevecz (1969, 1976) has observed oscillations
in the activity of a number of enzymes in synchronized mammalian cells and has
suggested in a general way that this might be an expression of the cellular clock.
Oscillations with periods of 3 or 4 hours were observed for a number of enzymes
that had no obligatory connection with other periodic events such as DNA synthesis. Furthermore, these oscillations involved protein synthesis and degradation as well as modulation, and the system displayed inertia in the sense that
inhibition of DNA and RNA synthesis damped, but did not obliterate, the enzyme oscillations. Other markers might also be repetitively expressed.

Does this mean, though, that the Klevecz fundamental cell cycle (G_q) is "the clock'' in the sense that it underlies and generates *circadian oscillations?* Does this mean that since $G_q \simeq$ 3–4 hours (for mammalian cells), then 6–8 (G_q) = 24 hours, or one circadian cycle? If so, then we are dealing with the achievement of relatively long 24-hour periods by a type of frequency demultiplication in which six to eight "ticks" create one "tock"; but what, if anything, counts the number of revolutions of G_q (Edmunds, 1978)?

B. NATURE OF THE CLOCK(S)

1. *Classes of Molecular Models for Circadian Clocks*

There are several different categories of models for an endogenous, self-sustaining circadian clock, which are neither mutually exclusive nor jointly exhaustive: (a) feedback loop models for oscillations in intermediary metabolism (e.g., glycolytic oscillations in yeast) in which longer periods are generated by cross-coupling among individual oscillators within a single cell or within a cell population (Pavlidis, 1969, 1971) that constitutes a network; (b) "tape-reading" transcription models, the most prominent of which is the chronon model of Ehret and Trucco (1967), in which the transcription of a long, polycistronic piece(s) of DNA with associated rate-limiting diffusion steps leading to translation meter time and generate circadian periods; and (c) membrane models (recently reviewed by Edmunds, 1980a), in which the transport activity and other properties of various membranes in the cell are intimately related to state transitions in the fluid mosaic membrane itself and which, in turn, affect membrane structure (e.g., Njus *et al.*, 1974; Schweiger and Schweiger, 1977). [For a general overview of these classes of models, see Chapter 8 in Palmer *et al.* (1976) on "Models and Mechanisms for Endogenous Timekeeping"; for a more comprehensive treatment, see the Dahlem Conference on "The Molecular Basis of Circadian Rhythms" (Hastings and Schweiger, eds., 1976).] In the following sections, we shall attempt to lay the speculative groundwork for molecular modeling of cell (division) cycles and circadian clocks, while drawing freely from elements of all three classes of models.

2. *Insertion and Deletion of Time Segments in Cell Cycles*

In Section II,A,4 the critical problem of variation in the length of individual cell division cycles was discussed, particularly in the case of those division cycles modulated by a circadian clock. The obvious question arises: How are the experimentally observed shortenings and lengthenings of individual cdc's generated by a master (circadian) clock at the biochemical or molecular level?

It would appear almost as if *Euglena* (and other microorganisms) has a programmable "clock for all seasons," or at least for a variety of sets of specified values for illumination, temperature, and nutritional conditions (K. A. Adams and L. N. Edmunds, (unpublished results). Although the term *programmable* smacks of deterministic sequences, one can couple various stochastic or probabilistic processes to sequential-type mechanisms, generating any desired degree of variance in the overall control system for the cdc (see Section II,A,1). It is perhaps a matter of personal preference (in our current state of knowledge) as to the degree which one focuses on a tight genetic control, as, for example, by sequential tapereading of a unique segment of DNA (e.g., the chronon of Ehret and Trucco, 1967) or RNA template; or, alternately, whether one assumes that the genetic program is relatively remote from the actual biochemistry of the cdc. In any case, the evidence reviewed in earlier sections formally demands that a clock of some sort (at least sometimes circadian in nature) predictably inserts time segments into, or deletes them from, the cell division cycle.

On the view that time "dilation" or "contraction" has an immediate molecular basis, one can envisage ways by which such time segments could be added or subtracted from the cdc by a master oscillator. On the one hand, an indeterminate, variable number of traverse of Klevecz's (1976) G_q subcycle could generate the necessary variance in g values. The fundamental period of 4 hours hypothesized for mammalian cell cultures in the ultradian growth mode would either have its analog in a circadian oscillator, which would then generate cdc's whose lengths would be integer *multiples of 24 hours,* or would be somehow transformed by a circadian clock into a longer-period subcycle. The former notion implicates a multiplicity of clocks (a "clockshop"); the latter requires frequency transformation by another control system or by an internal modification of a versatile, pliable oscillator. Unfortunately, the G_q quantal cycle thus far is only a descriptive notion without a molecular basis, although concommitant oscillations in enzyme activity have been observed (Klevecz, 1969) in mammalian cell cultures.

Alternatively, there is no reason to suppose that only one G_q subcycle is possible and that the duration of G_1 is to be accounted for solely by summation of a variable number of rounds of G_q. A variety of alternative subcycles of different lengths and functional roles might exist from which the cell could choose. For example, one way that the cdc might be programmed would be for a collection of timing loops of different lengths to couple together in various combinations to form a flexible timer, the *cytochron* (Edmunds and Adams, 1981), as diagrammed in Fig. 47. This scheme is sufficiently generalized to apply to any eukaryotic cell cycle, although it was originally devised to summarize experimental findings in *Euglena*. In this model the physiology of the cell is held in a steady state by feedback reactions, except for perturbations induced by environmental fluctuations, and the cell grows and accumulates reserves, or performs some steady state functional role, in an indeterminate sequence. The cytochron, however, is to be regarded as a separate entity—a clock, or timer—that meters time with the primary function of giving order and temporal separation to the key events (black bars) that trigger or initiate (small arrows) the determinate phases

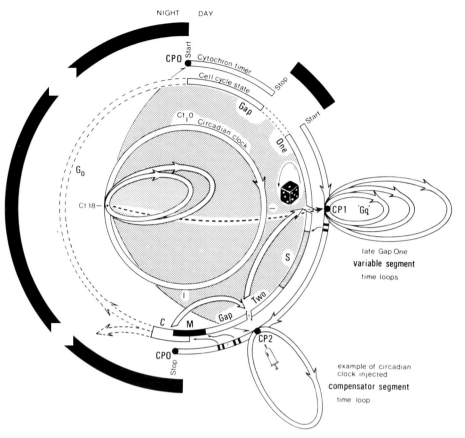

Fig. 47. A generalized model for the insertion and deletion of time segments in the track of the cytochron (cell-cycle clock) hypothesized to program the events of the cell division cycle of *Euglena*. Cytochrons in each cell start up synchronously at dawn and meter time (unless interrupted by a dark pulse) around to the control point (CP0), triggering sequentially (black bars) the events leading to chromatin replication (S), mitosis (M), and cytokinesis (C). In photoautotrophic cultures in LD regimes they stop (noncyclic mode) at CP0 in the dark, having triggered M and C, leaving cells in an untimed G_0 state until dawn restarts the cycle. In LL, or on certain organic substrates, however, the cytochron is cyclic and runs on through CP0 (i.e., there is no G_0). At CP1, the addition of one or more variable-segment time loops by a random selector (dice symbol) generates variability in the duration of Gap One and disperses (desynchronizes) subsequent cell divisions across the next dark period. A circadian clock, entrainable by LD *Zeitgeber*, can couple to the cytochron at a unique circadian time (*Ct* 18) and inject (syringe symbol) a determinate compensator-segment time loop into the cytochron track anywhere within the arc CP0 to CP2 (stippled), so that the cell cycle is extended until a subsequent circadian cycle. Thus divisions are phased in bursts or clusters at 24-hour intervals, the scatter within each cluster being generated by the Gap One variable-segment time loops, dawn synchronization of the cytochrons ensuring that each pulse is confined to the dark period in each LD cycle. The circadian clock can also apparently delete Gap One variable loops and reduce cell cycle variation under some regimes (dashed arrow). Finally, division is suppressed in cells approaching the infradian (stationary) phase of population increase, but multiple rounds of S raise the ploidy level. When conditions improve, ploidy is reduced by successive rounds of M and C. Anticlockwise arrows represent these temporary loop closures. (From Edmunds and Adams (1981) *Science* **211**, 1002–1013. Copyright 1981 by the American Association for the Advancement of Science.)

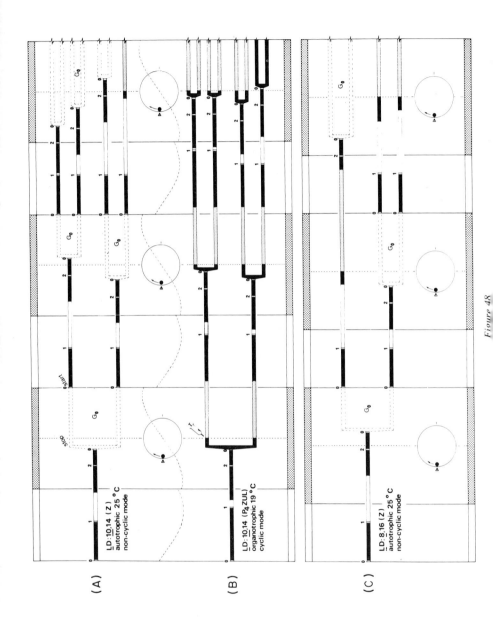

Figure 48

(S, M, and C) of the mitotic cycle, and possibly also the sequence of events leading to differentiation in multicells.

The cytochron (Fig. 47) is shown schematically to have a basic circular track which can be modified by the insertion (or deletion) of a variety of time loops of different lengths. These loops can be selected either at random (variable segments) to generate G_1 variability, or specifically (compensator segments) by the cell's circadian clock to manipulate the duration of cdc phases so that they coincide with appropriate time slots in the day–night cycle. The time track can be envisaged as being marked out in hours, such that if a *Euglena* cell followed the main track, bypassing all the lateral loops, its total traverse would require 8–10 hours, the minimum cycle time in its "best of all possible worlds" (organotrophic growth at 25°C; see Fig. 12A). In *Euglena* grown photoautotrophically in LD regimes, however, the cdc (and by implication the cytochron) clearly operates in a noncyclic mode (Fig. 48). Once mitosis and cytokinesis are completed, the daughter cells do not proceed to another cell division cycle unless illumination is maintained, and in darkness they appear to be held in an indeterminate or untimed G_0 state. At 25°C, as for *Chlamydomonas* (Spudich and Sager, 1980), a minimum of about 6 hours of light is required to prime a population for subsequent cell division in darkness under optimum conditions, a value which coincides with the shortest duration for G_1 (see Fig. 20), suggesting that the extra

---

Fig. 48. Generation of the two types of variance observed during synchronous or phased cell division by the cytochron and circadian clocks. Time tracks for cytochron clones in wild-type *Euglena* (strain Z) for photoautotrophic growth at 25°C in (A) LD: *10,*14 and (C) LD: *8,*16 cycles, and for the photosynthetically incapacitated mutant P_4ZUL (B) on organic medium in LD: *10,*14 cycles at 19°C. Open bands (0,1,2,) denote control points CP0, CP1, and CP2. Determinate segments CP0-to-CP1, S-to-CP2, and CP2-to-CP0 (black) are interspersed with variable segments (open track), selected randomly from a range modulated by the LD regime, and compensator segments (fine stipple), injected at circadian intervals by the entrained circadian clock (dials) into the cytochron tracks at the midnight point in each cycle. In photoautotrophic cells (A and C) cytochrons operate in the noncyclic mode, stopping in darkness at CP0 after initiating cell division, and restart in synchrony at dawn. Random choice of variable segments disperses cell divisions across each dark period, but the daughter cells are resynchronized as they collect in the untimed G_0 phase for a common start at the next dawn. Organotrophic growth (B) switches the cytochrons into the cyclic mode, in which they remain asynchronous and oblivious to LD cycles, but once a day the LD-entrained circadian clocks simultaneously inject an 8-hour compensator segment (time delay) into the tracks of all the cytochrons, irrespective of phase (except for the minority in the CP2-to-CP0 segment), giving rise to a stepped log–linear growth curve (see Fig.5b). In (A), synchronization of the cytochrons at dawn ensures that most cells reach CP2 by midnight, permitting a population doubling every cycle, since the circadian clock cannot couple to G_0 or post-CP2 segment cells. In (C), the shorter light periods increase the range of available variable segments, so that many of the cells fail to reach CP2 by midnight of the first cycle, and the circadian clocks inject 14-hour (at 25°C) compensator-segments that carry them over to complete division in the subsequent dark period (see Fig. 12). [From Edmunds and Adams (1981) *Science* **211**, 1002–1013. Copyright 1981 by the American Association for the Advancement of Science.]

variable segments, introduced into the cytochron tracks of other cells in the population to generate G_1 variability, may be put in at the end of a common light-driven segment.

It would appear, therefore, that temporal loci exist along the cytochron track at which decisions are made with respect to the addition or deletion of time loops, or to stop or start. To avoid the semantic problems involved in using terms such as "threshold" or "transition" point (often applied to temporal loci blocked by quite nonphysiological agents), these apparently natural branching points in the cdc program are referred to here as *control points* (CP). CP0 represents both the inception and termination locus for the cytochron timer, and the decision to set the timer running is made at this locus, for which illumination seems to be an obligate requirement in photoautotrophic *Euglena* populations, since the timer stops at and starts from CP0, on either side of a dark interval, only running in a truly cyclic mode in LL. If a carbon source such as ethanol (Edmunds, 1965a) is present, however, the cytochron immediately adopts the cyclic mode and programs successive mitotic cycles, apparently oblivious to any LD regime. As can be seen from Fig. 48, this stop–start mechanism serves to synchronize the cytochrons (and cdc inception) in LD-cycled populations dependent on chloroplast-fixed carbon.

A second control point, CP1, tentatively located at the inception of S in cells with minimal cycle times, is suggested to be the point at which variability is inserted into the Gap One segment and is probably the last locus along the cytochron track at which the sequence of events can be blocked or adjustments made before the cell irretrievably commits itself to DNA replication and chromatin duplication. Although it would appear that selection of individual variable segments is random, the range may be modulated, nevertheless, by environmental conditions, longer segments being made available in the infradian growth mode brought about by lower temperatures, low light intensities, or a poor nutrient status. Indeed, it seems probable that alternative time loops are also available for the CP0-to-CP1 and the CP1-to-CP2 segments to suit different environmental regimes. Certainly in *Euglena* additional options are available in the CP2-to-CP0 segments to permit the multiple rounds of S that occur in the stationary phase of batch cultures, while cell division is suppressed, and which lead to an increase in chromosome number from the usual 21 pairs to in excess of 80 pairs, this being followed on inoculation into fresh medium by successive rounds of M and C, which rapidly restore ploidy to the exponential-phase norm (K. Adams, unpublished).

Therefore, it begins to look as though the CP0-to-CP1, CP1-to-CP2, and CP2-to-CP0 segments may also be loops, and that the three control points may lie at the decision-making focus of a complex of loops (the cytochron). Such a state of affairs would also help to explain how it is that the circadian clock can apparently insert its delaying *compensator loops* (that under adverse conditions stretch the cdc from one circadian cycle to the next) at any locus along the cytochron–cdc track, with the exception of the post-CP2 segment (see Fig. 48). We have precisely located this last control point (CP2) at 1 hour (25°C) prior to mitosis; it may represent the locus at which the key events (four black bars) take place that trigger M and C. It could be that the cytochron runs only as far as CP2, and that the events of mitosis and cytokinesis, once initiated, are self-propelling, and run to completion of their own accord. The cytochron–cdc, however, can be blocked at this locus by a variety of environmental shifts; therefore, it must be distinct from CP0, although it must lie very close to it.

The CP0, CP1, and CP2 control points in the postulated *Euglena* cytochron are reminiscent of the G_1 and G_2 loci at which the cdc can be arrested in higher plant meristems and released by subsequent hormone action; by analogy, the equivalent of the CP0 control point in mammalian cells may represent the locus at which cycling and therefore unrestricted cell division is suppressed. Indeed, if the cytochron is fundamentally (primitively) a cyclic clock that can be stopped at CP0 by a variety of natural control mechanisms, it is perhaps hardly surprising that the transformation of mammalian cells can be accomplished by a single protein coded by a virus such as SV40.

3. *The Cytochron: A Molecular Model for Cell Cycle Clocks*

We have argued that the sequential events of the cdc, although in themselves a timer of sorts, cannot constitute the chronometer controlling the temporal spacing of these events by a simple stopping and restarting of the sequence, either in response to environmental shifts or to commands from an endogenous circadian oscillator. At least in an ordinary sense, the cell division cycle is not the clock. A separate programmable entity (which we term the cytochron) must interpret the environmental signals and insert, or delete, appropriate time loops of specified sign and duration into the cdc (selected from a finite library of available loops) and thereby control the time at which each change in cell cycle state is triggered. For example, at 19°C it appears that an 8-hour delay loop is inserted into the cdc program of *Euglena* in response to cues from a circadian clock, whereas at 25°C the loop length is 14 hours (see Fig. 48). In this example, loop length is apparently selected to compensate for a change in temperature—the higher the temperature, the longer the loop—so that the thermally induced acceleration of cellular metabolism does not advance the timing of cell division.

That the cytochron also must be functionally independent of the circadian clock (although not necessarily an entirely separate mechanism) is apparent from the way in which the two can be uncoupled in the P_4ZUL mutant (Fig. 19) when it is starved of certain sulfur compounds. Such cells continue to divide asynchronously, implying continued cycling of the cytochron, yet addition of cysteine, for example, reinstates insertion of a delay loop at circadian intervals in phase with a D/L transition mode several days prior to the addition. The circadian clock, therefore, must have been synchronized and have been cycling independently of the uncoupled cytochron. The persistence of circadian rhythmicity in stationary phase metabolism (infradian growth) in the absence of cell division also supports the concept of a functionally separate cdc.

We have yet to address ourselves to the identity of these time segments formally demanded by our data. Although it is obviously beyond the scope of this paper to assess all the evidence for the role of genes and their expression in the operation of the cdc and circadian chronometers, the concept of a discrete programmable cytochron, with optional time loops, and interacting with a resettable circadian clock, does suggest, however, a possible molecular basis of a kind not previously proposed. Given the complexity and apparent programmability of the cdc clock, it would appear that the information content required to operate the mechanism exceeds that available from any simple membrane-based oscillator (Njus *et al.*, 1974), and that perhaps some sort of direct readout of a nucleotide sequence is more likely as a basis for both the timing and programming functions of the proposed cytochron. Accordingly, Edmunds and Adams (1981) recently have hypothesized that a small segment of chromosomal DNA involving a hundred or so transcriptional units is folded into loops by bridging cross-links (of protein or RNA) at genetically defined loci to form a three-dimensional network of anastomosing loops (Fig. 49). The entire complex would constitute a giant functional gene of up to 3000 kb that is capable of metering periods as long as 24 hours by the incorporation of some 20 to 40 nucleotides per second, as originally proposed in the chronon model of Ehret and Trucco (1967).

This model differs radically from the chronon model, however, in proposing that (a) the transcription of one loop in the sequence triggers transcription of another specific loop somewhere else within the giant gene without the involvement of a translation step, and that (b) the sequence of transcription of the units is programmable. As a transcription complex nears the end of a transcription unit (Fig. 49), an initiator site on a loop joined to it by a cross-link opens up, and transcription of this second unit begins in response to a signal (possibly torsional) transmitted across the link. To permit programming of the network, each bridging cross-link (perhaps a protein dimer linked to the hairpins of two distant, inverted repeat-cruciforms) would be bistable, existing in a passive (flop) or an active (flip) mode. In the flop mode, a bridge cannot transmit a signal across to a distant loop; instead, as the transcription complex nears the end of a unit and approaches a link, an initiator site opens up on the next unit in tandem sequence along the DNA tape. When the bridge switches to the flip mode, however, the completion of transcription of a unit triggers transcription of the distant unit coupled to the other side of the bridge, instead of the one in genetic sequence. Modulation of the program by accessory proteins that change bridge-switching modes would permit time-metering loops (groups of transcriptional units) to be

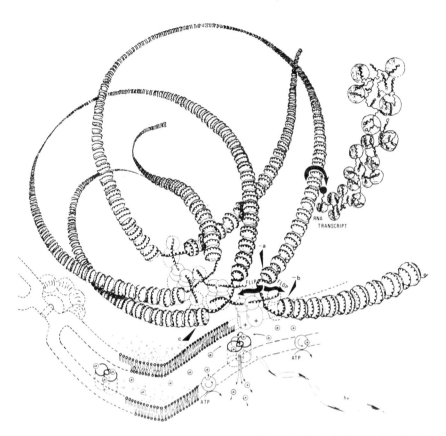

Fig. 49. Diagram of a chronogene segment, one possible molecular basis for metering time in longer-period cellular oscillations. Transcriptional units of chromosomal DNA, wound around their nucleosomes, loop out from protein complexes that anchor them by their inverted repeat cruciforms to the nuclear envelope and cross-link them to other units at paired genetic loci. At dawn, light-absorbing pigments in the membrane trigger the opening of ion gates that collapse a membrane potential accumulated by ATP-driven pumps; the resulting transient change in electric field reprograms the time-metering transcriptional sequence by switching all the protein links to the FLIP mode. Upon completion of transcription of the top loop (a), transcription is initiated on the adjacent unit (c) in response to a (torsional?) signal transmitted across the link rather than on the next unit (b) in the tandem sequence (as it would in the FLOP mode). As the membrane potential restabilizes, the FLIP mode decays so that the only effective switching is mediated by the link at the end of the transcriptional unit being actively transcribed at the time that the transitions between light and dark occur. In this way, long segments of tandemly arranged units are either inserted into, or are deleted from, a coupled transcriptional circuit at dusk or dawn to generate the advancing or delaying adjustments that serve to synchronize the clock with the earth's rotation. The genetically programmed siting of the links ensures that the precise loop lengths required to effect these phase shifts occur at the right places in the multiple-path transcriptional circuit. [From Edmunds and Adams (1981) *Science* **211,** 1002–1013. Copyright 1981 by the American Association for the Advancement of Science.]

added or deleted from the cyclic program, thus generating instant advances, or delays, or serving to compensate the cycle time for thermal effects on transcription rate. Some of the bridges may be attached to complexes embedded in the nuclear envelope and could be switched from the flop to the flip mode by the collapse of a membrane potential generated across the nuclear membrane (e.g., by charge accumulation in the perinuclear space). We suggest that this collapse could occur in response to the opening of ion gates linked to light-sensitive pigments (and temperature sensors) in the envelope (Njus *et al.*, 1974), the magnitude of the resulting advance or delay being dependent on the position of the time-metering transcriptional complexes in the transcription circuit at the time of the collapse and on the position of the next available bridge in the programmed topology of the network.

In this model the majority of the RNA transcripts have no coding function and are never capped or processed to serve as messages, being produced solely to meter time in between key structural genes, and that possibly derive from some of the highly mutated, "rusting hulks" of DNA that persist in eukaryotic genomes. Small RNA segments transcribed from critically spaced loops in the temporal program would serve to trigger events such as S, M, and C at the appropriate time, either by being processed and translated into an enzyme, or in a more subtle way, as for example, by forming the small, recently discovered snRNA's that are thought to hold the ends of intervening sequence loops in place so that splicing enzymes can process the mRNA precursors. Chronogenes may even be modulated by the products of successive cycles to generate long-term programs that control sequential development in multicellular organisms.

Such a molecular model originally developed to account for the cytochron might also account for the complex light- and temperature-resetting patterns (embodied in their phase response curves, or PRC's) of circadian clocks. In fact, transformation of PRC data onto a folded template map suggests a mechanism that could account for the many different PRC shapes, the gradual buildup of resetting amplitude with increasing irradiance, and the advancing transients (K. Adams, unpublished results). This type of model, invoking a discrete clock gene, or *chronogene,* similar to that proposed for the cytochron, is consistent with (a) observations that clock double mutants exhibit additivity and map at closely linked genetic loci (Bruce, 1974; Feldman and Atkinson, 1978); and (b) the demonstration of protein cross-linked, 30–90-kb (Paulson and Laemmli, 1977), apparently radially arranged (Marsden and Laemmli, 1979), DNA loops in eukaryotic chromosomes.

This class of transcriptional model may be in conflict with the observations of circadian rhythms in enucleated *Acetabularia* (see Schweiger and Schweiger, 1977), which indirectly support a nucleic acid-independent, membrane-based clock. As Scott and Gulline (1975) have pointed out, however, the possibility of membrane-based oscillators in some organisms does not preclude different mechanisms in others, or even in the same cell, because if rhythmicity confers a selective advantage on biological systems, various evolutionary strategies may have been adopted independently. One possibility here is that cytochrons may be based on chronogenes, and the circadian clocks on membrane-based devices requiring the nucleus only for a supply of parts.

On the other hand, the uncanny accuracy with which double mutants of *Neurospora* (Feldman and Atkinson, 1978) and *Chlamydomonas* (Bruce, 1974) add and subtract time segments from the circadian cycle, as if they were rails in a toy train circuit, is suggestive of a template clock read at a constant speed, to which a loop can be added and from which a loop can be deleted by, for example, mutations in a tandemly repeated family of pseudoallelic, but codominant, bridge protein genes. Although the cytochron and circadian clock of *Euglena* might or might not comprise, as we propose, a DNA or RNA template with a branching network of tracks and binary switching points (as in a rail freightyard), it would seem in principle that they could have a similar molecular basis, or at least that they could share certain elements or pathways in common, as has been suggested recently by Wille (1979). In fact, we suspect that there is but one programmable clock, a veritable on-board computer, that can plot midcourse corrections in response to real-time variables and ensure its survival against most possible odds. Until we can establish the existence of a single clock with a multiplicity of functions, however, any attempt to unravel its mechanism must take into account the evidence for two functionally separate cellular clocks, both clearly outside the passive sequence (or network) of metabolic states that we recognize as the cell division cycle.

At the moment, we simply do not know enough about the detailed mechanism underlying either circadian rhythms or the proposed cytochron in *Euglena* to be more mathematically precise (e.g., the modes and constraints of clock coupling, even assuming limit cycle dynamics of the circadian oscillator) or more explicit in molecular terms. (For example, we are ignorant as to whether our late-G_1, variable-segment time loops are quantized or generated instead by an entirely probabilistic mechanism.) Rather, we consider this to be a ''thought'' model having several possible solutions, arising from the need to explain the insertion and deletion of the finite *time* segments in cell division cycles formally demanded by their empirically observed variability.

C. Multiple Cellular Oscillators

1. *Intracellular Clockshops?*

A basic but unanswered question in the field of biological rhythms concerns the number of clocks that exist in a single cell. A large part of the difficulty in addressing this question lies in our ignorance of the mechanism of any clock.

Probably the majority opinion has it that there is but one central oscillator within the cell, and that the numerous overt circadian rhythms are merely "hands" of this driving entity. The observations of McMurry and Hastings (1972) that four separate rhythms in the luminescent marine dinoflagellate *Gonyaulax* all seem to have the same free-running period with similar temperature coefficients, and that they all respond identically when perturbed by a resetting dark pulse give rather strong support to this notion. On the other hand, some recent work (Laval-Martin *et al.*, 1979; Edmunds and Laval-Martin, 1981) with *Euglena* maintained in high frequency cycles (see Section II,C) has revealed that the phase relationship between the rhythm of photosynthetic capacity and that of chlorophyll content varied, suggesting the possibility of desynchronization among circadian rhythms in a multioscillator, unicellular organism. Clearly, the question must remain open until further experimental evidence is obtained.

2. Intercellular Communication—Coupled Oscillators?

The long-term persisting circadian rhythms of cell division (Section II,A), motility (Section II,B), and other variables raise another interesting question: Why does not the synchrony decay in the same manner that cell division synchrony decays in bacterial and mammalian cell cultures (usually within three or four cycles)? Assuming that no subtle geophysical factors are phasing the population, then we are left with two alternatives: (a) the free-running period of the oscillator must be almost improbably precise and almost identical with those of other cells; or (b) some sort of intercellular communication must occur that maintains synchrony within the population, or network, of self-sustaining oscillators (Edmunds and Funch, 1969b; Edmunds, 1971). With regard to the first alternative, much depends on the assumptions that are made as to the nature of the variance. Thus, by the "random walk" model (in which "fast-" or "slow running" cells do not necessarily transmit this property to their progeny, or even from cycle to cycle), the rate of dispersion could be quite slow, perhaps requiring several weeks before the peak of some rhythm became so spread out that it was obliterated. Indeed, recent work on the luminescence rhythm of *Gonyaulax* (D. Njus and J. W. Hastings, personal communication, 1980) suggests that gradual decay does occur over a timespan of 10–15 days. The alternative hypothesis of intercellular cross-talk has been examined theoretically (e.g., Goodwin, 1963, 1976; Winfree, 1967; Pavlidis, 1969) from both a mathematical and a biochemical standpoint. Experimental tests of this hypothesis however, have been inconclusive or negative. Thus, Brinkmann (1966, 1967) found no evidence for intercellular communication in *Euglena* when out-of-phase cultures were mixed and the resultant phase of the motility rhythm examined (Fig. 50). Similarly, mixing experiments with synchronously dividing populations of *Euglena* (Edmunds, 1971) did not support this notion. One must exercise some caution in generalizing these largely negative results, nevertheless, since the experimental protocol is critical (Edmunds, 1971).

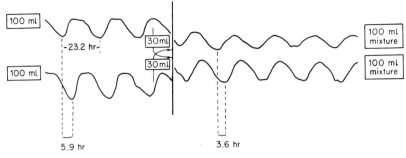

Fig. 50. Mixing experiment designed to test for intercellular communication in *E. gracilis.* Aliquots (30-ml) were taken simultaneously from two 100-ml autotrophic cultures 5.9 hours (90°) out-of-phase with respect to their free-running rhythms ($\tau = 23.2$ hours) of random motility and reciprocally mixed with the opposite culture. The resulting phase angle difference of 3.6 hours between the rhythms in the two mixed cultures was in close agreement with the theoretical prediction (3.1 hours) on the hypothesis that no cell–cell interaction occurred. (Courtesy of K. Brinkmann; modified from Brinkmann, 1967).

D. Chronopharmacology and Chronotherapy

Although *Euglena* is probably not the organism of choice for these studies, it is perhaps worth briefly noting that circadian and cell cycle clocks can play a profound role in the treatment of disease (see, e.g., Reinberg and Halberg, 1979). Perhaps the best illustrative example is that of cancer chemotherapy (reviewed by Edmunds, 1978), in which the goal is to learn how to select drugs, dosages, and intervals between doses or courses that will kill disseminated cancer cells (a) faster than they are being replaced, (b) for a sufficient time to reduce their number to zero, but (c) without killing the patient. By utilizing appropriate synchronizing agents and techniques and by employing various combinations of cdc stage-specific cytostatic drugs, therefore, one can maximize the chance for survival and cure by applying the minimum dose of the drug necessary to kill the phased cells at the time in their cdc when they are most susceptible.

An important part of this strategy, and one largely overlooked, is the role of circadian clocks in (a) modulating the cell division cycle itself and (b) regulating the overall susceptibility of the malignancy or of the host to cytostatic drugs. Halberg (1975) and co-workers particularly emphasized this question of the timing of treatment and have demonstrated that the tolerance of leukemic mice to cytosine arabinoside is a function of the circadian tolerance rhythm of the host. In fact, if drug administration schedules reflected these temporal parameters, there was a significant increase in both survival time and cure rate, as compared to the reference control animals receiving equal doses of the drug throughout the day. It appeared that the success of these "sinusoidal" treatments was due to an optimization of both host resistance to the drug coupled with greatest tumor susceptibility to it, each being regulated by a circadian clock (Haus *et al.*, 1974). Cell systems such as *Euglena,* therefore, may help elucidate basic cell cycle and clock mechanisms and hence have applicability to chronotherapy.

E. LIFE-CYCLE CLOCKS AND AGING

It is perhaps fitting that we close this review with a brief mention of the implications of cell cycle clocks and circadian rhythms toward senescence and aging. The viewpoint that aging can be viewed as the loss of temporal organization has been treated in a general fashion by Samis and Capobianco (1978). The problem is to distinguish between correlation and causality: Does the breakdown of circadian organization, for example, lead to aging or does it merely accompany it? There is some evidence that organisms function most effectively when, as an innately periodic system, they are driven at frequencies close to their own natural circadian frequency, i.e., when the organism is in "resonance" with its environment (see Edmunds, 1978), and they suffer deleterious effects and even decrease in longevity when their internal temporal organization is disrupted. Furthermore, among the many cellular theories of aging [see Finch and Hayflick (1977) for an exhaustive treatment], the hypothesis that aging is genetically programmed in the sense that there is some positive event leading to senescence and the organism's own destruction raises the possibility of a "life cycle clock." Such a timer could count cell divisions, on the one hand, or measure some recurrent or sequential cellular process, on the other. At a molecular level, this could occur by successive increments of gene depreciation, such as by the loss of a terminal segment of DNA with each replication or by repeated sequential base modification (Holliday, 1975). It is not clear at all whether a unicellular organism such as *Euglena* possesses a life cycle clock, or even whether it actually senesces in stationary culture, but perhaps critical experimental studies are warranted.

Acknowledgments

This work was supported by National Science Foundation research grant PCM78-05832. I thank the various publishers and authors, particularly Drs. K. A. Adams, K. Brinkmann, and D. L. Laval-Martin, upon whose work I have drawn heavily, for permission to reproduce their figures (credits are given in the legends) and to quote recent, unpublished results, and the many students and collaborators of my laboratory over the years whose research made much of this story possible. I am grateful to Dr. D. L. Laval-Martin and L. Misa for their assistance in preparing the manuscript.

References

Aschoff, J. (1965). "Circadian Clocks." North-Holland Publ., Amsterdam.
Aschoff, J., Klotter, K., and Wever, R. (1965). *In* "Circadian Clocks" (J. Aschoff, ed.), pp. x–xix. North-Holland Publ., Amsterdam.

Brinkmann, K. (1966). *Planta* **70**, 344–389.

Brinkmann, K. (1967). *Nachr. Akad. Wiss. Goettingen, Math. Phys. Kl. 2B:Biol. Physiol. Chem. Abt.* 138–140.

Brinkmann, K. (1971). *In* "Biochronometry" (M. Menaker, ed.), pp. 567–593. Nat. Acad. Sci., Washington, D.C.

Brinkmann, K. (1976a). *J. Interdiscipl. Cycle Res.* **7**, 149–170.

Brinkmann, K. (1976b). *Planta* **129**, 221–227.

Bruce, F. G. (1960). *Cold Spring Harbor Symp. Quant. Biol.* **25**, 29–48.

Bruce, V. G. (1965). *In* "Circadian Clocks" (J. Aschoff, ed.), pp. 125–138. North-Holland Publ., Amsterdam.

Bruce, V. G. (1974). *Genetics* **77**, 221–229.

Bruce, V. G., and Pittendrigh, C. S. (1956). *Proc. Nat. Acad. Sci. USA* **42**, 676–682.

Bruce, V. G., and Pittendrigh, C. S. (1957). *Am. Nat.* **91**, 179–196.

Bruce, V. G., and Pittendrigh, C. S. (1958). *Am. Nat.* **92**, 294–306.

Bruce, V. G., and Pittendrigh, C. S. (1960). *J. Cell. Comp. Physiol.* **56**, 25–31.

Buetow, D. E., ed. (1968). "The Biology of *Euglena*" Vols. I and II. Academic Press, New York.

Bünning, E. (1973). "The Physiological Clock." Springer-Verlag, Berlin and New York.

Campbell, A. (1957). *Bacteriol. Rev.* **21**, 263–272.

Campbell, A. (1964). *In* "Synchrony in Cell Division and Growth" (E. Zeuthen, ed.), pp. 469–484. Wiley, New York.

Codd, G. A., and Merrett, M. J. (1971). *Plant Physiol.* **47**, 635–639.

Cook, J. R. (1961). *Biol. Bull. (Woods Hole, Mass.)* **121**, 277–289.

Cook, J. R. (1966). *Plant Physiol.* **41**, 821–825.

Cook, J. R., and Cook, B. (1962). *Exp. Cell Res.* **28**, 524–530.

Cook, J. R., and James, T. W. (1960). *Exp. Cell Res.* **21**, 583–589.

Donachie, W. D., and Masters, M. (1969). *In* "The Cell Cycle" (G. M. Padilla, G. Whitson, and I. L. Cameron, eds.), pp. 37–76. Academic Press, New York.

Edmunds, L. N., Jr. (1964). *Science* **145**, 266–268.

Edmunds, L. N., Jr. (1965a). *J. Cell. Comp. Physiol.* **66**, 147–158.

Edmunds, L. N., Jr. (1965b). *J. Cell. Comp. Physiol.* **66**, 159–182.

Edmunds, L. N., Jr. (1966). *J. Cell Physiol.* **67**, 35–44.

Edmunds, L. N., Jr. (1971). *In* "Biochronometry" (M. Menaker, ed.), pp. 594–611. Nat. Acad. Sci., Washington, D.C.

Edmunds, L. N., Jr. (1974a). *Exp. Cell Res.* **83**, 367–379.

Edmunds, L. N., Jr. (1974b). *In* "Mechanisms of Regulation of Plant Growth" (R. Bieleski, A. Ferguson, and M. Cresswell, eds.), Bull. No. 12, pp. 287–297. Royal Society of New Zealand, Wellington.

Edmunds, L. N., Jr., (1975). *In* "Les Cycles Cellulaires et leur Blocage chez Plusieurs Protistes" (M. Lefort-Tran and R. Valencia, eds.), Colloque No. 240, pp. 53–67. Centre National de la Recherche Scientifique, Paris.

Edmunds, L. N., Jr. (1977). *In* "Proceedings of the XII International Conference of the International Society for Chronobiology, Washington, D.C., 1975," pp. 571–577. Il Ponte, Milan.

Edmunds, L. N., Jr. (1978). *In* "Aging and Biological Rhythms" (H. V. Semis, Jr. and S. Capobianco, eds.), pp. 125–184. Plenum, New York.

Edmunds, L. N. Jr. (1980a). *In* "Chronobiology: Principles and Applications to Shifts in Schedules" (L. E. Scheving and F. Halberg, eds.), pp. 205–228. Sijthoff and Noordhoff, Alphen aan den Rijn, Netherlands.

Edmunds, L. N., Jr. (1980b). *In* "Endocytobiology: Endosymbiosis Cell Biology, A Synthesis of Recent Research" (W. Schwemmler and H. E. A. Schenk, eds.) pp. 685–702. de Gruyter, Berlin.

Edmunds, L. N., Jr. (1981). *In* "International Cell Biology 1980–1981." (H. G. Schweiger, ed.), pp. 831–845. Springer-Verlag, Berlin and New York.

Edmunds, L. N., Jr., and Adams, K. (1981). *Science* **211**, 1002–1013.

Edmunds, L. N., Jr., and Cirillo, V. P. (1974). *Int. J. Chronobiol.* **2**, 233–246.

Edmunds, L. N., Jr., and Funch, R. (1969a). *Science* **165**, 500–503.

Edmunds, L. N., Jr., and Funch, R. (1969b). *Planta* **87**, 134–163.

Edmunds, L. N., Jr., and Halberg, F. (1981). *In* "Neoplasms—Comparative Pathology of Growth in Animals, Plants and Man" (H. E. Kaiser, ed.), pp. 105–134. William and Wilkins, Baltimore, Maryland.

Edmunds, L. N., Jr., and Laval-Martin, D. L. (1981). *In* "Photosynthesis," Vol. VI, *Proceedings of the 5th International Congress on Photosynthesis, September 7–13, 1980, Halkidiki, Greece.* (G. Akoyunoglou, ed.), pp. 313–322. International Science Services, Jerusalem.

Edmunds, L. N., Jr., Chuang, L., Jarrett, R. M., and Terry, O. W. (1971). *J. Interdisc. Cycle Res.* **2**, 121–132.

Edmunds, L. N., Jr., Sulzman, F. M., and Walther, W. G. (1974). *In* "Chronobiology" (L. E. Scheving, F. Halberg, and J. Pauly, eds.), pp. 61–66. Igaku Shoin, Tokyo.

Edmunds, L. N., Jr., Jay, M. E., Kohlmann, A., Liu, S. C., Merriam, V. H., and Sternberg, H. (1976). *Arch. Microbiol.* **108**, 1–8.

Edmunds, L. N., Jr., Tay, D. E., and Laval-Martin, D. L. (1982). *In* "Proceedings of the XV International Conference of the International Society for Chronobiology, 1981, Minneapolis (in press).

Ehret, C. F. (1974). *Adv. Biol. Med. Phys.* **15**, 47–77.

Ehret, C. F., and Dobra, K. W. (1977). *In* "Proceedings of the XII International Conference of the International Society for Chronobiology, Washington, D.C.," pp. 563–570. I. Ponte, Milan.

Ehret, C. F., and Trucco, E. (1967). *J. Theor. Biol.* **15**, 240–262.

Ehret, C. F., and Wille, J. J. (1970). In "Photobiology of Microorganisms" (P. Halldal, ed.), pp. 369–416. Wiley (Interscience), New York.

Ehret, C. F., Meinert, J. C., Groh, K. R., Dobra, K. W., and Antipa, G. (1977). *In* "Growth Kinetics and Biochemical Regulation of Normal and Malignant Cells" (B. Drewinko and R. Humphrey, eds.), pp. 49–76. Williams and Wilkins, Baltimore, Maryland.

Engelberg, J. (1964). *Exp. Cell Res.* **36**, 647–662.

Enright, J. T. (1971). *Z. Vgl. Physiol.* **72**, 1–16.

Feldman, J. F. (1967). *Proc. Natl. Acad. Sci. USA* **57**, 1080–1087.

Feldman, J. F. (1968). *Science* **160**, 1454–1456.

Feldman, J. F., and Atkinson, C. A. (1978). *Genetics* **88**, 255–265.

Feldman, J. F., and Bruce, V. G. (1972). *J. Protozool.* **19**, 370–373.

Finch, C. E., and Hayflick, L., eds. (1977). "Handbook of the Biology of Aging." Van Nostrand-Reinhold, Princeton, New Jersey.

Frisch, L., ed. (1960). "Biological Clocks," Vol. 25, *Cold Spring Harbor Symp. Quant. Biol.,* Cold Spring Harbor Laboratory, Cold Spring Harbor, New York.

Gilbert, D. A. (1978). *Biosystems* **10**, 235–240.

Goodwin, B. C. (1963). "Temporal Organization in Cells." Academic Press, New York.

Goodwin, B. C. (1976). "Analytical Physiology of Cells and Developing Organisms." Academic Press, New York.

Halberg, F. (1975). *Indian J. Cancer* **12**, 1–20.

Halberg, F., Tong, Y. L., and Johnson, E. A. (1967). *In* "The Cellular Aspects of Biorhythms" (H. V. Mayersbach, ed.), pp. 20–48. Springer-Verlag, Berlin and New York.

Halberg, F., Johnson, E. A., Nelson, W., Runge, W., and Sothern, R. (1972). *Physiol. Teacher* **1**, 1–11.

Halberg, F., Katinas, G. S., Chiba, Y., Garcia-Sainz, M., Kovacs, T. G., Künkel, H., Montalbetti, N., Reinberg, A., Scharf, R., and Simpson, H. (1973). *Int. J. Chronobiol.* **1**, 31–63.

Hastings, J. W. (1959). *Annu. Rev. Microbiol.* **13,** 297–312.

Hastings, J. W., and Schweiger, H.-G., eds. (1976). "The Molecular Basis of Circadian Rhythms." Dahlem Konferenzen, Berlin.

Haus, E., Halberg, F., Kühl, J. F. W., and Lakatua, D. J. (1974). *Chronobiologia Suppl. 1,* **1,** 122–156.

Hellebust, J. S., Terborgh, J., and McLeod, G. C. (1967). *Biol. Bull.* **133,** 670–678.

Hoffmans, M., and Brinkmann, K. (1979). *Chronobiologia* **6,** 111.

Holliday, R. (1975). *Fed. Proc., Fed. Am. Soc. Exp. Biol.* **34,** 51–55.

Jarrett, R. M., and Edmunds, L. N., Jr. (1970). *Science* **167,** 1730–1733.

Kauffman, S., and Wille, J. J. (1975). *J. Theor. Biol.* **55,** 47–93.

Kauffman, S. A., and Wille, J. J. (1976). *In* "The Molecular Basis of Circadian Rhythms" (J. W. Hastings and H.-G Schweiger, eds.), pp. 421–431. Dahlem Konferenzen, Berlin.

Kiefner, G., Schliessmann, F., and Engelmann, W. (1974). *Intl. J. Chronobiol.* **2,** 189–195.

Kirschstein, M. (1969). *Planta* **85,** 126–134.

Klevecz, R. R. (1969). *J. Cell Biol.* **43,** 207–219.

Klevecz, R. R. (1976). *Proc. Nat. Acad. Sci. USA* **73,** 4012–4016.

Klevecz, R. R. (1978). *In* "Cell Reproduction" (E. R. Dirksen, D. M. Prescott, and C. F. Fox, eds.), pp. 139–146. Academic Press, New York.

Klevecz, R. R., King, G. A., and Shymko, R. M. (1980). *J. Supramolec. Struct.* **14,** 329–342.

Klevecz, R. R., Kros, J., and Gross, S. D. (1978). *Exp. Cell Res.* **116,** 285–290.

Kreuels, T., and Brinkmann, K. (1979). *Chronobiologia* **6,** 121.

Kreuels, T., Martin, W., and Brinkmann, K. (1979). Proceedings of the Discussion Meeting on the Kinetics of Physiochemical Oscillations, Aachen, 1979, Vol. 1, pp. 51–60.

Krieger, D. T., ed. (1979). "Endocrine Rhythms." Raven, New York.

Laval-Martin, D., Shuch, D., and Edmunds, L. N., Jr. (1979). *Plant Physiol.* **63,** 495–502.

Ledoigt, G., and Calvayrac, R. (1979). *J. Protozool.* **26,** 632–643.

Lonergan, T. A., and Sargent, M. L. (1978). *Plant Physiol.* **61,** 150–153.

Lonergan, T. A., and Sargent, M. L. (1979). *Plant Physiol.* **64,** 99–103.

Lövlie, A., and Farfaglio, G. (1965). *Exp. Cell Res.* **39,** 418–434.

Marsden, M. P. F., and Laemmli, U. K. (1979). *Cell* **17,** 849–858.

McMurry, L., and Hastings, J. W. (1972). *Science* **175,** 1137–1139.

Menaker, M. (1971). "Biochronometry." Nat. Acad. Sci., Washington, D.C.

Mitchell, J. L. A. (1971). *Planta* **100,** 244–257.

Mitchison, J. (1971). "The Biology of the Cell Cycle." Cambridge Univ. Press, London and New York.

Mitchison, J. M. (1974). *In* "Cell Cycle Controls" (G. M. Padilla, I. L. Cameron, and A. Zimmerman, eds.), pp. 125–142. Academic Press, New York.

Nelson, W., Tong, Y. L., Lee, J.-K., and Halberg, F. (1979). *Chronobiologia* **6,** 305–323.

Njus, D., Sulzman, F., and Hastings, J. W. (1974). *Nature (London)* **248,** 116–120.

Palmer, J. D., Brown, F. A., Jr., and Edmunds, L. N., Jr. (1976). "Introduction to Biological Rhythms." Academic Press, New York.

Paulson, J. R., and Laemmli, U. K. (1977). *Cell* **12,** 817–828.

Pavlidis, T. (1969). *J. Theor. Biol.* **22,** 418–436.

Pavlidis, T. (1971). *J. Theor. Biol.* **33,** 319–338.

Pavlidis, T. (1973), "Biological Oscillators: Their Mathematical Analysis." Academic Press, New York and London.

Pohl, R. (1948). *Z. Naturforsch.* **3b,** 367–374.

Prézelin, B. B., and Sweeney, B. M. (1977). *Plant Physiol.* **60,** 388–392.

Proceedings XIII, XIV, and XV Int. Conf. Int. Soc. Chronobiol., Pavia (Italy) 1977, Hannover 1979, and Minneapolis 1981. In Press.

Reinberg, A., and Halberg, F., eds. (1979). "Chronopharmacology." Pergamon, Oxford.

Rusch, H. P., Sachsenmaier, W., Behrens, K., and Gruter, V. (1966). *J. Cell Biol.* **31**, 204-209.
Sachsenmaier, W. (1976). *In* "The Molecular Basis of Circadian Rhythms" (J. W. Hastings and H.-G. Schweiger, eds.), pp. 410-420. Dahlem Konferenzen, Berlin.
Sachsenmaier, W., Remy, U., and Plattner-Schobel, R. (1972). *Exp. Cell Res.* **73**, 41-48.
Samis, H. V., Jr., and Capobianco, S., eds. (1978). "Aging and Biological Rhythms." Plenum, New York.
Sargent, M. L. (1976). Group Report: The Role of Genes and Their Expression. *In* "The Molecular Basis of Circadian Rhythms" (J. W. Hastings and H.-G. Schweiger, eds.), pp. 295-310. Dahlem Konferenzen, Berlin.
Saunders, D. S. (1976). "Insect Clocks." Pergamon, Oxford.
Scheving, L., and Halberg, F., eds. (1980). Proc. NATO Advanced Study Institute on Chronobiology, Hannover 1979. A. W. Sijthoff, Alphen aan den Rijn, Netherlands.
Scheving, L., Halberg, F., and Pauly, J., eds. (1974). "Chronobiology." Igaku Shoin, Tokyo.
Schnabel, G. (1968). *Planta* **81**, 49-63.
Scott, B. I. H., and Gulline, H. F. (1975). *Nature (London)* **254**, 69-70.
Schweiger, H.-G., and Schweiger, M. (1977). *Int. Rev. Cytol.* **51**, 315-342.
Senger, H. (1975). *In* "Les Cycles Cellulaires et leur Blocage chez Plusieurs Protistes" (M. Lefort-Tran and R. Valencia, eds.), Colloque No. 240, pp. 101-108. Centre National de la Recherche Scientifique, Paris.
Smith, J. A., and Martin, L. (1973). *Proc. Nat. Acad. Sci. USA* **70**, 1263-1267.
Spudich, J. L., and Sager, R. (1980). *J. Cell Biol.* **85**, 136-146.
Sulzman, F. M., and Edmunds, L. N., Jr. (1972). *Biochem. Biophys. Res. Commun.* **47**, 1338-1344.
Sulzman, F. M., and Edmunds, L. N., Jr. (1973). *Biochim. Biophys. Acta* **320**, 594-609.
Sweeney, B. M. (1960). *Cold Spring Harbor Symp. Quant. Biol.* **25**, 145-148.
Sweeney, B. M. (1965). *In* "Circadian Clocks" (J. Aschoff, ed.), pp. 190-194. North-Holland Publ., Amsterdam.
Sweeney, B. M. (1969a). "Rhythmic Phenomena in Plants." Academic Press, New York.
Sweeney, B. M. (1969b). *Can. J. Bot.* **47**, 299-308.
Sweeney, B. M. (1972). Proc. Int. Symp. Circadian Rhythmicity, Wageningen, 1971, pp. 137-156.
Sweeney, B. M., Prézelin, B. B., Wong, D., and Govindjee (1979). *Photochem. Photobiol.* **30**, 309-311.
Terry, O., and Edmunds, L. N., Jr. (1969). *Biotechnol. Bioeng.* **11**, 745-756.
Terry, O. W., and Edmunds, L. N., Jr. (1970a). *Planta* **93**, 106-127.
Terry, O. W., and Edmunds, L. N., Jr. (1970b). *Planta* **93**, 128-142.
Tyson, J., and Kauffman, S. (1975). *J. Math. Biol.* **1**, 289-310.
Tyson, J., and Sachsenmaier, W. (1978). *J. Theor. Biol.* **73**, 723-738.
Vanden Driessche, T. (1970). *J. Interdisc. Cycle Res.* **1**, 21-42.
Vanden Driessche, T., Dujardin, E., Magnusson, A. and Sironval, C. (1976). *Int. J. Chronobiol.* **4**, 111-124.
Walther, W. G., and Edmunds, L. N., Jr. (1970). *J. Cell Biol.* **46**, 613-617.
Walther, W. G., and Edmunds, L. N., Jr. (1973). *Plant Physiol.* **51**, 250-258.
Wever, R. A. (1979). "The Circadian System of Man." Springer-Verlag, Berlin and New York.
Wille, J. J., Jr. (1979). *In* "Biochemistry and Physiology of Protozoa" (M. Levandowsky and S. H. Hutner, eds.), 2nd ed., Vol. 2, pp. 67-149. Academic Press, New York.
Winfree, A. (1967). *J. Theoret. Biol.* **16**, 15-42.
Wolken, J. J. (1967). "Euglena: An Experimental Organism for Biochemical and Biophysical Studies," 2nd ed. Appleton, New York.

CHAPTER 4

MOVEMENT AND LOCOMOTION OF *EUGLENA*

Eugene C. Bovee

I. Introduction

As noted in our earlier review (Jahn and Bovee, 1968a), *Euglena viridis* may have been one of the first swimming microorganisms seen by Leeuwenhoek in 1674 (as cited by Dobell, 1932). Nevertheless, the study of the movements and locomotion of *Euglena* has lagged far behind in the recent rush of research directed at its biochemistry, its growth (with associated rhythms), and its fine structure, all mainly in the single species, *Euglena gracilis*.

THE BIOLOGY OF *EUGLENA*, VOL. 3

It is encouraging that the new research on movements of *Euglena*, limited as it is, aims to determine the mechanisms within the organism that produce movement rather than to describe the movement or to "tinker" with external treatments in order to cause the organism to react.

Our earlier review (Jahn and Bovee, 1968a) devoted much space to a recounting and reassessment of the older literature of the previous century since Engelmann's early experiments (Engelmann, 1869). We shall not repeat that effort here (except to cite some of the older literature where it is particularly pertinent). Instead, we shall concentrate on the newer, though limited, literature and examine it critically.

II. Types of Movements by *Euglena*

The major kinds of movements demonstrated by *Euglena* and other euglenoids are (1) swimming, (2) contraction, (3) crawling, and (4) gliding. Some species perform all four types, for example, *Euglena spirogyra* (Leedale *et al.*, 1965). Others, with short or no discernible flagella, such as *Euglena mutabilis*, contract and crawl but do not swim, and rarely glide (Lackey, 1938). Yet others swim slowly, rarely contract, and do not glide, such as *Euglena acus*. Still others swim well, but are of rigid form and cannot contract, and have not been known to glide, such as *Euglena tripterus*. It is therefore difficult to generalize about any kinds of movements.

Besides the individual movements observed in a single species, *Euglena*, as other microorganisms, show phenomena of mass movement, particularly aggregation, and, in dense numbers, swimming in patterns. These latter movements will also be considered in this chapter.

III. Swimming

A. THE PROPULSIVE FORCE

To swim *Euglena* must develop a force that thrusts against the water and thereby propels it through the water. That is no mean feat! The mass of the body, even of a larger species of *Euglena*, is so small that it cannot move through the water without continually applying force. Calculations of the inertial force of its body in motion relative to the viscous force (drag) of the water show that without the application of propulsive force the *Euglena* cannot move. It has so little inertial force that the drag of the water stops it, almost instantly, in a fraction (less than ⅛) of the length of its body, as soon as the propulsive force ceases (Holwill, 1966b; Jahn and Bovee, 1968a, b).

The only organelle that produces such force is the flagellum. Most *Euglena* spp. have two flagella. However, one is short and found within the anterior pocket (reservoir) and serves only as a prop at the base of the longer, externally protruding flagellum. Furthermore, the longer, locomotory flagellum is also subject to the laws of hydrodynamics and physics and has so little mass that an undulation cannot progress along it by the inertia of a thrust generated at one end. It must, and does exert force sequentially and continuously along its length from one end to the other to produce a traveling wave (Gray, 1928; Taylor, 1951; Machin, 1958; Miles and Holwill, 1971; Rikmenspoel, 1971; Brokaw, 1972b).

B. The Swimming Path

The pathway that *Euglena* spp. follow as they swim is always a helix of variable diameter (Lowndes, 1943, 1944; Holwill, 1966a). The diameter may be relatively wide for some species, such as the small *Euglena pisciformis,* narrow for a slender, long, more massive species with a short flagellum, such as *E. acus.* The path is not a spiral, as some have called it (a spiral is a pathway traced around a cone; a helix is a pathway traced around a cylinder).

C. The Locomotory Apparatus (The Flagellum)

By electron microscopy, the anatomy of the flagellum* of various species of *Euglena* has been shown to consist of the flagellar membrane that surrounds the axoneme, the paraflagellar rod, and photoreceptor. Fibers include the mastigonemes that extend from and are attached to the membrane (and perhaps to the paraflagellar rod within) and the felted, fibrillar coat that covers the flagellar membrane.

1. The Flagellar Membrane

The flagellar membrane is a tubular extension of the surface membrane of the cell, and, like it, shows the bilaminar structure of a eukaryotic cell's membrane. There is excellent evidence that it is sensitive to stimuli (Holwill, 1966b; Leedale, 1967; Mikolajczyk, 1973, and Mikolajczyk and Diehn, 1976), and it is probably the principal sensory structure of the cell, other than the photoreceptor. It responds to electrical stimuli (Bancroft, 1915), chemical stimuli (Mikolajczk, 1973), mechanical stimuli (Mikolajczyk and Diehn, 1976), or combinations of these (Bancroft, 1915). These stimuli can be relayed by the flagellum to the cell, either by way of the membrane (Mikolajczyk, 1972) or internally, perhaps by the flagellar rod (Piccinni et al., 1975), which is in contact with the externally projecting mastigonemes (Leedale et al., 1965), or by both.

*See Figs. 1–4, 11–13, and 22 of Chapter 1 and Figs. 12–15 of Chapter 2 in this volume.

2. The Axoneme

As in other flagella, the axoneme is composed of nine peripheral doublet microtubules, each with a pair of so-called "arms" attached to the α-microtubule of each doublet. There are two separate, centrally located microtubules that run the length of the flagellum. The peripheral doublets are held in position by so-called "radial spokes" that are attached to a central sheath that surrounds the two central tubules at their inner ends and to the doublets peripherally (Reger and Beams, 1954). This complex proximally arises from a basal plate above (and attached to) a basal body (kinetosome or blepharoplast) of centriolar configuration, imbedded in the cytoplasm at the base of the anterior reservoir.

3. The Paraflagellar Rod

This organelle is a cylindrical bundle of longitudinal, solid fibrils that are cross-linked abundantly along their lengths, so that any cross section resembles a grid. It lies adjacent to the axoneme within the flagellar membrane and extends from the base to the tip (or nearly so) of the flagellum (Leedale et al., 1965; Mignot, 1965; Piccinni and Albergoni, 1973).

4. The Photoreceptor

Near the base of the flagellum, between the axoneme and the paraflagellar rod, lies a crystalloid, conical- to lens-shaped photoreceptor that is attached to the paraflagellar rod. Its fine structure is variously described by electron microscopists, but it appears to be composed of rodlets interwoven or twined so that in cross section the periodicity of a crystalloid is evident (Mignot, 1966; Walne and Arnott, 1967; Kivic and Vesk, 1972; Wolken, 1977). This lens-like crystalloid rests on a fibrillar mat that in turn lies against (or is attached to?) the paraflagellar rod and is partly surrounded by the basal end of this rod (Kivic and Vesk, 1972).

5. The Mastigonemes

The mastigonemes are proteinaceous fibrils that protrude through the membrane in a single row along the side of the flagellum that borders the paraflagellar rod. They are flexible and elastic. Those of E. gracilis are each about 1.5 to 4.0 μm long, about 0.2 to 0.3 μm in diameter, and set about 0.1 μm apart. Each one is composed of two or three subfibrils (DeFlandre, 1934; Pitelka and Schooley, 1955; Leedale et al., 1965; Mignot, 1966).

All species of Euglena have more or less distinct spiral striations on the surface of the body.

D. THE MECHANISM OF SWIMMING

1. *The Role of the Flagellum*

Since swimming requires the continual expenditure of energy, propulsion requires that force be applied against the water by the flagellum. Thus, most of the research on the mechanism of propulsion has been directed at the flagellum.

It is now confirmed that the axoneme contains structural elements—the peripheral microtubular doublets and their dynein arms—that use energy derived from adenosine triphosphate (ATP) to produce local and sequential contractions of the dynein arms. These bend and counterslide local regions of the microtubules to produce an undulating wave that progresses along the flagellum.

Visual observations (Holwill, 1966a,b; Jahn and Bovee, 1968a) and mathematical calculations (Holwill, 1966b; Brokaw, 1971; Rikmenspoel, 1971) suggest that the flagellum is stiff but also elastic wherever it is bent (Brokaw, 1971; Rikmenspoel, 1971; Miles and Holwill, 1971; Lindemann *et al.*, 1973).

2. *The Mechanochemical Mechanism*

The energy source is assumed to diffuse into the flagellum from the cell body (Carlson, 1962; Lin, 1972; Nevo and Rikmenspoel, 1970), being dephosphorylated by the ATPase activity of the dynein arms (Miles and Holwill, 1971; Bovee and Jahn, 1972; Brokaw, 1972a,b). Two ATPases (other than the potassium-activated enzyme of the membrane), one activated by Mg^{2+} and the second by Ca^{2+}, are present in the flagellum (Piccinni and Albergoni, 1973). As the ATP of a region of the flagellum is degraded to ADP, that region undergoes rigor and remains so until the ATP is restored by replacement or by reconstruction by the action of a kinase (Lin, 1972) or transphosphorylase (Acuña, 1970).

The physical couple between the dynein arms and the β-microtubule that precedes contraction of the dynein and bending of the tubule requires Mg^{2+}, as does the actin–myosin linkage in muscle (Wolken, 1967; Nichols and Rikmenspoel, 1978). Removal of Mg^{2+} from the external medium paralyzes the flagellum of *Euglena* (Wolken, 1967) as does chelation of internal Mg^{2+} by microinjection of EGTA (Nichols and Rikmenspoel, 1978), with Mn^{2+} then substituting for Mg^{2+} in restoring the linkage (Nichols and Rikmenspoel, 1978). Ca^{2+} may also be involved in the physical coupling (Wolken, 1967; Piccinni and Albergoni, 1973).

The availability of ATP regulates the rate of undulation of the flagellum and therefore the rate of swimming (Wolken, 1967; Brokaw, 1972a), although the amplitude of the waves is not thereby increased (Acuña and Bovee, 1979). Extraneous ATP added to the medium can be taken up by the flagellum and used either when the flagellum is broken free of the cell (Mahenda *et al.*, 1967) or by the intact cell (Bovee *et al.*, 1969; Bovee and Acuña, 1970; Acuña and Bovee, 1979). When $1 \times 10^{-5} M$ K_2ATP, dissolved in Chalkley's solution (Chalkley, 1930), was used as the medium for testing, frequency of flagellar undulation increased almost immediately from 20 pulses/second–47 pulses/second. In the same brief period of time, swimming rate accelerated from 142 μm/second–190 μm/second; after 3 hours it reached 260 μm/second, with subsequent return to normal after 24 hours. At $1 \times 10^{-4} M$ K_2ATP, swimming rate increased less rapidly initially, but reached 255 μm/second in 3 hours and continued to increase to 282 μm/second at 24 hours. Greater concentrations of ATP ($1 \times 10^{-3} M$ or more) inhibited swimming (Acuña and Bovee, 1979).

3. The Form of the Undulating Wave

Early critical observers noted that the undulatory waves along the flagellum are helical (Uhlela, 1911; Bancroft, 1915; Lowndes, 1944). High-speed cinematography has confirmed the helical nature of each wave (Holwill, 1966a; Jahn and Bovee, 1968a; Votta *et al.*, 1972). However, the waves are not continuously sequential, but progress intermittently, so that for species that have flagella about as long or slightly longer than the body, such as *E. viridis* and *E. gracilis,* only two waves (separated by a rigid, nearly-straight region of the flagellum) progress along the flagellum simultaneously. For species with a shorter flagellum, such as *E. acus* or *E. ehrenbergi,* one wave occurs at a time and for species with a flagellum twice the length (or more) of the body, such as *E. inflata*, three or four waves occur at the same time. Jahn and Bovee (1968a) stated that the beat is initially planar as it begins at the base of the flagellum in the anterior reservoir and is converted to helical at the bend of the flagellum (around the anterior end into the trailing position), which is also where the unilateral row of mastigonemes begins (Leedale *et al.*, 1965).

How the beat originates in the flagellum is not known, but most theories for base-to-tip undulations of flagella assume the impulse originates in the basal body of the flagellum (Lowndes, 1936; van Herpen and Rikmenspoel, 1968; Lubliner and Blum, 1971; Bovee and Jahn, 1972) and is relayed to the axoneme, perhaps being coordinated by cross fibrils at the basal level (Lang, 1963; Costello, 1973).

A planar beat would require that groups of subfibrils on opposite sides of the flagellum be activated alternately (Costello, 1973). A helical wave would require a minimum of three subfibrils to be activated out of phase and sequentially, or at least three groups to be so activated (Jahn and Landman, 1965).

Since the waves of *Euglena* begin as planar waves, it has been proposed that their conversion to helical is due to torque generated by the viscous drag of water on the flagellum, causing it to bend as it extends from the opening of the reservoir. This added viscous drag of the water on the unilateral row of mastigonemes adds to the torque (Holwill, 1966a). For species of *Euglena* having a short flagellum that projects forward, e.g., *E. acus,* the singly generated helical waves act as a rotating propeller ("like a spinning lasso," Leedale *et al.,* 1965), pulling the body forward through the water, as can be seen and demonstrated by models (Lowndes, 1943; Brown and Cox, 1954; Leedale *et al.,* 1965; Jahn *et al.,* 1979).

How the impulses that initiate and promulgate the waves travel along the flagellum is not known. The membrane can relay an impulse and alter the beat if it is stimulated (Holwill, 1966b; Mikojajczyk and Diehn, 1976), but such impulses are superimposed on the normal pulsing.

Bovee and Jahn (1972) suggest that alterations of cation flux from segment to segment of the quasicrystalline fibers of the flagellum, driven by piezo electric currents generated in the bending and straightening of the segments in sequence, constitute a possible mechanism of impulse transfer. They do not identify the segments, but it can be assumed that they are represented by the paired dynein arms, since the latter occur at regular intervals along each α-subfibril of the axoneme. Bovee and Jahn do not specify which fibrils are bent, but presumably they are the microtubules, either central, peripheral, or both. The initial impulse, they assume, is generated as a flow of electrons from intermittent pulses due to oxidation–reduction transfers, perhaps from mitochondrial activities or basal-body ATPases. They further suggested that changes in the ratios of cations as they oscillate due to the piezo-electric current changes, provide the necessary quantity of Mg^{2+} and Ca^{2+} at the sites required to establish the mechanochemical couple; that in turn promotes dynein–tubulin attachment and interaction and produces the sequential, localized bendings, resulting in the propagation of the wave. Bovee and Jahn presented no mathematical computations to support their assumptions, but did cite much of the known literature about flagellar movements of *Euglena* and of other flagella in support of their assumptions. Brokaw (1972a) also indicates that localized and sequential bending of the flagellum regulates the active undulation, and Lubliner and Blum (1971, 1972) suggest that an active sequence is started only after an initial, passive bending stimulates it, further sequential bending being an energy-using and -distributing process.

4. The Role of Mastigonemes

The unilateral row of mastigonemes on the flagellum of *Euglena* plays some role in the efficiency of flagellar propulsion since these structures are known to whip against the water. The added viscous drag of the water on them probably helps change the initially planar waves of the axoneme to helical waves as a result of increased torque on the flagellar shaft (Holwill, 1966a; Jahn and Bovee,

1968a). They move and bend as the flagellum undulates, but not independently as Fischer (1894) thought, thus indicating that they are elastic (DeFlandre, 1934; Leedale *et al.*, 1965).

If the flagellar waves were continuously sequential, they would wrap the mastigonemes around the flagellum; however, in the interrupted helical beating of the flagellum, they remain extended and are whipped against the water by each helical wave, probably adding thrust to that of the flagellar shaft (Jahn and Bovee, 1968a).

The surface of the flagellum is covered by a felted layer of microfilaments (Leedale *et al.*, 1965). Since a rough-surfaced cylinder develops more usable thrust at a micro-level than a smooth one (Taylor, 1951), the felted surface of the flagellum may also increase the thrusting efficiency of the flagellum.

5. *The Role of the Paraflagellar Rod*

This organelle has usually been ignored in discussions of the movements of *Euglena*. Bovee and Jahn (1972), however, mention a possible role in the conduction of cations as part of the relay system for promoting the progress of the undulating wave. Piccinni and Albergoni (1973) noted that it contains ATPase activity and also speculated that it may conduct impulses. Its physical contact with the photoreceptor also suggests a function in conducting a stimulus.

The presence of ATPase suggests that the paraflagellar rod may also be contractile (Piccinni *et al.*, 1975). Its potential contractility may be involved in throwing the flagellum sideways and/or forward during the well-known "shock response," perhaps so violently as to autotomize it, as often occurs in that response. Its stiffness, as suggested by its fibrillar nature, may be augmented by ADP, which accumulates after degradation of ATP during the passage of an undulatory wave. The stiffness could damp out some undulatory waves of the axoneme, whereas intermittent ATPase activity along the paraflagellar rod could augment other waves of the axoneme. The result could well be the intermittent helical undulations along the flagellum that are characteristic of swimming *Euglena*.

6. *The Role of the Photoreceptor*

The recent research of Tollin and Diehn and their co-workers in the 1960s and 1970s, as well as others, showed clearly that the photoreceptor is responsible for orientation of the body toward weak light and away from strong light. Much experimental evidence and electron microscopic studies suggest that the photoreceptor is a lattice-like, quasicrystalline organelle that acts as a capacitor, relaying electrons to the paraflagellar rod, which stiffens, causing the flagellum to be swung sideways, with waves of large amplitude coursing along the flagellum. These turn the body away from a source of light if it is stronger than a critical level (above 0.2 kW/m² of white light) or toward the light if it is weaker than another critical level [below 0.1 kW/m² of white light (Diehn *et al.*, 1975)]. The photoreceptor is particularly sensitive to blue light (420–490 nm) from the side (Gössel, 1957). The photosensitive molecules are probably flavinoid (Tollin and Robinson, 1969; Diehn, 1972; Wolken, 1977) and there may be two distinct types (Mikolajczyk and Diehn, 1976). Details of the activity of the photoreceptor are discussed more fully in Chapter 5 in this volume.

7. The Role of the Stigma

The stigma or "eyespot" was long-believed to be the photosensitive organelle (e.g., Künstler, 1886; Francé, 1893; Steuer, 1904; Magenot, 1926; Wolken, 1961). However, the stigma has been shown to absorb certain wavelengths of light, thus lowering their intensities so that, as the swimming *Euglena* gyrates along its helical pathway, the true photoreceptor is only intermittently stimulated (e.g., Gössel, 1957; Leedale, 1967; Jahn and Bovee, 1968a). *Euglena* that lack the stigma, but have the photoreceptor, are photonegative, whereas those lacking both are insensitive to light (Gössel, 1957; Vavra and Aaronson, 1962). Principal pigments of the stigma are carotenoids in lipid droplets and one to several droplets can be found within a membrane (Chadefaud, 1937; Gibbs, 1960; Krinsky and Goldsmith, 1960; Wolken, 1977). There may also be one or more flavins present (Sperling-Pagni *et al.*, 1976; Wolken, 1969). In *E. granulata* there appears to be a central crystalloid body in the center of the stigma (Walne and Arnott, 1967).

In strong intensities of light, the stigma cannot absorb enough quanta to prevent the photophobic response, but at lower intensities, it absorbs enough quanta to permit this response. This suggests that the photoreceptor has both an "on" and an "off" response (Diehn, 1972; Acuña and Bovee, 1979).

8. Role of the Cell Body

The body of various species of *Euglena* is surface-striated (β-helical) and in some is conspicuously ridged. In others, it is flattened and/or twisted; a combination of all of these types can occur. There has been much speculation and some observation about the role of the body in producing the thrust needed to promote swimming. Dangeard (1890) thought the rotation of the body in swimming was due to the spiral striations of the pellicle, an opinion reiterated recently by Guttman and Ziegler (1974), who likened them to the "rifling" of a gun barrel. Pochmann (1953) held a similar opinion. Others have indicated that the striations alone, even if they do affect rotation of the body, would not add to the forward component of movement but rather serve mainly to help keep the cell on a regular path (Holwill, 1966a). The shape of the body has more often been cited as contributing to forward motion. Holwill's computations (Holwill, 1966a) suggest that the spindle shape of some *Eugena* spp., such as *E. viridis,* adds no forward component to swimming but may reduce viscous drag. A spherical organism, such as *E. inflata,* would then be more subject to the effects of a viscous drag.

Lowndes (1943, 1944), Holwill (1966a), Jahn and Bovee (1968a), and Votta *et al.* (1972) suggest that body shape determines the length of the flagellum as well as the manner of its beating needed for propulsion. Gray (1953) and Brenner and Winet (1977) have demonstrated mathematically that the progression of helical waves along a flagellum develops a torque about its longitudinal axis such that material of the flagellum and cell body both rotate with the same angular velocity. This causes an equal and opposite torque to develop in the body of the cell, and the cell then gyrates. Since the flagellum of *E. gracilis* trails, the undulations pull upon the cell so that it is inclined by the drag on the flagellum (countered, partly, by the drag on the cell body), and the cell gyrates as an inclined plane. A spherical body would present more resistance (effective drag surface) than a spindle-shaped one, but neither, as Holwill (1966a) points out, would add an effective forward component to the progression of the cell. However, a flattened and/or twisted body would act as an inclined plane. If flattened, it adds a screw-like forward component of thrust and, if twisted, it would be even more effective, since the entire body would serve as a screw propeller (Lowndes, 1943; Jahn and Bovee, 1968a; Votta and Jahn, 1971; Votta *et al.,* 1972). Therefore, a short flagellum undulating at either right angles to or rotating while held forward of the flattened and/or twisted body would cause the latter to rotate and gyrate and move forward as its own screw propeller (Lowndes, 1943, 1944; Votta *et al.,* 1972). In the case of *E. acus,* the complicated "pretzel-like" beat of the short flagellum acts as the screw propeller, dragging the spindle-shaped body through the water (Jahn, 1974) with rotation, but little, if any, gyration, except at the slightly curved front portion.

Holwill's (1966a) mathematical computations would seem to deny the role of the body in developing the forward component of thrust. High-speed cinephotomicrographs, however, show that this does occur in euglenoids that have the flattened and/or twisted shape (Votta *et al.,* 1972) as Lowndes (1944) proposed. Holwill (1966a) also demonstrated by mathematical computation that a ridged and twisted species, such as *E. tripterus,* on which the ridges are twisted in a direction opposite to the helical path of the undulating waves along the flagellum, will develop a counter torque and thrust, which augments that produced by the flagellum. This thereby increases the efficiency of swimming.

Since the intermittent beating of the flagellum produces a pulsating torque, a swimming *Euglena* gyrates in a "wobbly" pattern (as noted by Lowndes, 1944) as shown photographically (Holwill, 1966a; Jahn and Bovee, 1968a; Votta *et al.,* 1972).

9. The Role of Mucus Secretion

Although mucocysts are known to be distributed along the striations of *Euglena*, e.g., in *E. spirogyra* (Leedale *et al.*, 1965), no comment has been made previously that mucus secretions may assist swimming. Jahn *et al.* (1965) showed photographically that ciliates, such as *Paramecium* and *Tetrahymena*, secret a "slip-stream" of mucus as they swim, leaving a thin but distinct trail of mucus behind. *Euglena* may also continuously secrete mucus, but no tests have yet established this likely occurrence. If this does occur, then the mucus, acting as a lubricant between the body and the water, would improve the efficiency of swimming. Brokaw (1971) has shown that increasing the viscosity of the medium increases the viscous drag on both the flagellum and the body, slowing the rate of flagellar undulations, gyration of the body, and forward motion. Spoon *et al.* (1977) found that polyethylene oxide, added as a viscosity-increasing material to water, progressively slowed the forward motion of *Euglena*, and that the decrease in motion was a linear function of the increasing viscosity so that at the level of viscosity which stops swimming progress, the cell is drawn backward by the elastic recoil of the viscous material. A spindle-shaped organism, such as *E. gracilis*, is the most efficient in reducing viscous drag. It follows, then, that any substance that reduces viscous drag by lubrication of the body surface will, as postulated above for mucus, increase the efficiency of propulsion and the swimming rate. This conclusion is predicted by Winet's (1976) calculations for the lubricating effect of mucus on an object passing through a tube of small diameter. In the case of *Euglena*, the "small diameter tube" would be the surrounding water.

10. The Efficiency (Inefficiency?) of Energy Use

Brokaw (1961) calculated that less than 10% of the energy used by a flagellum is used to propel the cell. The other 90% or more is required to overcome viscous and elastic forces within the flagellum, external viscous forces that oppose motion, and the inertia of the cell mass. When Reynold's number (a dimensionless number expressing the ratio of the inertial force of the body moving in a fluid to the viscous forces acting on its surface) is small, and especially if under 1.0 as it is in *E. viridis* (2×10^{-2}), the inertial force of either the flagellum or the body mass is negligible (Gittelson, 1974). The longer the flagellum, the greater is its R number, and the more efficient its propulsive mechanism. Therefore, the spindle-shaped species which have the longer flagella and the greater size swim more rapidly than other species.

11. Effects of Chemical and Physical Forces on Swimming

Our previous review (Jahn and Bovee, 1968a) covered the many experiments by researchers during the past century in which *Euglena* spp. were treated with a variety of chemicals and physical forces. Questions remain as to what these treatments do to the functioning of the motile machinery. When no chemical or physical alterations of conditions are imposed, *Euglena* swims steadily along a helical pathway that passes around a line through the center of the cylinder described by that path (Holwill, 1966a).

Since the flagellum is the only propulsive "muscle" *Euglena* has for swimming, any chemical or physical force that alters its action will affect the swimming rate and/or direction (Jahn and Bovee, 1968a).

Since body shape is related to the amount of viscous drag on the surface, and for flattened, twisted species is related to the generation of propulsive force, then any chemical or physical condition that alters body shape also will affect the efficiency of swimming and/or its direction. This will be the case whether or not the flagellum is also affected.

The observations that excesses of acetate (Danforth, 1953; Wilson *et al.*, 1959; Bates and Hurlbert, 1970), succinate (Bates and Hurlbert, 1970), or other organic materials (Bovee, 1965; Szabodós, 1936) reduce mobility of *Euglena* can be attributed to "overloading" the tricarboxylic acid cycle and thereby to blocking an efficient production of ATP by the mitochondria. As indicated previously, any agent that affects availability of ATP will affect the efficiency of swimming. Similarly, ATP overloading ($1 \times 10^3 M$ or more) interferes with locomotor efficiency (Acuña and Bovee, 1979) by its "plasticizing" effect (Lindemann *et al.*, 1973).

Any agent that alters the permeability of the flagellar membrane or the number and ratio of cations attached to the membrane (the Jahn ratio; Czarska, 1964) or their flux through the membrane (Eckert and Naitoh, 1972) also will alter the oscillatory movements of cations to and from their required sites for the regular stimulus of the motile mechanism in the axoneme and the paraflagellar rod. This will alter the position of the flagellum relative to the cell body, as well as its amplitude and frequency of beat.

For example, Mg^{2+} is required to promote flagellar motion (Wolken, 1977; Nichols and Rikmenspoel, 1978). This requirement is due to the role of the cation in physically attaching the dynein of the α-microtubule of one peripheral doublet to the proper sites on both ATP and the tubulin of β-microtubule, and perhaps also to its activity in triggering a functional ATPase. Mn^{2+} would be able to acts as a substitute because its physical, ionic symmetry is similar to that of Mg^{2+}. The requirement for certain trace elements for motility (Mo^{6+}, Cu^{2+}, Co^{2+}, Zn^{2+}, Mn^{2+}, Bo^{2+}: Wolken, 1967) can be explained by their cofactor actions in enzyme systems related to cellular oxidation-reduction mechanisms that lead to ATP production.

The effects of electrical current, mechanical stimuli, light, deuterium, hydrostatic pressure, pH, and other stimuli on swimming of *Euglena* cannot depend solely on the properties of the stimulus or on the ratio of cations in the solution to the square root of those attached to the surface of the flagellar membrane (the Gibbs-Donnan or Jahn ratio). Also to be considered are the effects that these physical and chemical stimuli have on energy flow during motility as well as any effects they may have on the structural elements of the motile machinery. More experiments such as those of Nichols and Rikmenspoel (1978) are needed, which microinject *Euglena* with inhibitors of known biochemical reactions or with membrane-active substances that alter the availability and flux of ions to their C sites or critical sites (Ling, 1962). Also needed are a micro-level application of physical forces with, for example, ion-selective microelectrodes and other sophisticated modern equipment.

E. The Flagellum as a Sense Organ

The role of the photoreceptor in the flagellum of *Euglena* as a light-sensitive organelle is now well-known; however, other sensory roles of the flagellum related to its motile behavior have been observed but not investigated. Mechanical stimulus of the flagellum, by pressure with a microneedle, will cause undulations to progress in opposite directions from the point of stimulus (Holwill, 1966b). Jennings (1904) and Lowndes (1936) noted that when *Euglena* "bumps into" an object, it retreats by contracting its body and swinging the flagellum forward. Then the base-to-tip waves drive the cell backward. The same reverse response occurs when *Euglena* encounters an electric field (Bancroft, 1915), an intense beam of light (Engelmann, 1882), a change in viscosity (Spoon et al., 1977), a sudden change of pH (Alexander, 1931), a change in osmotic concentration (Hofler and Hofler, 1952), or a change in organic substances (Jahn and McKibben, 1937). Often the flagellum is broken off by the reaction. The eliciting of this "shock" response by almost any sudden chemical or physical change suggests that the flagellum is highly sensory. No studies have been done to determine how the flagellum relays impulses. As mentioned earlier, though, the mastigonemes, the membrane, and the paraflagellar rod have all been suggested as being involved, either singly or together.

F. Aggregation

It is well-known that swimming *Euglena* tend to aggregate under certain circumstances. De Wildeman (1893) noted that *E. viridis* aggregates at the warmer end of a water-filled glass tube, whether that end is lighted or shaded, but it aggregates at the lighted end in a tube of even temperature. Aggregation toward the light source has often been observed (the "phototactic" response discussed by Diehn et al. in Chapter 5). In bright sunlight, *E. sanguinea* clusters in ball-like clumps (Heidt, 1934), perhaps an expression of "photophobotaxis" (also discussed in Chapter 5). *E. gracilis* aggregates in red light (Checucci et al., 1974; Colombetti and Diehn, 1978). Although the photoreceptor is not sensitive to red light, chloroplasts absorb it in photosynthesis. Therefore, this aggregation in red light is attributed to the presence of O_2, a chemotactic substance, which results from photosynthetic activity. Checucci *et al.* (1974) and Colombetti and Deihn (1978) assume that aggregation of *E. gracilis* around air bubbles is due to O_2, an observation made by Aderhold (1888) nearly a century ago. Colombetti and Diehn (1978) contend that hydrodynamic factors are not involved in such aggregations.

G. PATTERN SWIMMING

In dense populations in shallow containers, swimming *Euglena* spp. form patterns of equilateral triangles (Robbins, 1952). These patterns were earlier described by Wager (1910, 1911) who said they were due to a negative geotaxis. An old centrifugation experiment by Schwarz (1884) indicates the existence of a negative geotaxis: *E. viridis* swam centripetally until the centrifugal force exceeded eight times the gravity, perhaps because of the location of the center of gravity toward the rear (Brinkmann, 1968). Gravity is involved because weightlessness destroys the patterns (Jahn *et al.,* 1962); however, a negative geotaxis does not explain the formation of the patterns themselves. Dead organisms join the patterns at the surface and both living and dead ones fall together in columns at junctions where streams intersect. The dead ones are then left at the bottom of the container as living ones swim upward (Wager, 1911).

Several researchers cite hydrodynamic forces (capillary forces, Brinkmann, 1968; or cohesive forces, Wager, 1911) as being responsible for pattern formation among crowded organisms (Jahn and Brown, 1961; Jahn and Bovee, 1967, 1968a; Brinkmann, 1968; Nultsch, 1973). Most of those cited do not attempt to explain how the hydrodynamic forces cause the patterns to develop. The explanation advanced by Jahn (Jahn and Bovee, 1968a) and modified by Winet and Jahn (1974) suggests that for organisms such as *Euglena* and *Tetrahymena,* the calculated mean free path for the organisms swimming in near contact is such that as they approach one another at random (at a rate of 7% per second within a solid posterior angle of 90°), an end to end hydrodynamic linkage between organisms develops. This linkage is due to viscous drag forces and a pseudo-turbulent vortex behind each organism, the latter caused by flagellar (or ciliary) beating. Any organism entering the vortical ring is tied, hydrodynamically, to the one that develops the vortex and must follow it. Extraneous particles and dead euglenas should therefore enter the pattern, as they have been observed to do (Wager, 1911).

When streams of crowded organisms collide, they lose swimming space and motility, and they sink at the junctions where they collide (Wager, 1911). They fall faster than a single organism alone can fall, and are trapped in the falling column, despite their negative geotaxis, and escape only when that part of the falling column that they occupy reaches bottom (Wager, 1911). Bradley (1966) theorized that a column of descending particles, each with a density greater than that of an equal volume of water, creates a columnar density current that is contained by a shear zone and will continue to fall, at rates up to 50 times that at which a single particle will fall, as long as the density of the column is greater than that of water. This will continue until the column reaches bottom and disperses. Gittleson and Jahn (1968) used Bradley's theory to explain vertical column formation by *Polytomella,* and it appears to apply equally well to *Euglena.* Other, more complicated explanations for pattern swimming have been suggested (Nettleton *et al.,* 1953; Platt, 1961; Wille and Ehret, 1968) and some are set forth with extensive mathematical calculations (Winet and Jahn, 1974; Plesset and Winet, 1974; Childress *et al.,* 1975; Levandowsky *et al.,* 1975).

Since swimming motion is involved in pattern formation, anything that affects swimming and/or flagellar motion will alter the patterns.

IV. Contractile Movements of the Body (Metaboly)

The contractile movements of various *Euglena* spp. have been observed and described by many investigators and have been given a variety of names including "spasmodie" (Dangeard, 1902), "ameboid" movements (Haase, 1910; Bracher, 1937; Gojdics, 1953), peristaltic movements (Chadefaud and Provasoli, 1939; Kamiya, 1939; Hollandé, 1942; Hein, 1953; Kuźnicki and Mikolajczyk, 1973), "euglenoid" movements (Jahn, 1934), body movements (Mikolajczyk and Diehn, 1976), and "inchworm" movements (Hein, 1953). None of these names has come into general use, although the term *metaboly* ("metabolie"; Perty, 1852) has been the most often employed.

A. Complexity of Body Movements

According to Pringsheim (1948), the problem of describing body contractions is that the contractions are "complex and assume manifold forms," and that the kinds of contractions shown even by a single species of *Euglena* are many and varied.

B. The Shock Reaction

The classic shock reaction (Engelmann, 1882) by *Euglena* to a strong stimulus is a contraction of the entire body by shortening and rounding-up. A violent shock may cause so sudden and severe a reaction as to autotomize the flagellum at the opening to reservoir (Tchákotiné, 1936; Hilmbauer, 1954; Bates and

Hurlbert, 1970; Mikołajczyk and Diehn, 1976) and to contract the body, end to end, as a plaque (Dangeard, 1902) or a disc, with only "head" and "tail" projecting (Kuźnicki and Mikołajczyk, 1973; Acuña and Bovee, 1979).

C. OTHER TYPES OF CONTRACTIONS

Besides the classic shock reaction, many other forms of contraction occur in the genus *Euglena*. Pringsheim (1948) listed (1) slight curvatures, (2) unilateral bulging, (3) end-to-end contraction with increase in width, (4) local bulging and distension, and (5) local, pseudopodial, surface protuberances. Arnott and Walne (1966) noted for *E. granulata* simple surface fluctuations, anterior to posterior conicalization, and axial deformation. Others mention twisting (for *E. spirogyra:* Leedale *et al.*, 1965); swaying movements (for *E. deses:* Mast, 1941); stretching and contracting "inchworm" movements (for *E. mutabilis:* Hein, 1953); "earthworm-like" movements, (for *E. obtusa:* Palmer, 1967); unilateral bending to reverse direction (for *E. gracilis:* Mikołajczyk, 1972); oscillatory, peristaltic waves along the body (for *E. deses:* Kamiya, 1939; Hollandé, 1942); burrowing movements (for *E. limosa:* Bracher, 1937); creeping movements (for *E. deses:* Uhlela, 1911); coiling, twisting, and rotary movements if attached only at the rear (for *E. mutabilis:* Hein, 1953); and "ameboid" movements (Haase, 1910; Bracher, 1937; Pringsheim, 1948; Gojdics, 1953). Dangeard (1902) perhaps said it best by stating that *E. geniculata* "crawls and deforms in all sorts of fashions."

D. THE MECHANISM OF CONTRACTION

1. Assumptions for Contractile Fibrils

As early as a century ago, it was theorized that the pellicle and/or its striations might be the site of the contractile mechanism (Stein, 1878; Khawkine, 1887). Some modern electron microscropists, theorizing about fibrils observed in fixed slices, come to similar conclusions (e.g., Schwelitz *et al.*, 1970). Contractile longitudinal fibrils and circular fibrils (myonemes) were early assumed to exist, either in or under the pellicle, and to account, respectively, for longitudinal contractions and peristaltic waves (Stein, 1878; Khawkine, 1887; Chadefaud, 1937; Pringsheim and Hovasse, 1950). Electron microscopic studies have found such fibrils not clearly in the pellicle itself, but in the immediately adjacent cytoplasmic layer (Leedale *et al.*, 1965; Mignot, 1966; Arnott and Walne, 1966). Kuźnicki and Mikołajczyk (1973) cite a striated layer of fibrils under the cell surface as being responsible for the contractions. Arnott and Walne (1966) state that the subpellicular fibrils resemble smooth muscle fibrils, connecting the thicker edge of one strip of the pellicle to the thinner edge of the next. They also note polygonic longitudinal fibrils under the pellicle and they consider these fibrils to be longitudinally contractile. Any contractile nature for these fibrils remains to be proved.

2. The Role of the Pellicle

The pellicle* is composed of tapered, proteinaceous, elastic strips (Pigon, 1947) that lie directly beneath the cell membrane (Mignot, 1965, 1966; Leedale *et al.*, 1965; Arnott and Walne, 1966; Guttman and Ziegler, 1974) with grooves between them (the striations). Parallelling the striations and under the higher edge of each pellicular strip are longitudinal microtubules, one (sometimes two to five) for each strip. The pellicular strips vary in thickness from one species to another (Pringsheim, 1948; Leedale, 1967). Those with thick strips (that in some species interlock by projections) show little metaboly, whereas others with thinner strips are much more contractile with a large variety of body movements (Pringsheim and Hovasse, 1950). *E. spirogyra* having thick strips (Leedale *et al.*, 1965) can only twist; *E. deses* and *E. mutabilis*, each with thin strips, show a variety of movements (Dangeard, 1902; Lackey, 1938; Hein, 1953). The visible distinctness of striation is not related to the degree of contractile movement since some species with distinct striations are scarcely contractile and others with light striation are contractile (Pringsheim and Hovasse, 1950; Gojdics, 1953).

Rather than being the source of contractions, the pellicular strips and the parallelling microtubules tend to restrict contraction (Pringsheim, 1948; Gojdics, 1953; Mignot, 1965). The pellicular strips and the microtubules are now considered to give the body its shape, elasticity, and/or rigidity; their function in metaboly is to return the body to its normal shape during relaxation (Leedale *et al.*, 1965; Arnott and Walne, 1966; Guttman and Ziegler, 1974).

The shock reaction and the contractions that accompany it serve to direct the swimming *Euglena* away from the source of a strong stimulus. The stimulus may be visible light (Englemann, 1882; Jahn and Bovee, 1968a), ultraviolet radiation (Tchákotiné, 1936), heat (Hall, 1931), X-rays (Wichterman, 1955), osmotic shock (Hall, 1931; Höfler and Höfler, 1952; Hilmbauer, 1954), mechanical shock (Kuźnicki and Mikołajczyk, 1973), contact with an object (Lowndes, 1936; Mikołajczyk, 1972), entry into a highly viscous medium (Mikołajczyk, 1972), electrical current (Votta and Jahn, 1972), chemical shock by acidic substances (Alexander, 1931; LeFévre, 1931), especially organic ones (Bates and Hurlbert, 1970; Szabados, 1936; Jahn and McKibben, 1937; Votta and Jahn, 1971), or cationic changes in solution, especially K^+, Na^+, Ca^{2+}, by addition (Mikołajczyk, 1973) or deletion (Bancroft, 1915; Mikołajczyk, 1973). Alexander (1931) perhaps stated the situation accurately by noting that contractions of

*This topic is discussed further in Chapters 1 and 2 of this volume.

the body of swimming *Euglena* are the response to any sudden change in the environmental conditions.

Preceding or accompanying contractions that occur as a result of a sudden environmental change is autotomy of the flagellum (which also occurs for many species during cell division; Leedale, 1967). Flagellar autotomy plus contraction to a rounded form permits a euglena to sink slowly away from the source of shock to the bottom surface where it can crawl. The loss of the flagellum appears to activate the crawling mechanism, for example, by *E. deses* (Uhlehla, 1911), *E. limosa* (Conrad, 1940), and *E. mutabilis* (Lackey, 1938).

The contractions of metaboly also are used in other ways by other euglenas and euglenoids. *Euglena viridis,* when exiting from its cyst, breaks the cyst open with pressures exerted by peristaltic movements (Bovee, unpublished). Osmotic shock causes *Trachelomonas volvocina* to leave its test by peristaltic action (Hilmbauer, 1954). *Peranema trichophorum*, in order to scavenge the tissues of dead rotifers, uses peristaltic movements to wiggle in or out through small cracks in the rotifer's chitinous shell (Bovee, unpublished). Parasitic euglenoids of the genus *Astasia* use both the flagellum and body movements to penetrate the eggs of copepods (Michajlow, 1972).

3. *Control of Contractile Movements*

Little research or theory is available to explain how contractile movements of *Euglena* spp. are controlled and coordinated. Jahn and Bovee (1968a) have implied that contractile as well as flagellar movements may be stimulated and regulated by exchanges of cations at the surface and by fluxes of cations within the cell. Contractile movements occur when the cell surface is stimulated. The flagellum relays a stimulus to the body by way of the photoreceptor and/or the axoneme and paraflagellar rod, because the presence of light or contact of the flagellum with an object results in contractions (e.g., Lowndes, 1936; Lozina-Lozinska and Zaar, 1963). Autotomy of the flagellum that accompanies a shock response is accompanied and followed by contractions; however, whether the contractions are stimulated by the shock of the stimulus or the shock of flagellar autotomy, or both, is not clear. The contractions may be continued long *after* flagellar autotomy, as in *E. deses* and *E. mutabilis,* which then crawl. ATP is the source of energy and cation-flux, a source of stimulus. Increasing the external concentrations of cations induces contractions (Mikolajczyk, 1973) and adding ATP to the surrounding water at $1 \times 10^{-5} M$ augments the contractile, photophobic response (Acuña and Bovee, 1979) for *E. gracilis*. Also, dinitrophenol, which abolishes ATP synthesis, eliminiates contractions (Mikolajczyk, 1973). Minor contact of the flagellum or the body with a surface (wall of a glass tube, Mikolajczyk, 1972) can cause unilateral rippling waves of contraction. This indicates a differential routing of a localized stimulus along the surface of and/or within the body, whether the stimulus is relayed by the flagellum or by the body surface. Such a unilateral ripple causes *E. gracilis* to bend the body under itself and reverse its direction of swimming (Mikołajczyk, 1973).

Excess Ca^{2+} in the solution causes the shock reaction (Mikołajczyk, 1973). Therefore it is likely that the excessive Ca^{2+}, entering simultaneously through the entire surface, sets off the entire contractile machinery and causes the shock reaction. Also in excess in the surrounding water, K^+ or Na^+ could also stimulate contractions by displacing Ca^{2+} at the cell's surface, so that some of it could move into the cell where it would stimulate the contractions. The impulse of a shock reaction, relayed from the flagellum into the body, could cause an internal, sequential movement of cations from front to rear. This cation movement would set off the contractile response in the longitudinal fibers causing them to shorten. Oscillating peristaltic waves would occur as the cations flow back, thus stimulating circular fibrils sequentially while the long fibers relax. The unilateral bending, mentioned above for *E. gracilis,* could result from the sequential bending of the short transverse fibers that underlie each pellicular strip, so that the strips are bent inward along the stimulated side of the body, i.e., generating a partial peristaltic wave. Swaying movements in *E. deses* (Mast, 1941) could be derived by alternate stimulation of sides of the body as the photoreceptor swings back and forth in the path of the light stimulus.

Obviously, much remains to be done to explain the coordination of the contractile movements of *Euglena.*

V. Crawling Movements

Some euglenas crawl actively on a surface by contractile movements, usually by alternating extensions and contractions of the body. These movements commonly occur in *Euglena* spp. that have a short flagellum or one that is easily autotomized, e.g., in *E. deses* (Dangeard, 1902; Uhlehla, 1911; Bracher, 1919), or have a flagellum so short that it scarcely extends beyond the opening of the anterior reservoir, e.g., in *E. mutabilis* (Johnson, 1944; Gojdics, 1953).

The crawling in such euglenas was described by Hein (1953) in the case of *E. mutabilis:* "The anterior end snakes forward and extends the body. . . ." Bracher (1919) gave a description for *E. deses* by stating that the cell contracts forward as a ball, then extends forward full-length. Again for *E. mutabilis,* a bulge forms at the anterior end, which runs to the posterior end, and then back to the anterior end as the body contracts forward (Hollandé, 1942).

Apparently, contact of the anterior bulge stimulates a series or waves of contractions from front to rear, whereas lifting of the rear end stimulates the reverse wave. The result is sort of an "on–off" series of responses. If its rear end does not detach, the euglena erects itself and repeatedly coils and relaxes (Hein,

1953), a movement which is like that of a hemosporidian (Jahn and Bovee, 1968b).

Again, nothing is known of the control mechanism(s) for crawling.

VI. Gliding Movements

It has long been known that both swimming and crawling species of *Euglena* can glide over a moist surface. Gliding was studied by Günther (1927), who noted that *Euglena* secretes a mucus track on which it glides, but that it will also follow previously existing slime trails. Euglenas that inhabit marine or freshwater mud flats glide out of mucus-lined tunnels in the mud, at daytime low tides, and back into the tunnels before the tide rises again (Bracher, 1937; Palmer and Round, 1965; Palmer, 1967).

The mucus is laid down in lines that parallel the pellicular striations (Diskus, 1955; Leedale, 1967) and comes from mucocysts. The pores of mucocysts* appear as "warts" along the striating of *E. spirogyra* (Leedale *et al.,* 1965) and other species (Conrad, 1940). The existence of acid phosphatases in the endoplasmic reticulum around vesicles adjacent to mucocyst pores suggests the secretory activity of these structures (Sommer and Blum, 1965).

The few theories advanced about the basis of gliding movements assume that some interface-active, energy-using mechanism interacts with the mucus trail at the surface of the cell [first suggested by Max Schultz (1865) and revived a century later by Jarosch (1964)]. Jarosch postulated a fibrillar worm-gear mechanism involving multiple, rotating, protein helices. Other theories suggest gliding by surface undulations (Günther, 1927), gliding with the cell body as a helix (Hein, 1953) (a form that also occurs with sporozoan trophozoites; Jahn and Bovee, 1968b), and gliding by peristaltic movements (Chadefaud, 1937).

Helical gliding is probably caused by torsion of the body and its pellicle. The cell appears to be propelled by a force active between the cell surface and the mucus and sufficient to propel the body through the mucus sheath past a single point of mucus attachment to the surface or over a trail of it (Jahn and Bovee, 1968b). Visible undulations occur along the surface (Mikołajczyk, 1973), and these might act as a propulsive force. However, in some cases, surface undulations might occur at the submicroscopic level (Jarosch, 1964) or might result from submicroscopic shifting of polymeric linkages in a "ratchet fashion" mediated by membrane ATPases.

More experiments are needed to determine the energy source and the physical nature of the propulsive mechanism underlying gliding.

---

*These structures are considered further in this volume in Chapter 1.

VII. Recapitulation and Summary

Flagellar propulsion used by *Euglena* in swimming consists of mechanical mechanisms that employ hydrodynamic principles in conformance to the laws of a mass in motion. The result is the successful propulsion of the cell through a medium (water), which is viscous enough to halt movement instantly when the prodigious energy expenditure by the flagellum is severely reduced. For some species, such as the spindle-shaped *E. viridis* or *E. gracilis,* the flagellum is the only means of propulsive force. Rapid swimming is accomplished by intermittent helical waves that traverse the long flagellum from base to tip. The flagellum trails beside the cell body, thus producing a torque that gyrates and tilts it. This enables the cell to swim in a more-or-less wobbling fashion along a helical path. In the case of *E. acus,* a short flagellum beats anteriorly in a "figure-eight" (pretzel) pattern, thus pulling the cell slowly through the water with rotation, but with little gyration. Other *Euglena* species have a flattened, twisted shape and a short flagellum that whirls more-or-less at right angles. In this case, the cell-body acts as a gyrating inclined plane, essentially in the form of a screw, and provides the propulsive force, while the flagellum develops the torque required.

The energy-supply for flagellar propulsion is ATP. The energy-using machine is the flagellar axoneme, perhaps assisted actively by the paraflagellar rod, which also contains an ATPase. Flagellar undulation is probably initiated in or near the flagellar basal body and may be modulated by torque on the flagellum due to drag on its mastigonemes. The latter converts an initially planar beat to a helical one. The intermittent helical waves may be due (1) to an intermittent stimulus from the cell *via* the flagellar basal body or (2) to the stiffness of the paraflagellar rod and the intermittent waves along it that augment some flagellar undulations, while damping out others. The intermittent waves permit the maximal use of mastigonemal surfaces to augment the thrust.

The photoreceptor–paraflagellar rod system enables *Euglena* to sense and respond to light and directs it to the area of optimal light intensity for photosynthesis. Chemosensory mechanisms alter flagellar undulations to direct *Euglena* to regions of optimal pH and O_2 or CO_2 tension and away from osmotically incompatible situations.

In dense populations *Euglena* is doomed by hydrodynamics and gravity to swim in patterns until dispersed by outside forces.

Contractile movements enable *Euglena* to retreat from shock forces, crawl upon surfaces, change direction, glide, wiggle through narrow spaces, and (for parasitic forms) to invade cells of other organisms.

The mechanism of contraction is not clearly defined. It is apparently fibrillar and at least two-dimensional, is calcium-dependent, uses ATP as the energy source, i.e., it is quasi-muscular, and probably lies just underneath the pellicular strips. The latter are elastic but not contractile and, along with long microtubules that parallel them, resist contraction and restore the shape of the cell body during relaxation.

Crawling is apparently accomplished by alternate linear and circumferential contractions (often as peristaltic waves), the former shortening the body, and the latter lengthening it. How crawling is coordinated is not known.

Gliding occurs over a mucus track either self-secreted by the gliding cell or already existing. The mechanism of gliding is unknown.

It is unfortunate that there is a dearth of critical investigations on the movements of *Euglena*. This is especially so since the number of species is large and the cells are durable and easily grown in the laboratory and are large enough for observation and experimentation. Barker (1943) remarked, nearly 40 years ago, that more studies were required on the protein chemistry, thermodynamics, and biophysics of *Euglena*. Today, more study is still needed on the biochemistry of its motile machinery and the hydrodynamic principles that govern its swimming. In 1934, DeFlandre insisted that motion in the genus *Euglena* should be studied in as many species as possible. We agree and would like to add that it also needs to be studied in as many of its close relatives as possible. Although *Euglena* has been widely used as an experimental organism for many investigations, the study of its movements and locomotion have been sadly neglected.

References

Acuña, M. L. (1970). Master's Thesis. University of Kansas, Lawrence, Kansas.
Acuña, M. L., and Bovee, E. C. (1979). *Univ. Kans. Sci. Bull.* **51,** 669.
Aderhold, R. (1888). *Jena Z. Med. Naturwiss.* **22,** 310.
Alexander, G. (1931). *Biol. Bull. (Woods Hole, Mass.)* **61,** 165.
Arnott, H. J., and Walne, P. L. (1966). *J. Phycol. (Abstr.)* **2**(Suppl.), 4.
Bancroft, F. W. (1915). *Stud. Rockefeller Inst. Med.* **20,** 384.
Barker, D. (1943). *New Phytol.* **42,** 49.
Bates, R. C., and Hurlbert, R. E. (1970). *J. Protozool.* **17,** 134.
Bovee, E. C. (1965). *Hydrobiologia* **25,** 69.
Bovee, E. C., and Acuña, M. L. (1970). *J. Protozool. (Abstr.)* **17**(Suppl.), 14.
Bovee, E. C., and Jahn, T. L. (1972). *J. Theoret. Biol.* **35,** 259.
Bovee, E. C., Collins, S., and Acuña, M. L. (1969). *J. Protozool. (Abstr.)* **16**(Suppl.), 14.
Bracher, R. (1919). *Ann. Bot. (London)* **33,** 93.
Bracher, R. (1937). *Proc. Linn. Soc. (London)* **149,** 65.
Bradley, D. E. (1966). *Exp. Cell Res.* **41,** 162.
Brennen, C., and Winet, H. (1977). *Annu. Rev. Fluid Mech.* **9,** 339.
Brinkmann, K. (1968). *Z. Pflanzenphysiol.* **59,** 364.
Brokaw, C. J. (1961). *Exp. Cell Res.* **22,** 151.
Brokaw, C. J. (1971). *J. Exp. Biol.* **55,** 289.
Brokaw, C. J. (1972a). *Science* **178,** 455.
Brokaw, C. J. (1972b). *Biophys. J.* **12,** 564.
Brown, H. P., and Cox, A. (1954). *Am. Midl. Nat.* **52,** 106.

Carlson, F. D. (1962). *In* "Spermatozoan Motility" (D. W. Bishop, ed.), p. 137. Am. Assoc. Advanc. Sci., Washington, D.C.

Chalkey, H. W. (1930). *Science* **71**, 442.

Chadefaud, M. (1937). *Botaniste* **28**, 86.

Chadefaud, M., and Provasoli, L. (1939). *Arch. Zool. Exp. Gen. Notes Rev.* **80**, 55.

Checcucci, A., Columbetti, G., del Carratore, G., Farrare, R., and Lenci, F. (1974). *Photochem. Photobiol.* **19**, 223.

Childress, S., Levandowsky, M., and Spiegel, E. A. (1975). *J. Fluid Mech.* **63**, 591.

Colombetti, G., and Diehn, B. (1978). *J. Protozool.* **25**, 211.

Conrad, W. (1940). *Bull. Mus. R. Hist. Nat. Belg.* **16**, 1.

Costello, D. P. (1973). *Biol. Bull. (Woods Hole, Mass.)* **145**, 279.

Czarska, L. (1964). *Acta Protozool.* **2**, 287.

Dangeard, P. A. (1890). *Botaniste* **1**, 1.

Dangeard, P. A. (1902). *Botaniste* **8**, 97.

Danforth, W. (1953). *Arch. Biochem. Biophys.* **46**, 164.

DeFlandre, G. (1934). *Ann. Protistol.* **4**, 31.

de Wildemann, E. (1893-4). *Bull. Soc. Belg. Microsc.* **20**, 245.

Diehn, B. (1972). *Acta Protozool.* **11**, 325.

Diehn, B., Fonseca, J. R., and Jahn, T. L. (1975). *J. Protozool.* **22**, 492.

Diskus, A. (1955). *Protoplasma* **45**, 460.

Eckert, R., and Naitoh, Y. (1972). *J. Protozool.* **19**, 237.

Engelmann, T. W. (1869). *Arch. Gesamte. Physiol. Menschen Tiere* **2**, 307.

Engelmann, T. W. (1882). *Arch. Gesamte. Physiol. Menschen Tiere* **29**, 387.

Fischer, A. (1894). *Jahrb. Wiss. Bot.* **26**, 187.

Francé, R. (1893). *Z. Wiss. Zool. Abt.* **56**, 138.

Gibbs, S. P. (1960). *J. Ultrastruct. Res.* **4**, 127.

Gittleson, S. M. (1974). *Trans. Am. Microsc. Soc.* **93**, 272.

Gittleson, S. M., and Jahn, T. L. (1968). *Exp. Cell Res.* **51**, 579.

Gojdics, M. (1953). "The Genus *Euglena*." Univ. of Wisconsin Press, Madison.

Gössel, J. (1957). *Arch. Mikrobiol.* **27**, 288.

Gray, J. (1928). "Ciliary Movement." Cambridge Univ. Press, London.

Gray, J. (1953). *Q. J. Microsc. Sci.* **94**, 551.

Günther, F. (1927). *Arch. Protistenkd.* **60**, 51.

Guttman, H. N., and Zeigler, H. (1974). *Cytobiologie* **9**, 10.

Haase, G. (1910). *Arch. Protistenkd.* **20**, 47.

Hall, S. R. (1931). *Biol. Bull. (Woods Hole, Mass.)* **60**, 397.

Heidt, K. (1934). *Ber. Deut. Bot. Ges.* **52**, 607.

Hein, G. (1953). *Arch. Hydrobiol.* **47**, 516.

Hilmbauer, K. (1954). *Protoplasma* **43**, 192.

Höfler, K., and Höfler, L. (1952). *Protoplasma* **41**, 76.

Hollandé, A. (1942). *Arch. Zool. Exp. Gen.* **83**, 1.

Holwill, M. E. J. (1966a). *J. Exp. Biol.* **44**, 579.

Holwill, M. E. J. (1966b). *Physiol. Rev.* **46**, 696.

Jahn, T. L. (1934). *Cold Spring Harbor Symp. Quant. Biol.* **2**, 167.

Jahn, T. L. (1974). *In* "Swimming and Flying in Nature" (T. Y. Wu, C. J. Brokaw, C. Brennen, eds.), Vol. I, pp. 324-338. Plenum Press, New York.

Jahn, T. L., and Bovee, E. C. (1967). *In* "Research in Protozoology" (T.-T. Chen, ed.), Vol. 1, p. 39. Pergamon, Oxford.

Jahn, T. L., and Bovee, E. C. (1968a). *In* "Biology of *Euglena*" (D. E. Buetow, ed.), Vol. I, p. 45. Academic Press, New York.

Jahn, T. L., and Bovee, E. C. (1968b). *In* "Infectious Blood Diseases of Man and Animals" (D. Weinman and M. Ristic, eds.), p. 303. Academic Press, New York.

Jahn, T. L., and Brown, M. (1961). *Am. Zool.* **1,** 454.

Jahn, T. L., and Landman, M. (1965). *Trans. Am. Microsc. Soc.* **84,** 395.

Jahn, T. L., and McKibben, W. R. (1937). *Trans. Am. Microsc. Soc.* **56,** 48.

Jahn, T. L., Brown, M., and Winet, H. (1962). *Excerpta Med. Int. Congr. Ser.* **48,** 638.

Jahn, T. L., Bovee, E. C., Dauber, M., Winet, H., and Brown, M. (1965). *Ann. N.Y. Acad. Sci.* **118,** 912.

Jahn, T. L., Bovee, E. C., and Jahn, F. F. (1979). "How to Know the Protozoa," 2nd ed. W. C. Brown, Dubuque, Iowa.

Jarosch, R. (1964). *In* "Primitive Motile Systems in Cell Biology" (R. D. Allen and N. Kamiya, eds.), p. 599. Academic Press, New York.

Jennings, H. S. (1904). *Carnegie Inst. Washington Publ.* **16,** 129.

Johnson, L. P. (1944). *Trans. Am. Microsc. Soc.* **63,** 97.

Kamiya, N. (1939). *Ber. Dtsch. Bot. Ges.* **57,** 231.

Khawkine, W. (1887). *Ann. Sci. Nat. Bot. Biol. Veg.* **1,** 319.

Kivic, P. A., and Vesk, M. (1972). *Planta* **105,** 1.

Krinsky, N. I., and Goldsmith, T. H. (1960). *Arch. Biochem. Biophys.* **91,** 271.

Künstler, J. (1886). *J. Microgr.* **10,** 493.

Kuźnicki, L., and Mikołajczyk, E. (1973). *Prog. Protozool.* **4,** 237.

Lackey, J. B. (1938). *U.S. Public Health Rep.* **53,** 1499.

Lang, N. J. (1963). *J. Cell Biol.* **19,** 631.

LeFévre, M. (1931). *Trav. Crypt. Dédiés L. Magnin.* p. 343.

Leedale, G. F. (1967). "The Euglenoid Flagellates." Prentice-Hall, Englewood Cliffs, New Jersey.

Leedale, G. F., Meeuse, B. J. D., and Pringsheim, E. G. (1965). *Arch. Mikrobiol.* **50,** 68, 135.

Leeuwenhoek, A. (1674). Cited in Dobell, C. (1932/1960). Reprint of "Anton van Leeuwenhoek and His Little Animals." Dover, New York.

Levandowsky, M., Childress, S., Spiegel, E. A., and Hutner, S. H. (1975). *J. Protozool.* **22,** 296.

Lin, S. H. (1972). *Biophysik* **8,** 264.

Lindemann, C. B., Rudd, W. G., and Rikmenspoel, R. (1973). *Biophys. J.* **18,** 437.

Ling, G. N. (1962). "A Physical Theory of the Living State: The Association-Induction Hypothesis." Ginn (Blaisdell), Boston, Massachusetts.

Lowndes, A. G. (1936). *Nature (London)* **155,** 210.

Lowndes, A. G. (1943). *Proc. Zool. Soc. London* **A113,** 99.

Lowndes, A. G. (1944). *Proc. Zool. Soc. London* **A114,** 325.

Lozina-Lozinska, L. K., and Zaar, E. I. (1963). *Tsitologiya* **5,** 263.

Lubliner, J., and Blum, J. J. (1971). *J. Mechanochem. Cell Motil.* **1,** 15.

Lubliner, J., and Blum, J. J. (1972). *J. Theoret. Biol.* **31,** 1.

Machin, K. E. (1958). *J. Exp. Biol.* **35,** 796.

Magenot, G. (1926). *C. R. Soc. Biol.* **94,** 577.

Mahenda, Z., Vora, R., Wolken, J. J., and Ahn, K. S. (1967). *J. Protozool. (Abstr.)* **14**(Suppl.), 17.

Mast, S. O. (1941). *In* "Protozoa in Biological Research" (G. N. Calkins and F. M. Summers, eds.), p. 271. Columbia Univ. Press, New York.

Michajlow, W. (1972). *Bull. Acad. Pol. Sci. Ser. Sci. Biol.* **18,** 573.

Mignot, J.-P. (1965). *Protistologica* **1,** 5.

Mignot, J.-P. (1966). *Protistologica* **2,** 51.

Mikołajczyk, E. (1972). *Acta Protozool.* **11,** 317.

Mikolajczyk, E. (1973). *Acta Protozool.* **12**, 133.

Milolajczyk, E., and Diehn, B. (1976). *J. Protozool.* **23**, 144.

Miles, C. A., and Holwill, M. E. J. (1971). *Biophys. J.* **11**, 851.

Nettleton, R. M., Mefferd, R. B., Jr., and Loefer, J. B. (1953). *Am. Nat.* **87**, 117.

Nevo, A. C., and Rikmenspoel, R. (1970). *J. Theoret. Biol.* **26**, 11.

Nichols, K. M., and Rikmenspoel, R. (1978). *Exp. Cell Res.* **116**, 333.

Nultsch, W. (1973). *Arch. Protistenkd.* **115**, 336.

Palmer, J. D. (1967). *Nat. Hist.* **76**, 60–64.

Palmer, J. D., and Round, F. E. (1965). *J. Mar. Biol. Assoc. U.K.* **45**, 567.

Perty, M. (1852). ''Zur Kenntnis kleiner Lebensformen nach Bau, Funktionen, Systematik, mit Spezialverzeichnis in der Schweis beobachtet.'' Verlag von Jent und Reinor, Bern.

Piccinni, E., and Albergoni, B. (1973). *J. Protozool.* **20**, 456.

Piccinni, E., Albergoni, V., and Copellotti, O. (1975). *J. Protozool.* **22**, 331.

Pigon, A. (1947). *Bull. Int. Acad. Pol. Sci. Lett. Cl. Sci. Math. Nath. Ser. B 2*, **1946**, 111.

Pitelka, D. R., and Schooley, C. N. (1955). *Univ. Calif. Publ. Zool.* **61**, 79.

Platt, J. R. (1961). *Science* **133**, 1766.

Plesset, M. S., and Winet, H. (1974). *Nature (London)* **248**, 441.

Pochmann, A. (1953). *Planta* **42**, 478.

Pringsheim, E. G. (1948). *Biol. Rev.* **23**, 46.

Pringsheim, E. G., and Hovasse, R. (1950). *Arch. Zool. Exp. Gen.* **86**, 499.

Reger, J. F., and Beams, H. W. (1954). *Proc. Iowa Acad. Sci.* **61**, 593.

Rikmenspoel, R. (1971). *Biophys. J.* **11**, 446.

Robbins, W. J. (1952). *Bull. Torrey Bot. Club* **79**, 107.

Schultze, M. (1865). *Arch. Microscop. Anat.* **1**, 376.

Schwarz, F. (1884). *Ber. Dtsh. Bot. Ges.* **2**, 51.

Schwelitz, F. D., Evans, W. R. Mollenhauer, H. H., and Dilly, R. A. (1970). *Protoplasma* **69**, 341.

Sommer, J. R., and Blum, J. J. (1965). *J. Cell Biol.* **24**, 235.

Sperling-Pagni, P. S., Walne, P. L., and Wehri, E. L. (1976). *Photochem. Photobiol.* **24**, 373.

Spoon, D. M., Feise, C. I., and Youn, R. S. (1977). *J. Protozool.* **24**, 471.

Stein, F. R. (1878). ''Der Organismus der Infusionsthiere Section III. Der Naturgeschichte der Flagellaten oder Geisselinfusiorien.'' Engelmann, Leipzig.

Steuer, A. (1904). *Arch. Protistenkd.* **3**, 126.

Szabadós, M. (1936). *Acta Biol. (Szeged)* **4**, 49.

Taylor, G. (1951). *Proc. R. Soc. London* **A209**, 447.

Tchákotiné, S. (1936). *C. R. Soc. Biol.* **121**, 1162.

Tollin, G., and Robinson, M. I. (1969). *Photochem. Photobiol.* **9**, 41.

Uhlehla, V. (1911). *Biol. Zentralbl.* **31**, 657.

van Herpen, G., and Rikmenspoel, R. (1968). *Biophys. J. (Abstr.)* **8**, A-121.

Vavra, J., and Aaronson, S. (1962). *J. Protozool. (Abstr.)* **9**(Suppl.), 28.

Votta, J. J., and Jahn, T. L. (1971). *J. Protozool. (Abstr.)* **18**(Suppl.), 20.

Votta, J. J., and Jahn, T. L. (1972). *J. Protozool. (Abstr.)* **19**(Suppl.), 43.

Votta, J. J., Jahn, T. L., Griffith, D. L., and Fonseca, J. R. (1971). *Trans. Am. Microsc. Soc.* **90**, 404.

Wager, H. (1910). *Proc. R. Soc. (London)* **B83**, 94.

Wager, H. (1911). *Philos. Trans. R. Soc. London Ser. B* **201**, 333.

Walne, P. L., and Arnott, H. J. (1967). *Planta* **77**, 325.

Wichterman, R. (1955). *Biol. Bull. (Woods Hole, Mass.)* **109**, 371.

Wille, J. J., and Ehret, C. F. (1968). *J. Protozool.* **15**, 789.

Wilson, B. W., Buetow, D. E., Jahn, T. L., and Levendahl, B. H. (1959). *Exp. Cell Res.* **18**, 545.

Winet, H. (1976). *J. Exp. Biol.* **64,** 283.

Winet, H., and Jahn, T. L. (1974). *J. Theoret. Biol.* **46,** 449.

Wolken, J. J. (1961). "Euglena." Rutgers Univ. Press, New Brunswick, New Jersey.

Wolken, J. J. (1967). "Euglena," 2nd ed. Appleton, New York.

Wolken, J. J. (1969). *J. Cell Biol. (Abstr.)* **43,** 159.

Wolken, J. J. (1977). *J. Protozool.* **24,** 518.

CHAPTER 5

RESPONSES TO PHOTIC, CHEMICAL, AND MECHANICAL STIMULI

*Giuliano Colombetti, Francesco Lenci,
and Bodo Diehn*

I. Introduction

The unicellular flagellate *Euglena gracilis,* similar to several other microorganisms, has developed the ability to respond to a variety of external stimuli and therefore to establish an active relationship with its environment (see, e.g., Haupt and Feinleib, 1979; Seitz, 1975; Nultsch and Häder, 1979; Nultsch, 1980; Diehn, 1979a, 1980a; Carlile, 1980, and references therein). "Establishing an active relationship" means that the cell is able to perceive variations in the physicochemical properties of the external environment and to change its behavior accordingly, generally with the ultimate aim of finding the best environmental conditions for its survival. Because of the relative simplicity of this biological system, the apparently complex different behavioral responses can be

THE BIOLOGY OF *EUGLENA*, VOL. 3
Copyright © 1982 by Academic Press, Inc.
All rights of reproduction in any form reserved.
ISBN 0-12-139903-6

traced to variations in its swimming trajectory and, in some cases, to modifications of its body shape, such as contractions and rounding-up. A general scheme of the functional machinery connecting the phenomena of stimulus perception and response is shown in Fig. 1. The stimulus is perceived by a receptor, processed and elaborated and then transduced to the effector, which eventually acts on the motor apparatus. Depending on the type of stimulus acting on the cell, different subcellular structures and physiological processes carry out the sensory transduction functions shown in this figure. However, the generalized scheme shown is applicable in all cases. As discussed later, different sensory responses to various stimuli can be processed through at least partially coincident transduction chains. This fact can, of course, constitute a complication, potentially introducing some ambiguity and confusion (see, for instance, Feinleib, 1978), but can also help greatly in analyzing and trying to understand the different steps of the sensory response processes.

In this chapter we will present and discuss the various aspects of sensory responses of *E. gracilis* to photic, chemical, and mechanical stimuli, trying to follow the schematic block diagram presented in Fig. 1 as a logical and methodological guideline and devoting, whenever possible, particular attention to the molecular aspects of the entire phenomenon. Any progress in understanding the primary elementary events in sensory transduction of aneural systems will be applicable to more than this class of organisms, since basic mechanisms are fundamentally conserved through the evolution of living systems. This idea of utilizing aneural microorganisms as "simple" model systems for elucidating the molecular aspects of sensory responses in higher organisms must, of course, be

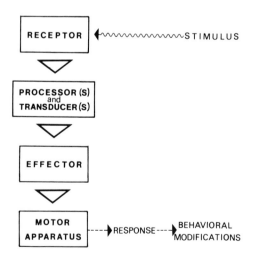

Fig. 1. Generalized scheme of sensory transduction.

applied with some discretion (Feinleib and Curry, 1974). The case of *Euglena* is of particular interest in its own right because of the richness of its sensory responses, the complexity of the molecular pathways through which these responses are elicited and, at the same time, the convergent character of some of the hypotheses that can be put forth about its sensory transduction chains (Lenci and Colombetti, 1978; Diehn, 1979a).

We will not discuss those interactions with the environment in which the responses are simply the consequences of modifications of the cellular metabolism (like photokinesis) and will limit ourselves to "pure" sensory perception processes.

II. Behavioral Responses to Photic Stimuli

Light-induced motor reactions are undoubtedly the most widely investigated sensory responses of *E. gracilis* (see, e.g., Doughty and Diehn, 1980, and references therein) and reasonably experimentally supported hypotheses can be put forward about the corresponding molecular photosensory physiology.

Euglena exhibits motor reactions in response to variations in external illumination conditions (for a discussion of possible behavioral responses of microorganisms, and of the appropriate terminology, see Diehn *et al.,* 1977): upon a sudden decrease (*step-down photophobic response*) or upon an increase, above a threshold of about 200 W/m², of light intensity (*step-up photophobic response*), the normally trailing flagellum reorients and beats more or less perpendicularly to the long axis of the cell, instead of beating approximately parallel to this axis, as shown in Fig. 2. The cell stops its forward motion and turns on the spot toward its stigma-bearing "dorsal side" (Diehn, 1969b, 1979a; Checcucci, 1976; Lenci and Colombetti, 1978, and references therein). This photomotile reaction ceases as soon as the previous illumination conditions are restored; otherwise the tumbling continues till adaptation occurs. The latter is always true for the step-down photophobic response, whereas light intensities above the step-up threshold can cause general damage to the cells (as evidenced by body contractions and immobilization). This can make it impossible or at least very difficult to measure adaptation to the new external conditions. A careful re-evaluation of the photic response of *Euglena* by Doughty and Diehn (1979) has recently revealed that the classical "tumbling" phase of the response described above is followed by a phase in which straight-line swimming is interspersed with short tumbling events. Thus, the photic response consists of a period of continuous flagellar reorientation (CFR), followed by a period of periodic flagellar reorientation (PFR). Both response phases are distinct and moreover exhibit differential dependence upon environmental parameters (Diehn, 1980b).

A detailed analysis of the step-up photophobic response, using high-speed

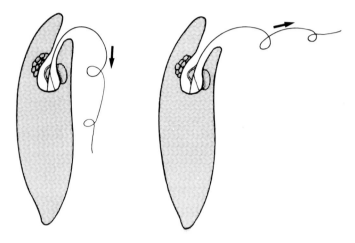

Fig. 2. Flagellar reorientation in the phobic response of *Euglena.*

cinemicrography, has been conducted by Diehn *et al.* (1975), who have deter-mined transduction times (100–400 milliseconds), flagellar beat frequencies (av-erage beat times 36 milliseconds in cells grown in a dark–light cycle and 41 milliseconds in continuously illuminated cells), and turning rates (1.15 seconds/turn). These responses of *Euglena* mediate the cell behavior which may result in accumulation in or dispersal from illuminated regions (see following paragraph). Because of the helical beating* of the flagellum and its asymmetrical thrust, the unstimulated *Euglena* moves forward, rotating around its longitudinal axis with a frequency of 1.8 Hz (Ascoli *et al.*, 1978). The cell body is inclined with respect to the direction of progress. Thus the cell moves in a helical path (Checcucci, 1976, and references therein).

Oriented movement of the cells toward (*positive phototaxis*) or away from (*negative phototaxis*) a light source probably utilizes this rotation (as we will see in the following), which allows detecting the source position by means of a two-instant mechanism (Feinleib, 1975). This type of movement-modulated de-tection mode compares light intensity in two different instants of time by means of a single photoreceptor (in other words *Euglena* is capable of detecting a temporal variation in light intensity). Positive and negative phototaxes, respec-tively mediated by step-down and step-up photophobic responses, are exhibited when the cells are oriented so as to have the photoreceptor permanently illumi-nated or shaded, respectively, as described by Diehn (1979a):

If light of an intensity such that its dimming will cause a stepdown phobic response (i.e., below 0.2 kW/m² of white light) is applied at right angles to the cell's path, then the shadow of the

*For further discussion of this topic, see this volume, Chapter 4.

stigma will fall upon the photoreceptor* (the paraflagellar swelling) once during each revolution, causing a phobic response that ceases when the stigma has turned past the shading position. This results in stepwise course corrections that lead to positive phototactic orientation.

In a test of this hypothesis, Diehn (1969a) illuminated a population of *Euglena* in the phototaxigraph with repetitive flashes of light and observed a clear resonance peak of photoaccumulation at a flash frequency that was equal to the rotation frequency of the cells.

At high intensities of lateral light (above 0.2 kW/m²), *Euglena* exhibits negative phototaxis as well as step-up responses upon application of the high-intensity stimulus. Diehn (1969a) has proposed that the two phenomena are causally related as follows (see also Diehn *et al.*, 1975):

> If the intensity of lateral illumination is above the threshold for the step-up photophobic response, the cell will commence a turning reaction upon stimulation. Since every shading event stops this response and resets the cell's sensory mechanism for a further step-up response upon reillumination, the only situation in which there will be no further phobic response is one in which the photoreceptor is permanently shaded. Permanent shading of the photoreceptor is best accomplished by the posterior end of the cell, and thus will lead to an orientation of the organism directly away from the light source.

Alternatively, according to Creutz and Diehn (1976), the cell would detect the position of the light source because of an orientation of the photoreceptor molecules such that their transition moments would be aligned perpendicularly to the cell's longitudinal axis. As an unpolarized stimulus is actually composed of perpendicular linearly polarized ones, cells moving transversely to the light beam will have only a part of their photoreceptor molecules with the transition moment parallel to one of the electrical vectors, and thus be able to absorb light, whereas in cells moving directly toward the light source, all the photoreceptor molecules will have their transition moments parallel to one of the polarizations of the beam. Since the mechanism controlling cell movement (the step-down phobic response) serves to maximize light absorption by the photoreceptor molecules, cells will orient toward the light source (Creutz and Diehn, 1976).

A. PHOTORECEPTIVE STRUCTURES AND PHOTORECEPTOR PIGMENTS

The schematic drawing of Fig. 3 shows in some detail the subcellular structures of *E. gracilis* involved in photobehavior. The stigma, or eyespot, is a cup-shaded organelle constituted of several granules of osmiophilic material; it is clearly visible in the transmission optical microscope, and its characteristic orange-red color is due to lipid-embedded carotenoids. These pigments have been identified and characterized in isolated stigma granules (Batra and Tollin,

---

*This structure is also discussed in this volume, Chapters 1 and 4.

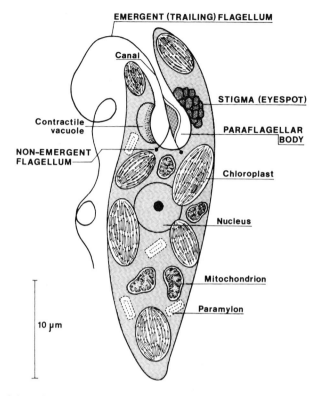

Fig. 3. Schematic representation of *E. gracilis,* including subcellular structures involved in photobehavior.

1964; Bartlett *et al.*, 1972; Sperling Pagni *et al.*, 1976; D. V. Heelis *et al.*, 1979a) and in intact cells by means of a microspectrophotometric technique (Benedetti *et al.*, 1976), which has also shown the absence of any dichroic absorption in these subcellular structures. The paraflagellar body (PFB), which is not visible under a transmission optical microscope, will be described in more detail in the following paragraph. Several kinds of experimental evidence suggest that the question whether the actual photoreceptor in *E. gracilis* is the stigma or the PFB is nearing a resolution.

The streptomycin-treated white mutants definitely lack chloroplasts and stigma granules, but still have the PFB (Kivic and Vesk, 1974; Ferrara and Banchetti, 1976; Benedetti and Checcucci, 1975) and do show photoinduced motile responses (Checcucci *et al.*, 1976a), whereas *Astasia*, a close relative of *Euglena* lacking the stigma as well as the PFB, does not exhibit any photodependent behavior (Goessel, 1957). More recently we have clearly observed, using a tunable dye laser coupled to a properly modified optical microscope, that flagellar reorientation in wild *E. gracilis* occurs only when the laser beam (0.8 μm in diameter) hits the emergent flagellum close to its base, at the same height of the stigma, whereas no photoinduced flagellar reorientation is observed when the same beam strikes the stigma or any other region of the cell body (Colombetti and Lenci, unpublished). It should, however, be noted that these two last experiments support the PFB-photoreceptor hypothesis only with respect to the step-up photophobic response. The streptomycin-treated cells, which have lost the stigma pigments, only show step-up photophobic responses under the microscope and photodispersal from light traps, down to the lowest light intensities used, whereas the laser beam excitation is a step-up experiment in itself. Even though no experimental evidence at present argues against the PFB as photoreceptor also for the step-down photophobic response, the problem still deserves some clarification. For example, Mikolajczyk and Diehn (1975) have suggested that two photoreceptor systems may exist, one for the step-up and one for the step-down response, and this hypothesis cannot be ruled out on the basis of the present knowledge of this subject. As we will see in the following, it seems, however, possible to exclude the notion that the pigments responsible for the two responses are different.

The morphological details of the PFB and of the stigma have been carefully investigated, and an up-to-date review can be found in Chapter 1 of this volume. Here we only want to recall that the PFB has a highly ordered, quasi-crystalline structure, the elementary cell parameters of which have been recently determined (Wolken, 1977; Piccinni and Mammi, 1978).

Concerning the identification of the photopigments, the best and definitely most unambiguous solution would be the isolation of the photoreceptive structure (the PFB). This problem has been tackled in the past by some research groups, but unfortunately without any significant success.

The first indications about the nature of the photopigments responsible for photobehavior of *Euglena* came, as is the case for most photosensitive microorganisms, from action spectra determinations. An action spectrum can be defined as the ratio between the magnitude of the response and the intensity of the photic stimulus causing this response, plotted as a function of the wavelength, provided that the response is a linear or a monotonous function of light intensity (Colombetti and Lenci, 1980a). In the simplest cases, a properly measured action spectrum will be proportional to the absorption spectrum of the pigments involved in the response. Nontrivial problems often arise from the presence of shading organelles that mediate the photomotile response and whose absorbing properties can be reflected in the structure of the action spectrum, as in the case of *Volvox* (Schletz, 1975). According to the hypotheses of Jennings (1906) and Mast (1911), true oriented movement of *E. gracilis* towards or away from the light source (see Diehn *et al.*, 1977) should be brought about by a periodic shading of the photoreceptor (the PFB) by the stigma. Therefore, also in the case of

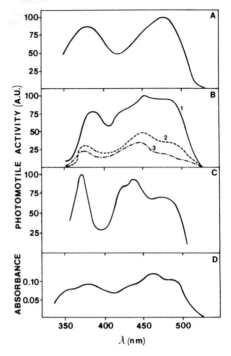

Fig. 4. Action spectrum of photomotile responses of *E. gracilis* (A–C): A, Accumulation in phototaxigraph (from Diehn, 1969a); B, photoaccumulation of wild-type (1), of dark-bleached cells (2), and photodispersal of streptomycin-bleached mutant (3) (from Checcucci *et al.*, 1976a); C, step-down photophobic response of individual cells (Bargighiani *et al.*, 1979a). D, Absorption spectrum of a flavo enzyme, D-amino acid oxidase (from Colombetti, *et al.*, 1975).

Euglena, the action spectrum could reflect the absorbing properties of both the photoreceptor and the stigma. As a matter of fact, all "modern" action spectra, from the very first ones extended to the near-UV region of the spectrum (Diehn, 1969a), are quite similar to each other and proportional to the optical absorption spectrum of flavin-type pigments (Fig. 4D). This holds true for the action spectrum of positive phototaxis (actually a photoaccumulation) of wild *Euglena* measured by means of the phototaxigraph (Diehn, 1969a, Fig. 4,A); for the action spectrum of photoaccumulation of wild-type (Fig. 4,B,1) and dark-bleached cells (Fig. 4,B,2) and of photodispersal of the streptomycin-treated mutant (Fig. 4,B,3), still measured on a cell population using a phototaxigraph (Checcucci *et al.*, 1976a); for the action spectrum of photodispersal of the wild-type determined by means of a phototaxigraph using a linearly polarized actinic beam (Diehn and Kint, 1970); and for the action spectrum of the step-down photophobic response of individual cells determined using a video recording system coupled to an optical microscope (Barghigiani *et al.*, 1979a; Fig. 4C). Even the dark-reversible photosuppression of photobehavior of *Euglena* has an action spectrum suggesting that the receptor pigment mediating the suppression response may be a flavin-type pigment (Tollin and Robinson, 1969). This agreement of the action spectra for the phobic responses and for photoaccumulation or photodispersal would appear puzzling until it is recognized that in the light trap of the "phototaxigraph" the accumulation–dispersal phenomena are mediated by phobic responses at the light–dark border of the trap (Diehn, 1973). There may in fact also be a contribution to accumulation by positive phototaxis toward light scattered by the cells in the illuminated region. Actually, in our experience, it is not easy to observe under the microscope a "taxis" of a cell population toward or away from the light source, possibly because for *Euglena* to exhibit a genuine taxis requires careful control of the experimental conditions of the actinic and observation systems and of the environmental conditions of these cells as well (C. Creutz, personal communication). If positive phototaxis were indeed a mechanism contributing to photoaccumulation, the question would arise as to why the stigma absorption spectrum does not influence the action spectrum. Perhaps phototaxis is indeed exhibited by *Euglena*, but the periodic shading mechanism of Jennings (1906) and Mast (1911) is not operating. It should be noted here that the Creutz–Diehn mechanism (1976) does not require stigma shading. Another reason for the independence of the action spectrum of stigma absorption could be that the stigma plays the role of a neutral gray filter (Checcucci *et al.*, 1976b).

Regardless of the type of cells examined and of the experimental procedure, and therefore of the particular photomotile reaction investigated, all action spectra reported up to the present time seem to point to a flavin-type pigment as responsible for light-induced behavioral responses in *Euglena*. This hypothesis, suggested on the basis of the structure of action spectra (Diehn, 1969a; Diehn and Kint, 1970; Checcucci *et al.*, 1976a; Barghigiani *et al.*, 1979a), has been confirmed by a series of *in vivo* measurements on the spectroscopic properties of the PFB pigments. Benedetti and Checcucci (1975) observed by fluorescence microscopy a green fluorescent organelle at the base of the flagellum, using an exciting light of wavelengths between 360 and 450 nm. Moreover, the relatively high degree of polarization of the fluorescence emitted seemed to be consistent with the highly ordered structure of the PFB described by Kivic and Vesk (1972), Piccinni and Omodeo (1975), and Piccinni and Mammi (1978). The PFB fluorescence emission spectrum has been recorded in intact wild *Euglena* (Benedetti *et al.*, 1975, 1976b), whereas the fluorescence excitation spectrum of this organelle has been determined, also in intact cells, by means of a tuneable dye-laser microspectroscopic apparatus (Ghetti, 1980; Colombetti *et al.*, 1981). Both sets of results confirm the presence of flavin-type pigments in the PFB,

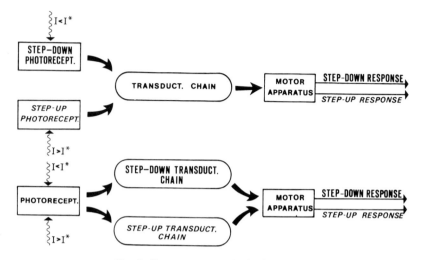

Fig. 5. Photosensory transduction in *Euglena*.

whereas no flavin-type fluorescence has ever been recorded in intact cells from the carotenoid-containing stigma, in the isolated granules of which, however, S. Pagni in 1976 observed a fluorescence emission around 520 nm, characteristic of flavin chromophores.

The previously mentioned hypothesis of the existence of two different photoreceptor systems for the step-up and the step-down responses in *Euglena* (Mikolajczyk and Diehn, 1975) seems to be further supported by the effects of cetyltrimethylammonium bromide (CTAB) on *Euglena* morphology and photobehavior (Mikolajczyk and Diehn, 1978). CTAB treatment selectively decreases the percentage of cells showing step-down photophobic responses, specifically destroys the reservoir and the flagellar membrane (which is further evidence that the photosensory transduction system for the step-down reaction of *Euglena* is located within the flagellar region of the cell), but has no effect on the step-up photophobic response.

On the basis of the foregoing results, Mikolajczyk and Diehn (1978) concluded that the step-down and step-up photophobic responses may employ physically distinct receptors, but could be mediated by the same transduction system. In conclusion, the two-photoreceptor/one-transduction-chain versus the one-photoreceptor/two-transduction-chain question (schematized in Fig. 5), is still open. However, the close similarity of the action spectra for photodispersal, for photoaccumulation, and for individual cell step-down response seems to indicate that, in any case (i.e., wherever they may be located), flavins are the photopigments responsible for *Euglena* photobehavior.

B. Photophysics and Photochemistry of the Photoreceptor Pigments

Even though the nature of the pigments involved in *Euglena* photobehavior seems to be ascertained, no reliable model for the *primary* molecular events occurring in the photoreceptor exist, although the biochemistry of subsequent transduction processes is beginning to emerge (see, e.g., Doughty and Diehn, 1980).

In vivo microspectrofluorometric measurements seem to indicate that the fluorescence emission intensity of the PFB is most probably not constant during excitation (λ_{exc} = 360–450 nm), but increases in intensity with irradiation time, reaching its maximum within a few seconds or tenths of a second, depending on the exciting beam intensity. Further prolonged irradiation causes a decrease of fluorescence intensity due to irreversible pigment photodegradation (Benedetti and Lenci, 1977; Colombetti and Lenci, unpublished). A possible tentative explanation of this time-dependent increase of fluorescence intensity might be that flavins in the PFB have a very low fluorescence quantum yield, probably because of the interactions with their molecular environment. Upon irradiation, a modification of the photoreceptor structure could occur, and this could be responsible for the observed increase in fluorescence intensity. It is interesting to observe that in some flavoproteins an increase in the flavin fluorescence at 520 nm can be detected following a blue light-induced release of the flavin moiety from the apoprotein as a final step of an interaction between the chromophore in an excited state (a triplet?) and the protein structure. This interaction does not damage either the coenzyme (chromophore)-binding site or the enzymatic activity of the flavoprotein, which can be fully restored upon addition of the flavin (Colombetti *et al.,* 1975). Of course, this photo-induced release of the flavin chromophore may have nothing to do with the molecular processes actually occurring in the photoreceptor, as is often true for molecular schemes and mechanisms based on results achieved in *in vitro* model systems. However, although these results cannot be immediately extrapolated to *in vivo* systems, interesting information can be obtained *in vitro* on some basic molecular reactions (see, e.g., Colombetti *et al.,* 1975; Lenci *et al.,* 1976; Heelis *et al.,* 1979b). Only in a few microorganisms, such as *Stentor coeruleus* (Song *et al.,* 1980), *Halobacterium halobium* (Hildebrand, 1978, 1980), and other procaryotes (Häder, 1980; Nultsch, 1980, and references therein), has the actual photoreceptive system been isolated and reliable hypotheses on the primary molecular photoevents put forward. We note that microorganisms for which the primary reactions occurring in the photoreceptor are at least partially known seem not to have specialized photoreceptor organelles. For some of them, for example *Halobacterium halobium* and *Phormidium uncinatum*, photomotile reactions are elicited through the same pigment systems (bacteriorhodopsin and the photosynthetic pigments, respectively), which control and regulate their metabolism (Lenci and Colombetti, 1978, and references therein).

The excitation of flavins to their first excited triplet state as a first step in the *Euglena* photoreception process has been suggested by Diehn and Kint (1970) and by Mikoḷajczyk and Diehn (1975) based on experiments of the KI effect on photomotile responses of *Euglena*. KI is known to be a quencher of excited states of flavins (Penzer and Radda, 1967) and, in particular, of triplet states (Song and Moore, 1974) and was found to specifically inhibit *Euglena* photoaccumulation and step-down photophobic responses, without, however, affecting the step-up photophobic responses (Diehn and Kint, 1970; Mikoḷajczyk and Diehn, 1975). The involvement of a flavin triplet in the early stages of photoreception of *Euglena* seems in agreement with the observation of *Phycomyces* phototropic and light growth responses under tuneable laser stimulation (575-630 nm), indicating a direct transition from the ground state to the lowest triplet state of riboflavin (Delbruck *et al.*, 1976). The first excited triplet state is a reasonable candidate for the primary reactive state in *Euglena* photobehavior, but the amount of experimental evidence is at present rather poor. Moreover, a much shorter-lived state, like the first excited singlet, could at least from a theoretical point of view be an even more efficient intermediary. In this context, the most important property of the triplet state is its rather long lifetime (10^{-6}-10^{-3} seconds) in comparison with that of the singlet state (10^{-9} seconds). Such a long life time may be a prerequisite for the occurrence of a secondary reaction mainly in diffusion-controlled processes. In relatively rigid systems (as most probably in the case of *Euglena* PFB), in which the diffusion of the reactant does not play a relevant role, the short-lived singlet might offer the advantage of being more energetic and, therefore, being able to initiate reactions that could not start from the less energetic triplet state. In the case of riboflavin, for example, $E(S_1)$ is about 2.3 eV, $E(T_1)$ about 2.0 eV, and E is therefore about 0.3 eV, an energy difference between the two states significantly larger than the KT at room temperature (0.025 eV). Moreover, the populations of the singlet is definitely much higher than that of the triplet.

Considering the presently available data, the suggestion that flavins may not be the photoreceptor pigments for the step-up photophobic response (Mikoḷajczyk and Diehn, 1975) seems to be, in some respects, a rash one (Diehn, 1979a; Colombetti and Lenci, 1980b). According to Diehn (1979a), KI could have different effects on the step-down and on the step-up responses by affecting differently two distinct, at least in part, sensory transduction chains, one operating for the step-down and the other for the step-up response.

C. PRIMARY MOLECULAR PHOTOREACTIONS

Starting with the hypothesis that inter- or intramolecular photochemical reactions of flavin chromophores may affect membrane ion pumps (Diehn and Kint,

1970), Froehlich and Diehn (1974) investigated light-induced effects in flavins embedded in lipid bilayer membranes (BLM). They proposed that upon illumination, the isoalloxazine moiety of riboflavin tetrapalmitate would become polarized, perhaps through an electron donor–acceptor reaction of an excited flavin (in a triplet state?) with another molecule of the lipid or the aqueous phase. The polarized chromophore might then interact electrostatically with water dipoles at the BLM surface, which had been previously oriented by an external electric field. The reorientation of the dipoles would correspond to a transient transport of charges against the applied potential. The possible relevance of these phenomena in *Euglena* photoreception processes is, necessarily, only hypothetical and, according to the authors (Froehlich and Diehn, 1974), even the almost identical dependence of membrane photoresponse and of *Euglena* step-up photoresponse on light intensity might be fortuitous. In any case, these considerations also would point to a flavin nature of the photoreceptor pigments for the step-up photophobic response.

A more direct approach to this problem has been followed by Creutz *et al.* (1978), who have investigated and measured the rate of *Euglena* photoaccumulation into light traps as a function of both light intensity within the trap, I, and the change of light intensity at the boundary of the trap, ΔI. From these experiments, it was found that the strength of the behavioral response in a single cell depends on I as well as on ΔI, and is moreover directly proportional to $\Delta I/I$. According to the authors' photoreceptor model, the strength of the step-down response would be proportional to the rate of change of the concentration of a photochemically active form, P, of the photoreceptor pigment, which, at a rate directly proportional to the light intensity, is converted to a photochemically quiescent form, Q. A mathematical analysis of this situation yields a relationship that is much the same relationship as found experimentally between the step-down response and $\Delta I/I$. The photodependent conversion of P to Q, at a rate $K_1 \cdot I \cdot P$, could cause a stoichiometric delivery of charges across the PFB membrane in one direction, whereas conversion from Q to P, at a light-independent rate $K_2 \cdot Q$, could cause charge transport in the opposite direction. The hypothesis that these charges might be protons was considered rather attractive because of the general importance and relevance of vectorial procticity in biological energy transduction phenomena, but seems to be ruled out by recent findings (Doughty and Diehn, 1982) excluding the intervention of proton translocation processes in *Euglena* photosensory transduction.

Before concluding this section, we would like to mention one of the still unsolved problems in *Euglena* photoreception, one which might have very interesting implications: the meaning (in terms of molecular processes and reactions) of the light intensity thresholds for the step-up and step-down responses. Photomotile activity of *Euglena* increases linearly with the logarithm of light intensity (Diehn and Tollin, 1966) up to about 10 W/m^2 of white light (Diehn, 1969b). Above this value photomotile activity (step-down responses) decreases and disappears at light intensities higher than 200 W/m^2, above which only step-up photophobic responses and photodispersal from light traps are the photomotile reactions exhibited by *Euglena*. This threshold value may vary in different cultures, depending on growth conditions, on the illumination conditions to which cells had previously been adapted, on temperature, and on age (Diehn *et al.*, 1975; Diehn, 1980b). The lowest light intensity capable of inducing a photomotile response in a cell population is of the order of 10^{-2} W/m^2 (Checcucci *et al.*, 1974a; Creutz and Diehn, 1976), whereas studies on individual cells show that a cell exhibits a step-down photophobic response with a 10% probability upon removal of a light stimulus of about 0.3×10^{-2} W/m^2 (Barghigiani *et al.*, 1979a; Doughty and Diehn, 1980). From these two values, which are in fairly good agreement with each other, it follows that a cell is capable of detecting light intensity variations corresponding roughly to 6.0×10^{15} photons/m^2, i.e., 6×10^3 photons/μm^2 (the cross section of the PFB is about 0.5 μm^2). This relatively low photosensing ability is approximately of the same order of magnitude as that observed in other microorganisms (Lenci and Colombetti, 1978, and references therein); for example, the minimum photoinduced electrophysiological signal detectable in *Stentor coeruleus* is elicited by a light intensity of about 2×10^{-2} W/m^2 (Wood, 1976). This relatively low sensitivity of microorganisms to light, in comparison with higher organisms, might be due to a low efficiency of amplification in their photosensory transduction chains. This appears reasonable also from an evolutionary point of view.

Keeping in mind that in *Euglena* step-down photophobic responses are elicited by a decrease in light intensity, whereas step-up photophobic responses are caused by an increase of light intensity above a threshold, a contribution to the understanding of the basic molecular mechanisms of photoreception could, undoubtedly, come from a comprehension of the factors affecting the thresholds and of their mechanisms. Interestingly, the transduction time for the step-up photophobic response was found to decrease upon repetition of stimulation. This might reflect a shift in the response threshold as a consequence of repeated stimulation (Diehn *et al.*, 1975). Lowering of the threshold for the step-up photophobic response was also systematically observed in CTAB-treated cells, in which the step-down photophobic response was selectively inhibited (Mikolajczyk and Diehn, 1978). We note that a streptomycin-treated mutant of *Euglena* only shows step-up photophobic responses, down to the lowest light intensity used (Checcucci *et al.*, 1976a).

D. Signal Processing and Transduction: The Black Box

The molecular pathways through which the signal from the photoreceptor is processed and transduced to the effector and eventually to the motor apparatus constitute a crucial problem in the sensory physiology of photomotile aneural organisms, which in most cases is still unsolved.

In the case of *Euglena* photobehavior, Tollin (1969, 1973) suggests that a change in light intensity falling on the photoreceptor might activate a switch controlling the flow of photosynthetically produced ATP to the locomotory flagellum, thus causing a variation of normal flagellar beating with a consequent change in the swimming direction. This hypothesis cannot, however, explain the fact that dark-bleached white cells, lacking chloroplast and photosynthetic capability, do show photomotile reactions (Checcucci *et al.*, 1976a). Moreover, there is no ultrastructural evidence for a connection between the chloroplasts and the flagellum.

Bovee and Jahn (1972), assuming that axoneme paraflagellar rod and PFB are piezoelectric, suggest that this piezoelectric activity displaces cations and drives them along the flagellum, thereby triggering sequential bending from base to tip. The photoreceptor (PFB) would play the role of a capacitor, which discharges when the light impinging on it varies. This charge released by the PFB upon light intensity changes would increase the effect of ion movements along the flagellum, thereby altering the position of the flagellum relative to the cell body. Notwithstanding its apparent completeness, this hypothesis has not received, up to the present, any experimental support.

Piccinni and Omodeo (1975) suggest that the signals from the photoreceptor (PFB) are proportional to the light-induced bleaching of the photoreceptor pigments. Variations in light intensity would be detected by a comparator (a component of the "processor" in the scheme described in Fig. 1), which would measure the difference between the real time signal from the photoreceptor with a signal received an instant before. The delay would be imposed by a synaptic-type junction in the contact area between the plasma membrane of the main flagellum and that of the short flagellum (see also Fig. 3). In this model the role of the effector for flagellum straightening would be played by the longitudinal fibrils assumed to be of an actomyosin nature, surrounding the canal of the cell, which would contract, modifying the flagellar position relative to the cell body, with consequent variation of the cell trajectory. The absence of any effect of both cytochalasins B and C (two drugs which are known to damage actomyosin systems) (Coppelotti *et al.*, 1979) on the step-down as well as the step-up photophobic response could be rationalized by postulating that while fibrils indeed act as the effector, in this case they may be of a type not sensitive to the action of the two cytochalasins; alternatively, one would have to grant that the fibrils are simply not involved in the process of photosensory transduction in *E. gracilis* (Coppellotti *et al.*, 1979). The localization by electron microscopy of ATPase activity in the paraflagellar rod (the highly ordered structure that extends along the flagellum) (Piccinni *et al.*, 1975) might support the hypothesis of the paraflagellar rod as the effector structure (Checcucci *et al.*, 1974b) and, in any case, suggests that the lattice structure of this organelle is related to the motility of the *Euglena* flagellum.

A systems analysis approach to the problem of photosensory transduction in

E. gracilis was used by Diehn (1973). This model of the signal processing system, however, could not at that time be described in terms of identified subcellular components and physiological processes. It might be interesting to reconsider this approach in the solution of the problem, utilizing all the information available at present.

Concerning possible direct interactions of the photomotile sensory systems of *Euglena* with the photosynthetic systems PS I and PS II, the electron transport chains, oxidative phosphorylation, and photophosphorylation, Barghigiani *et al.* (1979b) have shown that the step-down and the step-up photophobic responses of the cell are not affected by metabolic drugs such as 3(3',4'-dichlorophenyl)-1,1-dimethyl urea (DCMU), 2,4-dinitrophenol (DNP), and sodium azide (NaN$_3$), even at concentrations that severely impair cell motility and viability and induce serious morphological alterations. DCMU and NaN$_3$, while completely inhibiting O$_2$ light-induced evolution and damaging the chloroplast structure, do not have any effect on photophobic responses. This shows that in *Euglena* photosynthetic capability and photosensory transduction are not related, which is in agreement with the fact that dark-bleached nonphotosynthetic cells clearly show photomotile reactions (Checcucci *et al.*, 1976a). The delayed inhibition of photoreactions by DCMU reported by Diehn and Tollin (1967) might be due more to an effect on energy sources for photobehavior rather than on the actual photosensory transduction chain. The lack of any effect of NaN$_3$ on *Euglena* photobehavior may also suggest that membrane photopotentials of the type described for *Haematococcus pluvialis* (Litvin *et al.*, 1978) and assumed for *Chlamydomonas reinhardii* (Stavis and Hirschberg, 1973; Stavis, 1974; Marbach and Mayer, 1971; Schmidt and Eckert, 1976; Nultsch, 1979) are not involved in *Euglena* photobehavior. In conclusion, neither the photosynthetic process, in particular noncyclic photophosphorylation and electron transport, nor oxidative phosphorylation seems to play any direct role in *Euglena* photosensory transduction.

Ion transport and ion transport-dependent membrane phenomena are, however, involved in *Euglena* photosensory transduction, similarly to the situation in other microorganisms (see, for example, Colombetti and Lenci, 1980b, and references therein). As far as protons are concerned, the use of drugs affecting proton transport processes [such as gramicidin D and carbonyl cyanide *m*-chlorophenyl hydrazone (CCCP)] and varying the pH of the medium do not have any effect on the step-down and the step-up photophobic responses of *Euglena* (Doughty and Diehn, 1982). These results suggest that there is no direct linkage between proton transport processes and motile responses to photic stimuli, in agreement with the previous described ineffectiveness of DNP (Barghigiani *et al.*, 1979b).

Sound support for the ionic nature of photosensory transduction in *Euglena* has recently been provided by Doughty and Diehn (1979) and by Doughty *et al.*

(1980). The model is based on a great number of experimental findings, the most important of which are briefly mentioned in the following. *Euglena* exhibits optimum motility in the presence of Mg^{2+}, Ca^{2+} and K^+, whereas Ni^{2+} ions immobilize the cells. In increasing Ca^{2+}, Mn^{2+}, and Ba^{2+} concentrations, the frequency of directional change of the cells increases and the duration of the step-down photophobic response is enhanced in the presence of divalent cations in the order $Ca^{2+} > Ba^{2+} > Mn^{2+} > Co^{2+} > Mg^{2+} > Ni^{2+} = 0$. An increase of the duration of the photophobic responses is also caused by NaCl and by the drug ouabain (an inhibitor of Na^+–K^+ ion transport across membranes), whereas chloride (administered with the impermeant cation choline) reduces the duration of the photophobic responses. The NaCl-induced enhancement of photobehavior is reduced by KCl, and at constant Cl^- concentration, photobehavior is independent of KCl and NaCl relative concentrations. The Ca^{2+} conductance antagonist Verapamil and the Ca^{2+} conductance–active transport antagonist ruthenium red do not modify the effect of Ca^{2+}. In the presence of the Ca^{2+} ionophore A23187, in contrast, Ca^{2+} induces a specific light-independent but concentration-dependent discontinuous tumbling response of the cells. Finally, antagonists of voltage-dependent monovalent cation fluxes in membranes (tetrodotoxin, procaine, tetraethylammonium, 4-aminopyridine) do not have any effect on *Euglena* photobehavior. Figure 6 shows a strictly preliminary and entirely hypothetical

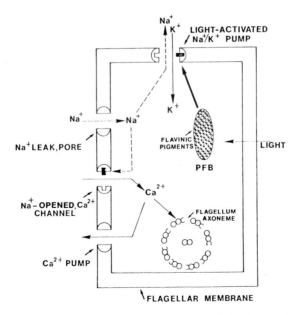

Fig. 6. Hypothetical scheme of the molecular processes mediating the step-down photophobic response of *Euglena*.

schematic representation of the overall process: flagellar reorientation, and therefore the photophobic response, is triggered by a transient increase of Ca^{2+} concentration in the intraflagellar space. Light, however, does not directly regulate Ca^{2+} permeability of the flagellar photoreceptor membrane. Modifications of *Euglena* photobehavior and photosensitivity are induced by monovalent and divalent cations, the fluxes of which could be controlled by a light-activated (blocked by ouabain) Na^+–K^+ pump (Fig. 6). The pump could also serve to accumulate K^+ in the cell, perhaps as a prerequisite for net K^+ efflux to drive flagellar reorientation. A comment may be in order concerning this stimulating model: The flagellar reorientation time is most probably related to the phenomenon of adaptation of the cells to the stimuli, so that the possibility should be carefully considered that the light-dependent Na^+–K^+ transport activity regulates the subsequent adaptation steps rather than the actual excitation steps in the photoreceptor system. In view of the success of the above-described neuropharmacological approach to elucidating the molecular mechanisms involved in the step-down response, a similar experimental approach to the step-up photophobic response would be of great interest and could help in settling the question relative to the two-photoreceptors/two transduction-chains.

III. Behavioral Responses to Chemical, Mechanical, and Gravitational Stimuli*

Like several other microorganisms, *E. gracilis* can use the same type of response when reacting to different external stimuli; thus the chemophobic response toward oxygen (Colombetti and Diehn, 1978) and the response to mechanical forces (Jahn and Bovee, 1968) are expressed through the same elementary motile reactions as the photophobic response, i.e., flagellar reorientation and subsequent turning toward the dorsal side* (see Fig. 2). In the case of mechanical stimulation, the situation is often complicated by body contractions, as we shall see in more detail in the following paragraph. In contrast to the case of photic stimuli, the effects of chemical, and, above all, mechanical and gravitational stimuli on *Euglena* behavior have not been extensively investigated.

A. Chemical Stimuli and Chemobehavior

A concentration gradient of oxygen induces behavioral modifications of *Euglena*. This is the only chemosensory response of *Euglena* that has been investigated in some detail. Checcucci *et al.* (1974a) and, independently, Diehn (1973) reported a red light-induced accumulation of green *Euglena*, whereas no

*This is also discussed in this volume, Chapter 4.

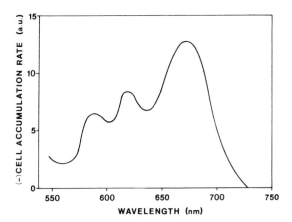

Fig. 7. Action spectrum of red light-induced photoaccumulation of *E. gracilis* (from Checcucci *et al.*, 1974a).

response to red light was observed in dark-bleached white cells (Checcucci *et al.*, 1974a). The action spectrum of the photoresponse (Fig. 7) as well as the complete inhibition of the red light response by DCMU and NH_2OH pointed to a chemoaccumulation of *Euglena* toward oxygen, photosynthetically evolved in the illuminated regions of the culture suspension (Checcucci *et al.*, 1974a). In agreement with these findings, Colombetti and Diehn (1978), in cultures containing 10^6 cells/ml, have observed ring pattern formations of green *Euglena* around air bubbles and spontaneous formation of dynamic ring patterns (100–300 μm ring width) in the dark, very similar to those observed in bacterial chemotaxis. Cells moving inward as well as outward from the ring show phobic responses, the ring defining the region between a step-down and step-up threshold concentration of oxygen. The step-up threshold concentration of oxygen seems lower than 0.2 mM (the concentration of O_2 in air-saturated water). Exactly the same mechanism, i.e., movements of the cells along the O_2 gradient, can explain the behavior of *Euglena* in the presence of air bubbles.

Spontaneous ring formation in the dark is markedly affected by irradiation with red light (575–650 nm; 5 W/m²): ring contraction or formation of the ring patterns in the dark area around the red light spot was clearly observed. This effect of red light was completely inhibited by DCMU (Colombetti and Diehn, 1978). The phenomenon of ring contraction can be explained as a chemodispersal in response to photosynthetically evolved O_2. Rings move inward because the O_2 concentration exceeds the step-up threshold value everywhere in the sample, except inside the ring, if the ring diameter is smaller than that of the light spot. For the same reasons, rings form in the dark area

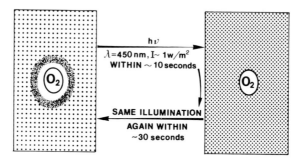

Fig. 8. Effect of blue light upon chemoaccumulation toward oxygen of *Euglena*.

around the red light spot, if the diameter of the original ring is larger than the lighted region.

B. CHEMORECEPTORS

Miller and Diehn (1978) have assayed the effect of various inhibitors on *Euglena* chemoaccumulation toward oxygen, monitoring spontaneous formation of ring accumulation patterns in the dark. They report a selective inhibition of ring formation by sodium cyanide (10^{-6}–$10^{-4}M$), without any effect on cell motility and viability. Control experiments also showed that neither sodium cyanide nor carbon monoxide had any effect on the photosensory responses of *Euglena*, thus indicating that the inhibitory effect of the two drugs is specifically at the level of the receptor and does not affect other parts of the sensory transduction chain. NaN_3 was found to have no effect on the chemoresponse nor on the photoresponse of *Euglena* (this last result confirmed also by the findings of Barghigiani *et al.*, 1979b). On the basis of these results, the authors conclude that the cytochrome a_3 moiety of cytochrome-*c* oxidase is the receptor molecule for oxygen-mediated chemosensory responses of *Euglena*, most probably located in the anterior part of the cell.

C. INTERACTION OF CHEMOSENSORY AND PHOTOSENSORY SYSTEMS

An interesting and promising approach to the problem of sensory transduction in microorganisms is the analysis of behavioral modification induced by the superimposition of different stimuli.

A linkage between the physiological processes responsible for chemoaccumulation and for step-down photophobic response of *Euglena* has been shown by Colombetti and Diehn (1978) on the basis of their observation that low intensity blue light causes dispersal of oxygen-mediated ring patterns (Fig. 8). This phenomenon has been studied and discussed in more detail by Diehn (1979a).

It is fairly easy to subject a population of *Euglena* to light and chemical stimuli at the same time. Cells in a chemoaccumulation pattern such as a ring around an oxygen bubble are kept within this pattern by constant chemophobic responses at the pattern boundaries. Photic stimuli can be added by varying the illumination of the cell suspension (ordinarily, chemoresponses are studied under yellow illumination of 577 nm, which is a wavelength to which the cell's photosensory system is not responsive).

If a photic stimulus is applied at an intensity below the step-up threshold, the cells will, as one would expect, not exhibit any photophobic responses. However, the cells immediately begin to leave the chemoaccumulation ring, apparently because they no longer respond with chemophobic reactions to the chemical stimulus gradients. As a consequence, the ring pattern disappears within approximately 10 seconds. This effect is transient: Upon continuing illumination, the pattern forms again with another 30 seconds or so (Fig. 8). If the light stimulus is removed at this time, the cells execute normal-appearing step-down photophobic responses. Simultaneously, the ring pattern boundaries sharpen perceptibly.

At the present time, any interpretation of this behavior must of necessity be speculative. Clearly, the photic and chemical stimuli do not simply compete for the flagellar response system. The light stimulus, when given by itself, would only cause a resetting of the photosensory transduction system, but no overt response whatever in the absence of the chemical stimulus.

One must conclude from the dispersal of the chemoaccumulation pattern that this photic stimulus causes a temporary erasure of chemosensory transduction. Apparently, control of the flagellar response system is transiently switched from the chemosensory to the photosensory transduction system. One might speculate that this occurs in anticipation of the step-down stimulus that would be presented if the light were to be turned off again. If this stimulus does not materialize within approximately 30 seconds, the chemosensory apparatus resumes control over the response system. Nevertheless, the *photic* transduction system remains reset and will again take control of the effector if the step-down stimulus is given now. At this time, the control is no longer absolute: The ring pattern sharpens during the photophobic response, indicating that chemophobic responses occur in addition to the photophobic responses, and moreover that the range between their step-up and step-down thresholds has been reduced.

The above conjectures call for further experimental tests. For instance, if it is the "resetting" characteristic of the stimulus that causes the cell to switch from one sensory system to another, then resetting for the step-up photophobic response should have the same effect as resetting for the step-down response. In other words, turning off high intensity illumination should have the same effect as turning on low intensity light.

There are two problems if one attempts such an experiment with ordinary light-grown *Euglena:* First, the cells are photosynthetic and will evolve oxygen upon high intensity illumination. This, in turn, will destroy the chemoaccumulation pattern by raising the stimulus concentration above the step-up reference level. Second, turning off the light will cause an albeit delayed and attenuated, step-down response (Diehn, 1969b). Both problems can be overcome by using dark-grown cells. Without a functioning photosynthetic system, these cells neither evolve oxygen, nor are they capable of step-down responses (Diehn and Tollin, 1967; Diehn, 1973). They do, however, form chemophobic accumulation patterns.

The results of experiments with etiolated cells were exactly as predicted: Increasing the illumination above the step-up reference level triggered photophobic responses and at the same time sharpened the chemophobic pattern. When the light was turned off, dissolution of the ring was indeed observed within a few seconds.

Present studies are concerned with the erasure of chemotransduction as a function of the magnitude of the resetting stimulus. These experiments should yield information on the characteristics of the *Euglena* signal processor. Other problems that will surely be investigated include the question of whether resetting for the chemoresponses in turn inactivates the phototransduction system, or whether the latter is assigned an overriding importance in the hierarchy of sensory systems.

D. MECHANICAL AND GRAVITATIONAL STIMULI

Motor responses of *Euglena* have also been observed upon stimulation by mechanical means (Mikołajczyk and Diehn, 1976, 1979) and, possibly, in response to the force of gravity (Creutz and Diehn, 1976). Since these responses are in part expressed through the flagellar apparatus, as are the photically induced responses, it is clear that mechano- and photosensory pathways must converge and probably interact, just as discussed above for the chemo- and photosensory systems.

Upon simultaneous photic and mechanical stimulation of a *Euglena* cell, flagellar phobic responses are followed by body contractions (Mikołajczyk and Diehn, 1976). These contractions are not induced by light or mechanical stimuli acting alone, even though these stimuli can independently trigger flagellar reorientation, i.e., photo- and mechanophobic responses of the cell.

For cells immersed in media of varying viscosity, the transduction time for light-induced body contractions was found to depend on the viscosity of the medium, with the shortest transduction times observed at the higher viscosities.

From 500–2000 centipoise (cP), during which the cells are capable both of body contraction and forward locomotion (only at 4000 cP and above does complete immobilization occur), a logarithmic dependence of both transduction time and forward velocity upon the viscosity was found. Within this limited viscosity range, each doubling of the viscosity resulted in a decrease of the transduction time to 60%, and of the forward velocity to 20% of the preceding value.

Although the transduction time for contraction depended on the intensity of the mechanical stimulus, it was found to be independent of the intensity of the light stimulus from threshold (at 0.10 kW/m^2) to 1.13 kW/m^2. This is a very intriguing situation: A Weber-Fechner relationship exists not between the response and triggering stimulus, but between the response and the mechanical stimulus that couples the body contraction system to the apparatus responsible for flagellar reorientation.

Since evidence is accumulating that all sensory functions of the cell reside in the flagellar and reservoir membranes (Mikolajczyk and Diehn, 1978, 1979), the above results may well be due to a mechanosensory stimulus exerted upon the reservoir region by the asymmetric thrust of the reoriented flagellum.

The evidence for a geosensory response of *Euglena* is indirect (Creutz and Diehn, 1976). These investigators have observed that in the absence of any experimenter-controlled stimulation, the cells preferentially assume an otherwise random orientation in the horizontal plane. Since, as described in the same report, the cells orient perpendicularly to the plane of polarized stimulating light, it is possible to establish an experimental situation in which photic and gravitational stimuli counteract each other. Such an experiment would yield information on the interaction of the two sensory systems, and on the relative degree of control over the motor apparatus exerted by either system.

Acknowledgments

G. C. and F. L. acknowledge the support of C. N. R. on the Cooperative Research Project, Italy–United States (Grant Numbers 104320/02/8102190 and 102060/02/8102191).

References

Ascoli, C., Barbi, M., Frediani, C., and Muré, A. (1978). Measurement of *Euglena* motion parameters by laser light scattering. *Biophys. J.* **24**, 585–599.

Barghigiani, C., Colombetti, G., Franchini, B., and Lenci, F. (1979a). Photobehavior of *Euglena gracilis:* Action spectrum for the step-down photophobic response of individual cells. *Photochem. Photobiol.* **29**, 1015–1019.

Barghigiani, C., Colombetti, G., Lenci, F., Banchetti, R., and Bizzarro, M. P. (1979b). Photosensory transduction in *Euglena gracilis:* Effect of some metabolic drugs on the photophobic response. *Arch. Microbiol.* **120**, 239–245.

Bartlett, C. M., Walne, P. L., Schwarz, O. J., and Brown, D. H. (1972). Large scale isolation and purification of eyespot granules from *Euglena gracilis*. *Plant Physiol.* **49**, 881–885.

Batra, P. P., and Tollin, G. (1964). Phototaxis in *Euglena*. I. Isolation of the eyespot granules and identification of the eyespot pigment. *Biochim. Biophys. Acta* **79**, 371–378.

Benedetti, P. A., and Checcucci, A. (1975). Paraflagellar body (PFB) pigments studied by fluorescence microscopy in *Euglena gracilis*. *Plant Sci. Lett.* **4**, 47–51.

Benedetti, P. A., and Lenci, F. (1977). *In vivo* microspectrofluorometry of photoreceptor pigments in *Euglena gracilis*. *Photochem. Photobiol.* **26**, 315–318.

Benedetti, P. A., Bianchini, G., and Grassi, S. (1975). An electrodynamic fast-scanning instrument for microspectroscopy. *Int. Biophys. (IUPAB) Congr., 5th, 1975.* Abstr. P/188.

Benedetti, P. A., Bianchini, G., Checcucci, A., Ferrara, R., Grassi, S., and Percyval, D. (1976).

Spectroscopic properties and related functions of the stigma measured in living cells of *Euglena gracilis*. *Arch. Microbiol.* **111**, 73–76.

Bovee, E. C., and Jahn, T. L. (1972). A theory of piezoelectric activity and ion-movements in the relation of flagellar structures and their movements to the phototaxis of *Euglena*. *J. Theor. Biol.* **35**, 259–276.

Carlile, M. J. (1980). Sensory transduction in aneural organisms. *In* "Photoreception and Sensory Transduction in Aneural Organisms" (F. Lenci and G. Colombetti, eds.), pp. 1–22. Plenum, New York.

Checcucci, A. (1976). Molecular sensory physiology of *Euglena*. *Naturwissenschaften* **63**, 412–417.

Checcucci, A., Colombetti, G., Del Carratore, G., Ferrara, R., and Lenci, F. (1974a). Red-light induced accumulation of *Euglena gracilis*. *Photochem. Photobiol.* **19**, 223–226.

Checcucci, A., Colombetti, G., Ferrara, R., and Lenci, F. (1974b). Modelli molecolari per la fototassi in *Euglena gracilis*. *Proc. Natl. Congr. Biophys. Cybernet., 3rd, 1974*, pp. 275–282.

Checcucci, A., Colombetti, G., Ferrara, R., and Lenci, F. (1976a). Action spectra for photoaccumulation of green and colorless *Euglena*: Evidence for identification of receptor pigments. *Photochem. Photobiol.* **23**, 51–54.

Checcucci, A., Colombetti, G., Ferrara, R., and Lenci, F. (1976b). Further analysis of the mass photoresponses of *Euglena gracilis* Klebs. *Monit. Zool. Ital.* [N.S.] **10**, 271–277.

Colombetti, G., and Diehn, B. (1978). Chemosensory responses toward oxygen in *Euglena gracilis*. *J. Protozool.* **25**, 211–217.

Colombetti, G., and Lenci, F. (1980a). Identification and spectroscopic characterization of photoreceptor pigments. *In* "Photoreception and Sensory Transduction in Aneural Organism" (F. Lenci and G. Colombetti, eds.), pp. 171–188. Plenum, New York.

Colombetti, G., and Lenci, F. (1980b). Photosensory transduction chains in Eukaryotes. *In* "Photoreception and Sensory Transduction in Aneural Organism" (F. Lenci and G. Colombetti, eds.), pp. 341–354. Plenum, New York.

Colombetti, G., Lenci, F., McKellar, J. F., and Phillips, G. O. (1975). Light-induced effects in a flavoprotein: D-amino acid oxidase. *Photochem. Photobiol.* **21**, 303–306.

Colombetti, G., Ghetti, F., Lenci, F., Polacco, E., and Quaglia, M. (1981). *In vivo* microspectrofluorometry of photoreceptor pigments. *J. Photochem.* **17**, 36.

Coppellotti, O., Piccinni, E., Colombetti, G., and Lenci. F. (1979). Responses of *Euglena gracilis* to cytochalasins B and D. *Boll. Zool.* **46**, 72–75.

Creutz, C., and Diehn, B. (1976). Motor responses to polarized light and gravity sensing in *Euglena gracilis*. *J. Protozool.* **23**, 552–556.

Creutz, C., Colombetti, G., and Diehn, B. (1978). Photophobic behavioral responses of *Euglena* in a light intensity gradient and the kinetics of photoreceptor pigment interconversion. *Photochem. Photobiol.* **27**, 611–616.

Delbruck, M., Katzir, A., and Presti, D. (1976). Responses of *Phycomyces* indicating optical excitation of the lowest triplet state of riboflavin. *Proc. Natl. Acad. Sci. U.S.A.* **73**, 1969–1973.

Diehn, B. (1969a). Action spectra of the phototactic responses in *Euglena*. *Biochim, Biophys. Acta* **177**, 136–143.

Diehn, B. (1969b). Phototactic response of *Euglena* to single and repetitive pulses of actinic light: Orientation time and mechanism. *Exp. Cell Res.* **56**, 375–381.

Diehn, B. (1973). The receptor–effector system of phototaxis in *Euglena*. *Science* **181**, 1009–1015.

Diehn, B. (1979a). Photic responses and sensory transduction in protists. *In* "Handbook of Sensory Physiology" (H. Autrum, ed.), Vol. 7, Part 6A, pp. 23–68. Springer-Verlag, Berlin and New York.

Diehn, B. (1979b). The interactions of photic and chemical stimulus/response systems in *Euglena gracilis*. *Acta Protozool.* **18**, 7–16.

Diehn, B. (1980a). Photomovement and photosensory transduction in microorganisms. *Photochem. Photobiol.* **31**, 641–644.

Diehn, B. (1980b). Experimental determination and measurement of photoresponses. *In* "Photoreception and Sensory Transduction in Aneural Organisms" (F. Lenci and C. Colombetti, eds.), pp. 107–125. Plenum, New York.

Diehn, B., and Kint, B. (1970). The flavin nature of the photoreceptor pigments for phototaxis in *Euglena. Physiol. Chem. Phys.* **2**, 483–488.

Diehn, B., and Tollin, G. (1966). Phototaxis in *Euglena*. II. Physical factors determining the rate of phototactic response. *Photochem. Photobiol.* **5**, 839–844.

Diehn, B., Fonseca, J. R., and Jahn, T. L. (1975). High speed cinemicrography of the direct photophobic response of *Euglena* and the mechanism of negative phototaxis. *J. Protozool.* **22**, 492–494.

Diehn, B., Feinleib, M. E., Haupt, W., Hildebrand, E., Lenci, F., and Nultsch, W. (1977). Terminology of behavioral responses of motile microorganism. *Photochem. Photobiol.* **26**, 559–560.

Diehn, B., and Tollin, G. (1967). Phototaxis in *Euglena*. IV. Effect of inhibitors of oxidative and photophosphorylation of the rate of phototaxis. *Arch. Biochem. Biophys.* **121**, 169–177.

Doughty, M. J., and Diehn, B. (1979). Photosensory transduction in the flagellated alga, *Euglena gracilis*. I. Action of divalent cations, Ca^{2+} antagonists and Ca-ionophore on motility and photobehavior. *Biochim. Biophys. Acta* **588**, 148–168.

Doughty, M. J., and Diehn, B. (1980). Flavins as photoreceptor pigments for behavioral responses in motile microorganisms, especially in the flagellated alga *Euglena* sp. A critique. *Struct. Bond.* **41**, (P. Hemmerich, ed.) pp. 45–70, Springer-Verlag, Berlin and New York.

Doughty, M. J., Grieser, R., and Diehn, B. (1980). Photosensory transduction in the flagellated alga *Euglena gracilis*. II. Evidence that blue light effects alteration in Na-K permeability of the photoreceptor membrane. *Biochim. Biophys. Acta* **602**, 10–23.

Doughty, M. J., and Diehn, B. (1982). In preparation.

Feinleib, M. E. (1975). Phototactic response of *Chlamydomonas* to flashes of light. I. Response of cell population. *Photochem. Photobiol.* **21**, 351–354.

Feinleib, M. E. (1978). Photomovement of microorganisms. *Photochem. Photobiol.* **27**, 849–854.

Feinleib, M. E., and Curry, G. M. (1974). The nature of the photoreceptor in phototaxis. *In* "Handbook of Sensory Physiology" (W. R. Lowenstein, ed.), Vol. 1, pp. 366–395. Springer-Verlag, Berlin and New York.

Ferrara, R., and Banchetti, R. (1976). Effect of streptomycin on the structure and function of the photoreceptor apparatus of *Euglena gracilis. J. Exp. Zool.* **198**, 393–402.

Froelich, O., and Diehn, B. (1974). Photoeffects in a flavin-containing lipid bilayer membrane and implications for algal phototaxis. *Nature* **248**, 802–804.

Ghetti, F. (1980). Microspettroscopia Laser su Pigmenti, Thesis, Physics Institute, University of Pisa.

Goessel, J. (1957). Uber das Actionsspektrum der Phototaxis chlorophyllfreier *Euglenen* und uber die Absorption des Augenflecks. *Arch. Mikrobiol.* **27**, 288–305.

Häder, D. P. (1980). Photosensory transduction chains in prokaryotes. *In* "Photoreception and Sensory Transduction in Aneural Organisms" (F. Lenci and G. Colombetti, eds.), pp. 355–372. Plenum, New York.

Haupt, W., and Feinleib, M. F., eds. (1979). "Encyclopedia of Plant Physiology," New Series, Vol. 7, Springer-Verlag, Berlin and New York.

Heelis, D. V., Kernick, W., Phillips, G. O., and Davies, K. (1979a). Separation and identification of the carotenoid pigments of stigmata isolated from light grown cells of *Euglena gracilis* strain Z. *Arch. Microbiol.* **121**, 207–211.

Heelis, P. F., Parsons, J. B., Phillips, G. O., Barghigiani, C., Colombetti, G., Lenci, F., and McKellar, J. F. (1979b). Flavin pigments embedded in lipid matrices: A spectroscopic and photochemical investigation. *Photochem. Photobiol.* **30**, 507–512.

Hildebrand, E. (1978). Light-controlled behaviour of bacteria. *In* "Receptors and Recognitions—Taxis and Behaviour" (G. L. Hazelbauer, ed.), pp. 1–68. Chapman & Hall, London.

Hildebrand, E. (1980). Comparative discussion of photoreception in lower and higher organisms. Structural and functional aspects. *In* "Photoreception and Sensory Transduction in Aneural Organisms" (F. Lenci and G. Colombetti, eds.), pp. 319–340. Plenum, New York.

Jahn, T. L., and Bovee, E. C. (1968). Locomotive and motile responses in *Euglena*. *In* "The Biology of *Euglena*" (D. E. Buetow, ed.), Vol. 1, pp. 45–108. Academic Press, New York.

Jennings, H. S. (1906). "Behavior of the Lower Organisms." Columbia University Press, New York.

Kivic, P. A., and Vesk, M. (1972). Structure and function in the euglenoid eyespot apparatus: The fine structure, and response to environmental changes. *Planta* **105**, 1–14.

Kivic, P. A., and Vesk, M. (1974). The structure of the eyespot apparatus in bleached strains of *Euglena gracilis*. *Cytobiologie* **10**, 88–101.

Lenci, F., and Colombetti, G. (1978). Photomovements of microorganisms: A biophysical approach. *Annu. Rev. Biophys. Bioeng.* **7**, 341–361.

Lenci, F., Colombetti, G., Del Carratore, G., and Gualtieri, P. (1976). Photomodifications of the flavoprotein D-amino-acid oxidase in the presence of a substrate competitive inhibitor. *Bull. Mol. Biol. Med.* **1**, 119–128.

Litvin, F., Simeshchekov, O., and Simeshchekov, V. A. (1978). Photoreceptor electric potential in the phototaxis of the alga *Haematococcus pluvialis. Nature (London)* **271**, 476–478.

Marbach, I., and Mayer, A. M. (1971). Effect of electric fields on the phototactic response of *Chlamydomonas reinhardii. Isr. J. Bot.* **20**, 96–100.

Mast, S. O. (1911). "Light and the Behavior of Organisms." Wiley, New York.

Mikołajczyk, E., and Diehn, B. (1975). The effect of potassium iodide on photophobic responses in *Euglena:* Evidence for two photoreceptor pigments. *Photochem. Photobiol.* **22**, 268–271.

Mikołajczyk, E., and Diehn, B. (1976). Light-induced body movement of *Euglena gracilis* coupled to flagellar photophobic responses by mechanical stimuli. *J. Protozool.* **23**, 144–147.

Mikołajczyk, E., and Diehn, B. (1978). Morphological alterations in *Euglena gracilis* induced by treatment with CTAB (cetyltrimethylammonium bromide) and Triton X-100: Correlations with effects on photophobic behavioral responses. *J. Protozool.* **25**, 461–470.

Mikołajczyk, E., and Diehn, B. (1979). Mechanosensory responses and mechanoreception in *Euglena gracilis. Acta Protozool.* **18**, 591–602.

Miller, S., and Diehn, B. (1978). Cytochrome *c* oxidase as the receptor molecule for chemoaccumulation (chemotaxis) of *Euglena* toward oxygen. *Science* **200**, 548–549.

Nultsch, W. (1979). Effect of external factors on phototaxis of *Chlamydomonas reinhardii.* III. Cations. *Arch. Microbiol.* **123**, 93–99.

Nultsch, W. (1980). Photomotile responses in gliding organisms and bacteria. *In* "Photoreception and Sensory Transduction in Aneural Organisms" (F. Lenci and G. Colombetti, eds.), pp. 69–87. Plenum, New York.

Nultsch, W., and Häder, D. P. (1979). Photomovements of motile microorganisms. *Photochem. Photobiol.* **29**, 423–437.

Penzer, G. R., and Radda, G. K. (1967). The chemistry and biological function of isoalloxazines (flavines). *Q. Rev. Chem. Soc.* **21**, 43–65.

Piccinni, E., and Mammi, M. (1978). Motor apparatus of *Euglena gracilis:* Ultrastructure of the basal portion of the flagellum and the paraflagellar body. *Boll. Zool.* **45**, 405–414.

Piccinni, E., and Omodeo, P. (1975). Photoreceptors and phototactic programs in protista. *Boll. Zool.* **42**, 57–79.

Piccinni, E., Albergoni, V., and Coppellotti, O. (1975). ATPase activity in flagella from *Euglena gracilis*. Localization of the enzyme and effects of detergents. *J. Protozool.* **22**, 331–335.

Schletz, K. (1975). Phototaxis bei Volvox: Pigmentsysteme der Lichtrichtungsperzeption. *Z. Pflanzenphysiol.* **11**, 189–211.

Schmidt, J. A., and Eckert, R. (1976). Calcium couples flagellar reversal to photostimulation in *Chlamydomonas reinhardii*. *Nature (London)* **262**, 713–715.

Seitz, K. (1975). Orientation in space: Plants. *In* "Marine Ecology" (O. Kinne, ed.), pp. 451–497. Wiley, New York.

Song, P. S., and Moore, T. A. (1974). On the photoreceptor pigment for phototropism and phototaxis: Is a carotenoid the most likely candidate? *Photochem. Photobiol.* **19**, 435–441.

Song, P. S., Walker, E. B., and Yoon, M. J. (1980). Molecular aspects of photoreceptor function in *Stentor coeruleus*. *In* "Photoreception and Sensory Transduction in Aneural Organisms" (F. Lenci and G. Colombetti, eds.), pp. 241–252. Plenum, New York.

Sperling Pagni, P. G., Walne, P. L., and Werhy, L. (1976). Fluorometric evidence for flavins in isolated eyespots of *Euglena gracilis* var. bacillaris. *Photochem. Photobiol.* **24**, 373–375.

Stavis, R. L. (1974). The effect of azide on phototaxis in *Chlamydomonas reinhardii*. *Proc. Natl. Acad. Sci. U.S.A.* **71**, 1824–1827.

Stavis, R. L., and Hirschberg, R. (1973). Phototaxis in *Chlamydomonas reinhardii*. *J. Cell Biol.* **59**, 367–377.

Tollin, G. (1969). Energy transduction in algal phototaxis. *Curr. Top. Bioenerg.* **3**, 417–446.

Tollin, G. (1973). Phototaxis in *Euglena*. II: Biochemical aspects. *In* "Behavior of Microorganisms" (A. Perez-Miravete, ed.), pp. 91–105. Plenum, New York.

Tollin, G., and Robinson, M. I. (1969). Phototaxis in *Euglena*. V. Photosuppression of phototactic activity by blue light. *Photochem. Photobiol.* **9**, 411–416.

Wolken, J. J. (1977). *Euglena*: The photoreceptor system for phototaxis. *J. Protozool.* **24**, 518–522.

Wood, D. C. (1976). Action spectrum and electrophysiological responses correlated with the photophobic responses of *Stentor coeruleus*. *Photochem. Photobiol.* **24**, 261–266.

CHAPTER 6

STIMULATION AND INHIBITION OF THE METABOLISM AND GROWTH OF *EUGLENA GRACILIS*

E. S. Kempner

I. Introduction and Scope

Almost all of the reports of *Euglena* research in recent years concern the species *Euglena gracilis;* the sporadic studies of other species are also included in this chapter, but are specifically stipulated as such. Because of the great similarity of *Astasia longa* to *Euglena,* studies of this organism are also considered, although it should be remembered that *Astasia* does differ significantly from *Euglena* and justifiably belongs in a separate genus (Blum *et al.,* 1965).

A great variety of substances affect the metabolism of *E. gracilis,* but only a few elements are needed for cell growth. The heterotrophic growth of *E. gracilis* strain Z requires H, C, N, O, Mg, P, S, Cl, K, Ca, Mn, Co and Zn; other elements, if needed at all, are only required at levels below $10^{-6} M$ (Kempner and

THE BIOLOGY OF *EUGLENA*, VOL. 3
ISBN 0-12-139903-6

Miller, 1972a). The chemical form of these elements may be crucial, as for example the requirements for vitamins B_1 and B_{12}. The main organic supply for cell growth can be carbon dioxide, ethanol, acetate, lactate, glucose, glutamic acid, or malic acid. Almost all *Euglena* cultivation media include one of this group exclusively as the carbon source. All other organic compounds can be considered gratuitous or supplementary.

Because of its photosynthetic capabilities, *Euglena* has long been a popular experimental organism. The large number of chloroplast-related chapters in these volumes attests to the continued interest in this aspect of *Euglena* metabolism. The reader is referred to those chapters for discussion of chemical modifiers of photosynthesis and chloroplast structure.

The second largest group of *Euglena* literature citations concerns the absolute requirement for vitamin B_{12}. As will be seen, there are dramatic and fascinating effects of vitamin B_{12} deprivation. In spite of all the studies, however, the fundamental biochemical role of the vitamin in *Euglena* remains obscure.

Other interesting phenomena exhibited by *Euglena*, such as phototaxis and swimming movements are influenced by culture composition and are covered in the appropriate chapters. The fundamental (i.e., nonspecialized) aspects of *Euglena* metabolism are similar to those of most organisms, but are combined in a package that is exceptionally convenient for research. Although *Euglena* can be grown as easily and reproducibly as *E. coli*, it has the distinct advantage of having a subcellular organelle complement similar to those of the higher organisms. *E. gracilis* appears to be an ideally designed cell for research in molecular biology and cell physiology.

The literature up to 1968 has been covered thoroughly in prior volumes of this series, especially Chapter 6, Volume I (Cook, 1968) and Chapter 8, Volume II (Hutner *et al.*, 1968). The present work discusses reports published between 1968 and January 1, 1980.

II. Growth Media

Both *E. gracilis* and *A. longa* have been cultivated in a variety of defined growth media. Some of the more popular and significant recipes are given in Table I. Many different organic sources are listed, and where none is listed, CO_2 was supplied for autotrophic growth. The inclusion of any given compound in these formulations should not be taken as evidence for its stimulatory effect on growth of *Euglena*. Clearly many compounds, especially some of the inorganic chemicals, are not required. Indeed, if stimulation of *Euglena* growth is measured by rapid doubling and yield of cells, it is clear that the most dilute medium (Kempner and Miller, 1972a) is one of the most stimulatory. The necessity for each component of this medium was tested both by omission and reduction. The inorganic components contain the only metals required for the growth of

Euglena, at least at levels of 1 μg/liter or higher. The list is surprisingly short: K, Mg, Ca, Mn, and Zn. The only anions present are phosphate, sulfate, and chloride. Glutamic acid in this medium is the sole supplier of carbon, nitrogen, and energy. These can be supplied by many other compounds as seen in the other recipes. The only unique organic requirements are the two vitamins, and no substitutions are known for these.

Four aspects of culture conditions have received attention: light, temperature, pH, and aeration rate. Light intensity, primarily with respect to photosynthesis and phototaxis, has been widely studied and is discussed elsewhere in these volumes. The temperature optimum for growth seems to be near 28°C and is relatively independent of culture composition. There have been several studies of the pH effects (Cook, 1968; 1971a,b). The acid pH used in most media generally leads to most rapid growth.

III. Inorganic Agents

A. Required Elements

1. *Oxygen and Ozone*

Several significant studies of the effects of oxygen and ozone on *E. gracilis* and *A. longa* have appeared recently. In all cases cells were exposed in culture media to bubbles of the gases.

Growth of *A. longa* on acetate is modified by changes in the gases bubbled through the culture. A mixture of 95% air–5% carbon dioxide was compared with 95% oxygen–5% carbon dioxide (Begin-Heick and Blum, 1967). The doubling time went from 8.5–16.5 hours when the oxygen was increased. There was no differential effect observed in cellular RNA, protein, or DNA, although there was a decrease in cellular RNA and protein throughout the exponential phase of growth in both cultures. Such unbalanced growth is well-known in *Euglena* and has been discussed in detail (Wilson and Levedahl, 1968). Respiration rate of cells was higher in air than in oxygen, no matter how the cells were originally grown. Isolated mitochondria from air-grown cells were incubated in air, in oxygen, or in oxygen at 2 atmospheres pressure; no effect on respiration was observed, but mitochondria from oxygen-grown cells had reduced activity of five enzymes. Both ergosterol-related and quinone-related compounds increased in mitochondria from oxygen-grown cells. The principal effects were observed on succinate dehydrogenase, ergosterol compounds, and rhodoquinone, and it was suggested that oxygen had multiple sites of action.

Oxygen toxicity and carbon starvation in *A. longa* were studied in cells grown on ethanol (Begin-Heick, 1970). Increased oxygen tension and carbon starvation had the same depressing effects on growth, cell composition, and the activity of certain enzymes. Other parameters were different, however. The effect of oxygen was found to be due principally to the inhibition of NADH: cytochrome *c*

Table I

SEVERAL POPULAR GROWTH MEDIA FOR *E. gracilis*[a]

	Cramer and Myers (1952)	Buetow (1965)	Blum (1965)	Greenblatt and Schiff (1959)	Hutner et al. (1956)	Hutner et al. (1956)	Hutner et al. (1966)	Kirk and Keylock (1967)	Price and Vallee (1962)	Raison and Smillie (1969)	Lyman and Siegelman (1967)	Kempner and Miller (1965a)	Kempner and Miller (1972a)
Sucrose	—	—	—	—	15 000	—	5 000	—	—	—	—	—	—
Glucose	—	—	—	—	—	10 000	—	—	—	—	—	—	—
Urea	—	—	—	—	—	—	400	—	—	—	—	3 000	3 000
NH$_4$ glutamate	—	—	—	5 000	—	—	—	—	—	—	—	—	—
l-Glutamic acid	—	—	—	—	3 000	—	5 000	—	3 000	—	—	—	—
dl-Aspartic acid	—	—	—	—	2 000	—	2 000	—	—	—	—	—	—
Ethanol	—	—	2 400	—	—	—	—	—	3 000	—	—	—	—
dl-Malic acid	—	—	—	2 000	1 000	—	5 000	—	270	—	—	—	—
Maleic anhydride	—	—	1 960	—	—	—	—	—	—	—	—	—	—
Glycine	—	—	—	—	2 500	—	2 500	—	—	—	—	—	—
Na acetate	—	5 000	—	—	—	—	—	2 000	—	—	—	—	—
Na succinate·6H$_2$O	—	—	—	—	—	—	100	—	—	16 200	—	—	—
NH$_4$ succinate	—	—	—	—	600	—	—	—	—	—	—	—	—
Na citrate	800	645	650	—	—	—	—	500	—	—	—	—	—
Na$_2$ EDTA	—	1 000	—	—	—	—	—	—	—	500	500	—	—
KH$_2$PO$_4$	1 000	—	2 720	400	300	200	400	766	195	300	300	44	44
K$_2$HPO$_4$	—	—	—	—	—	—	—	136	—	—	—	—	—
NH$_4$Cl	1000	1000	800	—	—	—	—	—	—	—	—	79	—
(NH$_4$)$_2$HPO$_4$	—	—	200	—	—	—	—	—	—	—	—	—	—
NH$_4$HCO$_3$	—	—	—	—	—	1 000	—	—	—	—	—	—	—
NH$_2$NO$_3$	—	—	—	—	—	—	—	300	—	—	—	—	—
(NH$_4$)$_2$SO$_4$	—	—	—	—	—	—	—	—	—	1 000	1 000	—	—
MgSO$_4$·7H$_2$O	200	200	200	500	400	400	—	300	405	500	500	39	11.7
MgSO$_4$·3H$_2$O	—	—	—	—	—	—	100	—	—	—	—	—	—
MgCO$_3$	—	—	—	—	—	—	400	—	—	—	—	—	—
CaCl$_2$·2H$_2$O	20	26.5	20	—	—	—	—	—	—	—	—	—	73.4 µg
CaCO$_3$	—	—	—	500	80	50	100	60	—	60	60	735	—
Ca(NO$_3$)$_2$·H$_2$O	—	—	—	—	—	—	—	60	33.1	—	—	—	—

Table (concentrations in mg/liter unless otherwise specified):

Component														
$FeCl_3 \cdot 6H_2O$	—	3	—	—	24.2	—	—	—	10	18.1	—	—	5	—
$Fe(SO_4)_3 \cdot xH_2O$	3	3	—	—	—	3	—	—	—	—	—	—	—	—
$FeSO_4 \cdot (NH_4)_2SO_4 \cdot 6H_2O$	—	—	—	—	—	21	3	21	—	14	70	70	—	—
$ZnCl_2$	—	—	—	—	—	—	—	—	—	—	—	—	—	4.2 µg
$Zn(C_2H_3O_2)_2 \cdot 2H_2O$	—	—	—	—	—	—	—	—	—	—	22	67.2	—	—
$ZnSO_4 \cdot 7H_2O$	0.4	0.4	0.4	88	88	1.5	1	11	3.5	4.4	4.4	72	—	24 µg
$MnCl_2 \cdot 4H_2O$	1.8	1.8	1.8	—	—	0.4	0.4	7.6	2.0	1.55	31	—	—	—
$MnSO_4 \cdot 4H_2O$	—	—	—	81.6	81.6	—	—	—	—	—	—	—	—	—
H_2MoO_4	0.2	0.2	0.2	20.2	20.2	—	—	—	0.2	—	—	20.2	—	—
$Na_2MoO_4 \cdot 2H_2O$	—	—	—	—	—	0.16	0.16	1.8	—	0.64	0.72	—	—	—
$(NH_4)_6Mo_7O_{24} \cdot 4H_2O$	1.3	1.3	1.3	1.6	1.6	0.12	—	0.24	0.2	0.48	—	1.62	—	—
$CoCl_2 \cdot 6H_2O$	—	—	—	—	—	—	—	—	—	—	2.4	—	—	—
$CoSO_4 \cdot 7H_2O$	—	—	—	—	—	—	—	0.5	0.04	0.31	4	0.6	—	—
$Co(NO_3)_6 \cdot 6H_2O$	1.3	1.3	—	—	—	0.08	—	—	—	—	—	—	—	—
$Cu(C_2H_3O_2) \cdot H_2O$	—	—	—	—	—	—	—	—	—	—	—	0.6	—	—
$CuSO_4$, anhydrous	—	—	—	—	—	—	—	0.5	0.04	—	4	—	—	—
$CuSO_4 \cdot 5H_2O$	0.02	0.02	0.02	0.78	0.31	0.14	—	0.29	0.25	0.31	—	0.57	0.57	—
H_3BO_3	—	—	—	0.57	0.57	—	—	—	1.0	0.57	0.57	0.57	—	—
NaI	—	—	—	24 µg	24 µg	—	—	—	—	—	—	24 µg	—	—
$Na_3VO_4 \cdot 16H_2O$	0.01	0.02	—	—	93 µg	23 µg	1.8	0.6	—	93 µg	0.46	—	93 µg	5 µg
Thiamine HCl	0.02	0.01	0.01	1.0	0.6	0.5	0.6	0.6	1.0	0.63	0.6	1.0	—	0.6
Vitamin B_{12}	0.5 µg	0.01	?	0.2 µg	var.	0.4 µg	0.5 µg	?	0.01	5 µg	5 µg	0.2 µg	0.2 µg	—
pH	?	6.8	25	3.3	3.6	3.2–3.6	3.1–3.4	?	30	3.9	3.5	3.4	3.1	—
T, °C	25	25	25	25	?	25–28	25–28	?	21–25	27	26	26	27.5	—
Euglena strain	Bacillaris / SM L1	SM L1	SM L1	and Z	and Z	Z	Z	Z	Z	Z	Z	Z	Z	Z
Doubling time (hours)	18?	25.5	24	?	?	?	?	?	?	?	?	16	13	—
Max. cells/ml $\times 10^{-6}$	20?	0.36	0.28	?	?	?	?	?	?	?	1	1–2	1–2	—

[a] Concentrations in mg/liter unless otherwise specified.

oxidoreductase activity and an increase in NAD-dependent alcohol dehydro-genase. The two effects keep NAD reduced and therefore limit all cellular reactions that require oxidized NAD.

Subsequently the studies were expanded to compare the effects of air and oxygen on *Astasia* cells grown on either ethanol or acetate as a primary carbon source (Morsoli and Begin-Heick, 1973). Ethanol metabolism was found to be blocked in oxygen-rich atmospheres, and acetic acid accumulated in the culture medium. In air, however, acetate was efficiently used for growth. The cellular NAD/NADH ratio in ethanol-grown cells was reduced to 25% of the normal value when 95% oxygen was substituted for 95% air, confirming the prediction in the previous report. In these three studies, the oxygen level in the experimental atmosphere was increased almost fivefold at the expense of nitrogen gas, which was reduced from 74% to zero. This profound change in nitrogen may have con-tributed significantly to the complex effects ascribed entirely to oxygen toxicity.

A vital enzyme of the glycolate pathway, glycolate dehydrogenase, is found in two forms in *Euglena*. One is localized in the mitochondrion and appears to be relatively insensitive to culture conditions. The other form, localized in micro-bodies, is under strict environmental control. In a very precise study, Yokota *et al.* (1978a) showed that this form of the enzyme displayed an absolute re-quirement for atmospheric oxygen and an independent demand for light.

Cultures of *E. gracilis* strain Z were grown in the light on a pH 3.5 lactate medium in the presence of 2.5×10^{-5} *M* DCMU (see Section IV,D,4) with a gas phase of air or nitrogen (Calvayrac and Ledoigt, 1976). The same generation time (12 hours) was observed in air with or without DCMU. In nitrogen, the doubling time slowed to 14 hours until the cell density reached 2×10^5/ml, and then slowed further to about 30 hours. The amount of chlorophyll per cell and the ratio of chloroplast/cytoplasmic RNA both dropped with continued growth in pure nitrogen, whereas in air both parameters reached a steady state value.

The phenomenon of chemoaccumulation in *Euglena* induced by oxygen was discovered by Colombetti and Diehn (1978), and the identification of cyto-chrome *c* oxidase as the chemoreceptor molecule has recently been reported (Miller and Diehn, 1978). These fascinating studies are discussed in Chapter 5 of this volume.

Exposure of *E. gracilis* cultures to gaseous ozone (de Koning and Jegier, 1969; 1970a) at 0.5–0.8 ppm had only small effects on NADH concentration, oxidative phosphorylation, or photosynthesis. The largest effect, a 12% decrease in the rate of NADH formation in cells in the light, may be significantly greater than statistical error or biological variability.

2. *Nitrogen*

Considerable information published previous to 1968 about nitrogen effects on *Euglena* growth and metabolism has been reviewed by Cook (1968). Since that time, only a few additional studies have appeared, but they have been very significant.

The metabolism of glycolate in mitochondria from light-grown *Euglena* was found to be under the control of exogenously supplied ammonium ion (Yokota *et al.*, 1978b). In the absence of medium nitrogen, the uptake of glycolate was inhibited. With continued ammonium deprivation, glycolate dehydrogenase was repressed. These effects severely curtail the incorporation of the glycolate products, glycine and serine, into protein.

An informative study of nitrogen effects on fatty acid and lipid biosynthesis was made by Pohl and Wagner (1972). Compared with ammonium chloride, potassium nitrate was found to be a poor nitrogen source for cell growth, but the choice or concentration had no effect on the distribution of fatty acids in dark-grown *Euglena*. When grown in the light, the fatty acid and lipid profile of *Euglena* was changed markedly. This shift was prevented in the light if the concentration of available nitrogen was too low. The scheme of fatty acid biosynthesis proposed by these and other authors correlate well with the distribution changes and suggest that light and nitrogen control an entire pathway (Fig. 1).

A third study concerns the fixation of CO_2 in the dark. This had already been

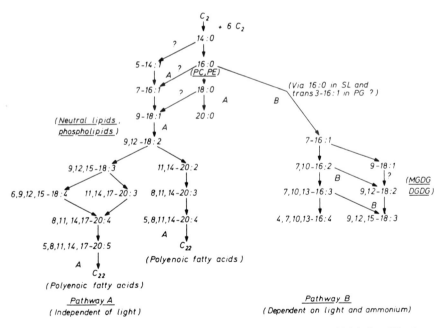

Fig. 1. Pathways of fatty acid biosynthesis in *E. gracilis*. PC, phosphatidylcholine; PE, phosphatidylethanolamine; SL, sulfolipid; PG, phosphatidylglycerol; MGDG, monogalactosyldiglyceride; DGDG, digalactosyldiglyceride. (From Pohl and Wagner, 1972.)

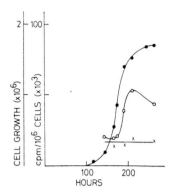

Fig. 2. Heterotrophic growth and CO_2 fixation by *Euglena* in the dark. Growth (●) and CO_2 fixation (○) in glucose growth medium; CO_2 fixation in phosphate buffer (x). (From Peak and Peak, 1977.)

the subject of several conflicting reports, some of which have now been explained by the work of Peak and Peak (1977). Ammonium is normally depleted from their culture medium during the exponential phase of growth. During the first 5–7 hours after this depletion, the cells become "sensitized" to subsequent replenishment of ammonium ion, which then stimulates dark CO_2 fixation. It can be observed in cells which are washed and resuspended in fresh medium, but resuspension in distilled water or phosphate buffer does not expose the cells to the required ammonium ion and no stimulation is seen (Fig. 2). Hopefully, the biochemical mechanism(s) of these effects will be resolved in subsequent studies.

Taken together, these reports outline a pattern of metabolic control dependent on nitrogen. In *Euglena,* ammonium ions can shift the flow of carbon from amino acid–protein pathways to lipids and can also change the pattern of fatty acids produced. If these observations are found to be specific for ammonium ion and are general in metabolism, a novel biochemical control of crucial activities and structures in cellular functions will be revealed.

One additional report should be noted. Schwelitz *et al.* (1978a) studied the inhibition of chlorophyll synthesis by glucose or ethanol. A 50% increase over the ammonium sulfate already present in the medium immediately relieved the ethanol inhibition. However, additional ammonium sulfate had no effect on glucose inhibition until the exogenous glucose was reduced to very low levels by metabolic activity.

3. *Sulfur*

An excellent review of the extensive literature on sulfur in *E. gracilis* has been presented (Cook, 1968). Only a few additional studies have appeared since that time.

Growth of *Euglena* at full rate and with undiminished cell yield requires 1.5 mg/liter sulfur and these cells are found to contain 3.1 mg sulfur/g dry cells (Kempner and Miller, 1972a). In a slightly different growth medium Buetow (1965) had earlier reported that the final yield of cells was constant when the sulfur concentration of the medium was varied between 10 and 680 mg/liter. However, at 1580 mg/liter sulfur the growth rate was reduced.

At limiting sulfur concentrations, ($2.2 \times 10^{-6} M = 0.07$ mg/liter sulfur) the macromolecular composition of *Euglena* changes, the cells have only six chloroplasts, and the UV survival curve is the same as with reduced phosphorus (Epstein and Allway, 1967). These authors suggest that the low sulfur leads to the same half-ploid state as shown in phosphorus-limitation studies. *A. longa* cells grown in a chemically defined medium in a chemostat were studied by Morimoto and James (1969a). The generation time of the culture population was controlled by the rate of addition of the limiting sulfate compound. The protein content of *Astasia* was independent, but the cellular RNA and DNA levels fell as generation time increased (Fig. 3). The decreased DNA per cell in slow-growing cultures could be due to the ploidy change suggested in phosphorus- or sulfur-limited *Euglena,* or to a change in polyteny (amount of DNA per chromosome). With unlimited nutrients, *Astasia* showed a generation time of 10 hours. When the sulfur level was reduced to 1 μM (.03 mg/liter sulfur), the cells grew very slowly, with a generation time of 56 hours. Morimoto and James (1969b) pulsed these slow-growing cultures with 5 μM sulfur (0.15 mg/liter), and observed a

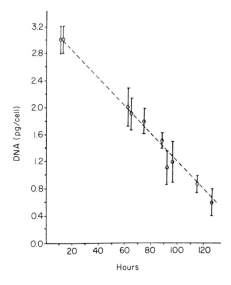

Fig. 3. DNA content of *Astasia longa* cells as a function of growth rate. Abscissa: generation time (hours); ordinate, DNA per cell (pg). The most rapidly growing cells have the highest DNA content. (From Morimoto and James, 1969a.)

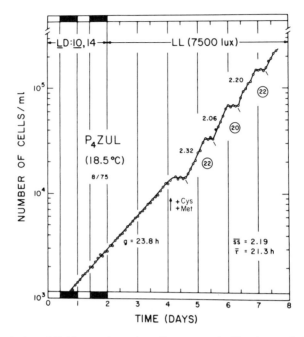

Fig. 4. Induction of division synchrony by sulfur compounds. The photosynthetically blocked mutant of *Euglena*, P_4ZUL, grew asynchronously for 2 days on a light–dark regime followed by 2 days in continuous light. Addition of cysteine plus methionine on the fourth day induces synchrony with divisions occurring during what would normally be the "night" period as is typical for *Euglena*. (From Edmunds *et al.*, 1976.)

synchronous burst of division 3 hours later. Between 3 and 6 hours, over 90% of the cells divided. During this 6-hour recovery process, cellular DNA was found to increase within the first hour; protein increased over the first three hours, while RNA displayed an increase just prior to the cell division. These changes are similar to the pattern found in temperature-synchronized *Astasia* cells.

Several photosynthetic mutants of *E. gracilis* were found to lose their potential for synchronization of cell division when cultured in the laboratory over a period of many months or years. Edmunds *et al.* (1976) were able to restore it by addition* to the culture medium of any of several sulfur-containing compounds at concentrations of 10^{-5}–10^{-4} *M*. The compounds include methionine, cysteine, dithiothreitol, and thioglycolic acid (Fig. 4). Equally effective were the inorganics, sodium thiosulfate or sodium sulfide. It was further shown that the deficiency was due to an uncoupling of cell division from the underlying clock mechanism, which had apparently continued to run undisturbed. The biochemi-

---

*This is discussed in detail in Chapter 3, this volume.

cal mechanism by which the cell division was reassociated with the clock is unknown.

There have been two recent studies concerned with the cellular toxicity of the environmental pollutant, SO_2. In both cases the photosynthetic rate of *Euglena* cells in buffer was observed. De Koning and Jegier (1970a) bubbled air containing various concentrations of SO_2 through the suspension. Photosynthesis was increased at low concentrations but was inhibited at concentrations higher than 3 ppm. There was no observable change in the concentrations of chlorophyll *a* or *b* during the 1-hour exposure. Using sodium bisulfite, which would be produced by the SO_2 in water, Wodzinski *et al.* (1977) could find no effect of 0.1 mM (6.4 mg/liter) on *Euglena* photosynthesis.

4. *Phosphorus*

The role of phosphorus and phosphates in metabolism and growth of *E. gracilis* has been the subject of a great many studies. Many strains and mutants have been examined in a variety of media in light and dark and also at different pH. Phosphorus is required for *Euglena* growth, but the requirement may be fulfilled over a wide range of phosphate concentrations (Doemel and Brooks, 1975). Buetow and Schuit (1968) reported that at pH 6.8, strain SM-L1 generation time was normal at phosphorous levels as high as 650 mg/liter (21 mM), but slowed at 1530 (49 mM). At the other extreme, they found the lowest phosphorous concentration at which the cell yield was maximal was 5 mg/liter (160 μM). Similarly, strain Z at pH 3.1 showed normal growth rate and cell yield when the phosphorus level was 10 mg/liter (320 μM), whereas at one-third of this value both parameters were altered (Kempner and Miller, 1965a). In all other reports, *Euglena* cultivation below these concentrations was found to be limited.

The total phosphorus content of *Euglena* measured by several authors was 5–18 mg/g dry cells (Buetow and Schuit, 1968). The total nucleic acids ranged from 18–49 $\mu g/10^6$ cells (Cook, 1968) or 28–76 mg nucleic acids/g dry cells, which would require 1.5–4 mg phosphorus/g dry cells. These data indicate clearly that the bulk of cellular phosphorus in *Euglena* is not in the form of RNA or DNA, a fact not generally recognized. Estimates of the different phosphate compounds can be obtained by ^{31}P-NMR of intact *Euglena* as shown in Fig. 5 (L. Jacobson and E. S. Kempner, unpublished observations). Nucleic acids are not seen as a discrete peak because of its excessively broadened linewidth. The major resonance at -22 ppm is assigned to the polyphosphates, and integration of this spectrum indicates that 80% of all cellular phosphorus is in this form, with only small amounts of other phosphorus compounds. The "hot TCA-soluble" fraction in the conventional extraction procedure for *Euglena* has been assumed to be principally composed of nucleic acids, although this may not always be correct (Edmunds, 1965; Fig. 6). Colorimetric determinations of

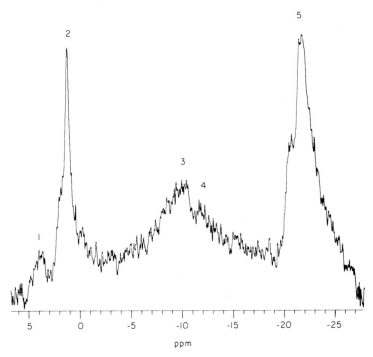

Fig. 5. ^{31}P-NMR spectrum (109.3 MHz) of *E. gracilis* cells. Cells from logarithmic phase of growth in minimal medium were washed twice and resuspended in phosphate-free medium. Spectrum recorded at 5°C. Tranients (10^4) were accumulated. Downfield shifts marked positive, referenced to external 85% H_3PO_4. (1) Phosphomonoesters (e.g., glycolytic intermediates, nucleotide mono-phosphates); (2) intracellular orthophosphate; (3) terminal phosphate of tripolyphosphate; (4) pyrophosphate; (5) middle phosphates of long-chain polyphosphates.

ribose and deoxyribose in this fraction lead to a calculated titer of nucleic acids that accounts for only one-half of the phosphorus actually observed (Kempner and Miller, 1965b). Large quantities of phosphorylated compounds are extracted together with nucleic acids (Edmunds, 1965; Speiss and Richter, 1970). Analysis of phosphorus kinetics during cell uptake, starvation, or "chase" experiments must be cautiously reconsidered. As shown by Edmunds (1965), interpretation of nucleic acid synthesis and degradation based on ^{32}P labeling in this fraction is especially suspect, unless these corrections have been properly performed.

It has long been known that *E. gracilis* contains significant amounts of polyphosphates (Cook, 1968), although only limited quantitative data has been available (Edmunds, 1965). Polyphosphates have been revealed as a major peak on MAK column chromatography of the nucleic acid fraction of ^{32}P-labeled *Euglena* (Speiss and Richter, 1970, 1971); 60% of the radioactivity was found to

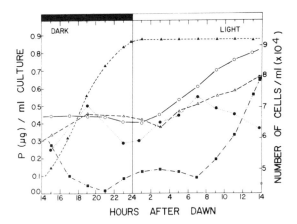

Fig. 6. Changes in cellular phosphate compounds during the cell cycle. Cell division occurs uniquely during the dark phase, whereas nucleic acid synthesis is seen only during the light period. Lipid phosphorus compounds show no cycle dependence. The acid-soluble phosphorus compounds and the acid-insoluble polyphosphates increase primarily in the light and can account for appreciable portions of the total cellular phosphorus. Key: ▲, Cell number; ○, nucleic-acid fraction phosphorus; △, acid (5% TCA) soluble phosphorus; ●, lipid phosphorus; ■, acid insoluble polyphosphates. (From Edmunds, 1965.)

be polyphosphate. The biochemical role and kinetics of this material are not understood. One study of the presence of inorganic phosphatases and phosphagen kinases in *Euglena* concluded that the polyphospates store inorganic phosphorus, whereas phosphoarginine and related compounds serve as energy donors (Piccinni and Coppellotti, 1977). However, chase experiments of ^{32}P-labeled polyphosphate showed no transfer of radioactivity to other cell fractions even under phosphate starvation; if storage is the role of polyphosphates in *Euglena,* it must be dead storage.

The kinetics of radiophosphorus uptake by *Euglena* has been examined in different growth conditions (Chisolm and Stross, 1976a,b). Total cellular radioactivity was measured during light–dark synchronized growth (Chisolm and Stross, 1976a). Edmunds' observation (1965) that phosphorus incorporation was limited to the light period was confirmed, although another report indicated that significant incorporation does take place in the dark (Cook, 1968). A lag phase was reported before ^{32}P uptake became linear with time (Chisolm and Stross, 1976a) although earlier data (Cook, 1968) did not reveal any lag. During the light phase, uptake kinetics of phosphorus over a wide concentration range was found to be complex (Chisolm and Stross, 1976a). No simple Michaelis–Menten analysis fits the data, and at least three separate sets of constants are indicated. Considering the 2-hour preincubation used in this study and the many possible pathways that phosphorus enters, it is not surprising to find complex results. In

Euglena cells grown in phosphate-limited medium until the stationary phase, Chisolm and Stross (1976b) reported that the phosphate uptake lag disappeared and the kinetics became simple. A single Michaelis–Menten constant was found. Furthermore, uptake in the dark was no longer inhibited, although somewhat lower than in light. In view of the stationary phase cultures used, it can be assumed that most biochemical pathways were depressed or completely inhibited. Reintroduction of phosphate might then temporarily activate one pathway selectively, resulting in the simpler relations found. These studies would have been much more understandable if they had been related to the wealth of earlier studies of ^{32}P uptake and phosphate content in the various cell fractions of *Euglena* (Edmunds, 1965; Kempner and Miller, 1965b; Cook, 1968, Buetow and Schuit, 1968; Parenti *et al.*, 1971, 1972).

With an adequate supply of phosphorus, *Euglena* cells maintain a constant cellular concentration of RNA and DNA (Kempner and Miller, 1965a; Epstein and Allaway, 1967; Parenti *et al.*, 1971, 1972) and protein (Kempner and Miller, 1965a) throughout the exponential and stationary phases of growth (Parenti *et al.*, 1971, 1972).

As with other required nutrients, a low level of phosphorus concentration in the culture medium which adversely affects *Euglena* growth is termed "deficiency" of the nutrient; when there is no available external supply for the cells, the convention is to refer to "starvation" conditions. Below 5 μg/ml phosphorus the growth rate of *Euglena* remains normal for a period but then slows gradually (Kempner and Miller, 1965a; Parenti *et al.*, 1971, 1972) and finally stops at lower final cell yields, which are proportional to the phosphate concentration available (Kempner and Miller, 1965a; Buetow and Schuit, 1968; Parenti *et al.*, 1971). When *Euglena* were grown photoautotrophically at 0.47 μg/ml phosphorus (Cook, 1971a) or mixotrophically (organic media in the light) at 0.14 μg/ml phosphorus (Davis and Epstein, 1971), the culture growth could not be synchronized by light–dark cycles. These are the results of profound effects on cellular metabolism by the limited phosphate supply.

In a remarkable paper, Epstein and Allaway (1967) studied the *bacillaris* strain grown mixotrophically in a chemostat with a reduced phosphorous level of 0.31 μg/ml. They reported RNA, chlorophyll, and chloroplast levels reduced to one-half or one-third of normal; cellular DNA was one-half of the normal value (Table II). These authors also determined that the number of chromosomes in normal and phosphate-deficient cells was 90 and 45, respectively. This discovery of a change in ploidy with external nutrient concentration has been important for subsequent research. The change in macromolecular composition in *Euglena* at reduced phosphate levels (1.8 μg/ml phosphorous) was subsequently confirmed in strain Z (Parenti *et al.*, 1971, 1972) and extended to protein also. They showed that the reduction was due to differential inhibition of macromolecular synthesis as compared to cell division, resulting in progressively lower cellular

Table II
GENERAL DATA ON CHEMOSTAT-GROWN *Euglena*[a]

	Normal cells	P/500 cells[b]
RNA (pg per cell)	30–40	10–15
DNA (pg per cell)	3.0	1.5
Chlorophyll (pg per cell)	15	5
Chloroplasts per cell	11–13	5–7

[a] From Epstein and Allaway (1967).
[b] Cells grown in media containing 1×10^{-5} M phosphate, which is 1/500 of the normal concentration.

concentrations throughout the exponential phase of growth. On subculturing in phosphate-deficient medium, they reported a rapid synthesis of macromolecules followed by a decreased rate as the exponential phase was repeated. Chlorophyll synthesis and plastid replication were depressed less than cell division, so that these cells became enriched with chlorophyll and chloroplasts throughout the growth of the culture.

In an especially clear set of experiments, Liedtke and Ohmann (1969) showed that at low levels of phosphorus (16 μg/ml) in the medium phosphatase activity was derepressed in *E. gracilis* strain Z. The increase was due to net enzyme synthesis as shown by blockage by antibiotics or parafluorophenylalanine. Replenishment of phosphorus led to a rapid decrease in enzyme activity, which was too rapid to be accounted for by dilution due to growth. There is an inactivation process that is stimulated by phosphate, and the inactive enzyme cannot be "reactivated" by a second decrease in external phosphorus concentration; new synthesis is required.

Other studies of the recovery of *Euglena* from phosphorus deficiency or starvation have been widely reported. Buetow and Schuit (1968) found that regrowth after phosphate replenishment was characterized by a lag, followed by a transient rapid increase in cell number with a generation time only 70% that of the control. Similar kinetics were observed for protein synthesis, but not for DNA or RNA synthesis (Parenti *et al.*, 1971). Plastid resynthesis occurred more slowly than cell division at first, so that the number of plastids per cell fell during the initial regrowth after replenishment (Parenti *et al.*, 1972). The most impressive study of phosphate recovery was published by Davis and Epstein (1971). In mixotrophic growth conditions, the initial lag in growth was confirmed. The resurgence of RNA synthesis was very rapid, whereas DNA and chlorophyll recovered slowly (Table III). From the studies of phosphate deficiency and recovery, it was concluded that organelle synthesis was greatly dependent on culture conditions (Davis and Epstein, 1971). Cells under optimum growth conditions could become polyploid in their plastids and nuclear chromosomes. If photosyn-

Table III

CHANGES IN WILD-TYPE MIXOTROPHIC CELLS FOLLOWING PHOSPHATE RESTORATION[a,b]

Hours after phosphate addition	Cells/ml ×10⁵	DNA/cell (pg)	RNA/cell (pg)	Chlorophyll/cell (pg)	Chloroplasts/cell	Chloroplasts/ml ×10⁶
0	2.3 ± 0.3	1.9 ± 0.1	14 ± 2	2.6 ± 0.2	5.0	1.2 ± 0.2
6					5.4	
9	2.4 ± 0.6				6.4	1.5 ± 0.5
17	2.6 ± 0.9				6.1	1.4 ± 0.5
20					6.5	
24	3.3 ± 0.4	2.5 ± 0.2	42 ± 3	2.6 ± 0.1	7.9	2.6 ± 0.4
34					8.4	
42					8.0	
48	7.2 ± 0.8	2.9 ± 0.1	56 ± 4	3.0 ± 0.1	8.2	5.9 ± 1.0
66					8.4	
72	19 ± 2	2.9 ± 0.1	56 ± 1	4.7 ± 0.2	9.1	17 ± 3
96	37 ± 6	3.2 ± 0.1	42 ± 4	6.6 ± 0.2	9.5	35 ± 7
Steady-state full-ploid		3.6 ± 0.1	49 ± 5	25 ± 1	= 10	

[a] Values are mean ± SEM.
[b] From Davis and Epstein (1971).

thesis was then prevented, the cells preferentially discarded extra plastids, whereas conditions which required photosynthesis resulted in cells discarding extra chromosomes rather than losing plastids.

5. *Manganese*

There is an absolute requirement for manganese for continuous *Euglena* growth. All common growth media include either the chloride or sulfate salt, although the amount varies by a factor of more than 1000. Normal photoautotrophic growth was observed in cultures containing 0.49 mg/liter manganese, but growth was sharply depressed if the manganese salt was omitted; some manganese was available as a contaminant in other salts (Constantopoulos, 1970). Autotrophic growth in the dark was observed to be unaffected by manganese levels of 0.013 mg/liter or greater. Kempner and Miller (1972a) found a continual incorporation of exogenous manganese throughout the exponential growth phase (Fig. 7), and in the steady state the cells contained 0.010 mg Mn/g dry cells. A manganese content of 0.0014 mg/g dry cells was also found for *Euglena* grown in a much more complex medium (Knezek and Maier, 1971).

Constantopoulos (1970) found that chlorophyll synthesis was not depressed by manganese deficiency, whereas the synthesis of galactosylglycerides was severely inhibited, even below the depression in cellular dry weight. Matson *et al.* (1972)

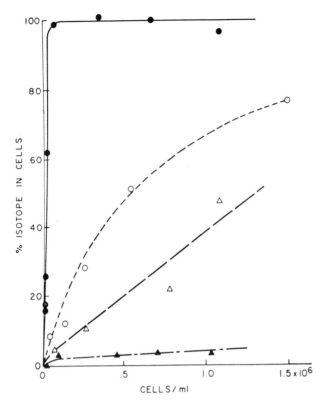

Fig. 7. Incorporation of radioisotopes by *Euglena.* Initial innoculum 300 cells/ml; cell population increased exponentially with time. Initial concentrations as given in Table I, except of chlorine, which was 13 times that of normal. Key: ●, Zinc; ○, manganese; △, calcium; ▲, chlorine. (From Kempner and Miller, 1972a.)

found the fatty acid composition of whole cells grown without organic carbon not greatly affected by manganese deficiency. However, if organic carbon sources were available, manganese deficiency would lead to severe depression in both chlorophyll and galactosyldiglycerides. In isolated chloroplasts high concentrations of manganese (550 mg/liter) decreased by 35% the synthesis of both galactosylmonoglycerides and galactosyldiglycerides.

Another role of manganese in enzymological reactions was elucidated by Falchuk *et al.* (1978). The base composition of RNA synthesized by normal *Euglena* RNA polymerases, as well as the bizarre polymerase found in zinc-deficient cells, was markedly altered by the presence of manganese in the incubation mixture; these authors suggest that manganese may affect the interaction between the enzymes and the bases of the template or the nucleotide substrates.

The wide spectrum of manganese effects in *Euglena* metabolism is also demonstrated in the unique experiments of Nichols and Rikmenspoel (1978). Microinjection of salt solutions into impaled *Euglena* showed that flagellar mobility required magnesium, and that manganese acted as an analog for magnesium in this system.

6. *Zinc*

The growth of *E. gracilis* requires zinc. However, growth was completely inhibited at zinc concentrations of 163 mg/liter as $ZnSO_4 \cdot 7 H_2O$ (Simeray and Delcourt, 1979). On the other hand, after a 3-hour exposure to as much as 5000 mg/liter zinc (also as the sulfate), at least some of the cells were still alive (Ruthven and Cairns, 1973). No studies of the mechanism of zinc toxicity in *Euglena* have been reported in recent years.

Normal growth of *Euglena* has been reported in growth media with zinc concentrations from as low as 0.014 mg/liter as $ZnCl_2$ to as high as 20 mg/liter as zinc acetate, with more rapid growth and slightly larger average cell volume at the lower concentration (Kempner and Miller, 1972a; Shehata and Kempner, 1977). At the lowest zinc concentration that supported rapid growth, radiozinc was rapidly removed from the medium (Fig. 7); more than a 100-fold increase in cell number occurred after the external zinc supply was exhausted (Kempner and Miller, 1972a). *Euglena viridis* also showed high avidity for external zinc, and exhibited optimum growth at 4.2 mg/liter as zinc sulfate (Coleman *et al.*, 1971).

Growth inhibition due to lack of zinc at levels below 0.010 mg/liter zinc has been widely reported (Prask and Plocke, 1971; Falchuk *et al.*, 1975a,b,c, 1978; Nakano *et al.*, 1978; but see Mills, 1976). After 7–10 days of growth in limited zinc medium, the 87 S ribosomes disappeared but were restored upon addition of zinc. It was suggested that defective ribosomes were synthesized under zinc-deficient conditions, and that these disintegrated when the zinc level dropped below a certain critical concentration (Prask and Plocke, 1971). The zinc requirement cannot be satisfied by cadmium (Falchuk *et al.*, 1975b; Nakano *et al.*, 1978). The latter authors did not find the extremely large cells in starved cultures that were reported by the previous authors. In a series of reports, Falchuk and his colleagues (1975a,b,c, 1978) showed that the dry weight of *Euglena* grown in 0.0065 mg/liter zinc was more than ten times greater than normal. There were large increases in the cellular content of other metals, roughly proportional to the increase in cell mass and volume. There was a decreased rate of RNA synthesis as measured by tritiated uridine incorporation, but organelle morphology was normal. The DNA content of zinc-deficient *Euglena* was greater than normal, and there was an increase in the number of cells in the S and G_2 phases of the cell cycle. The effect was reversible by addition of zinc to the culture. In spite of the decreased rate of RNA synthesis and the tenfold increase in cell dry weight (Falchuk *et al.*, 1975a), Falchuk *et al.* (1978) found no change in the amount of

RNA per cell compared to the normal zinc-sufficient cells. Unfortunately, the authors did not offer any explanation for this apparent inconsistency. A single, unusual zinc-dependent RNA polymerase was found in the deficient *Euglena* cells, and the messenger RNA synthesized under these conditions had a drastically altered base composition (Falchuk *et al.*, 1978).

The zinc metalloenzymes of *Euglena* have been reviewed by Cook (1968). The RNA polymerases found in normal *Euglena* cells have also been shown to be zinc-containing enzymes (Falchuk *et al.*, 1976). The unique single RNA polymerase found only in zinc-deficient cells is reputedly also a zinc-containing protein.

Zinc solutions have been introduced directly into *Euglena* cells. Impaled cells were injected with 9×10^{-14} liter of 0.1 M $ZnSO_4$. After dilution in the cell volume, the final internal zinc concentration was 2 mM. This resulted in a 50% decrease in the beating rate of the flagellum. Although Mg^{2+} caused a similar decrease, it was shown that zinc and magnesium do not act at the same site (Nichols and Rikmenspoel, 1978).

7. *Magnesium and Calcium*

E. gracilis cells contain 1.4 mg of magnesium per gram dry weight of cells when grown at an external concentration of 1.15 mg/liter magnesium as the sulfate heptahydrate. However, this is not the minimal required concentration of magnesium as shown by experiments with $MgCl_2$ and K_2SO_4 (Kempner and Miller, 1972a). Cells grown in 0.49 mg/liter magnesium showed exponential growth characteristics, but growth rate and final cell yield were reduced compared to *Euglena* in higher magnesium concentrations and were therefore called magnesium-deficient cells (Zielinski and Price, 1978). Growth rate and cell yield increased with external magnesium levels up to 39 mg/liter. In these studies, magnesium was supplied as the sulfate, but $KHSO_4$ was added to maintain a constant sulfate level. Protein and chlorophyll synthesis varied with magnesium concentration, but at low magnesium concentrations, the rates were always less than the generation time, yielding *Euglena* with reduced cellular concentrations of protein and chlorophyll. This was rapidly reversed by addition of magnesium to the culture. Because of the importance of magnesium in ribosomes, Zielinski and Price (1978) examined the polysomes in cells deficient in magnesium but observed no differences from that of normal cells.

Folkman and Wachs (1973) studied the removal of algae from pond water by flocculation with $Mg(OH)_2$. Magnesium was added as the chloride, and the pH was raised with NaOH. At pH 10.9, they were able to sediment a mixture of *Chlorella* and *Euglena* completely when magnesium concentration was 14–17 mg/liter.

Although the common growth media for *Euglena* contain calcium at many different concentrations, all supply the metal in vast excess. It was possible to reduce calcium level to as low as 73 μg/liter without inhibiting growth or cell yield. Under these conditions, the calcium was removed from the medium at a constant rate (Fig. 7) and the *Euglena* contained 5 μg Ca/g dry cells (Kempner and Miller, 1972a).

The only other recent study of calcium involving *Euglena* was that of Folkman and Wachs (1973), in which it was found that 500 mg/liter CaO raised the pH of pond water to 11.2 and effectively removed *Euglena* and *Chlorella* by flocculation.

B. GRATUITOUS ELEMENTS

1. *Cadmium and Mercury*

One of the major points of interest in studies of *Euglena* has been the possible role of cadmium as a substitute for zinc. However, it has been clearly shown that in the presence of zinc, cadmium is excluded from *Euglena*. Cells starved of the required zinc ion do allow cadmium to enter, but cadmium does not substitute for the missing element. Cadmium concentrations of the order of 1 mg/ml, in the presence of comparable amounts of zinc, had no effect on growth rate, cell dry weight, metal content, or morphology. However, higher molar ratios of cadmium to zinc (up to 10-fold) resulted in progressively slower growth rates (Falchuk *et al.*, 1975b; Nakano *et al.*, 1978; Fennikoh *et al.*, 1978). A 100-fold excess of cadmium to zinc was reported to be immediately lethal to *Euglena*, although it should be noted that the absolute concentration of cadmium was 5600 mg/liter (Bonaly *et al.*, 1978). These authors also found that at lower cadmium to zinc ratios (Cd 56 and Zn 40 mg/liter) there was a slow growth rate initially, followed, after 10 days incubation, by recovery due to the acquisition of cellular resistance to cadmium; after 16 days growth the cell yield was the same as the controls without cadmium. Very interesting aspects of *Euglena* metabolism will be implied if these unique findings can be confirmed.

In the absence of zinc, which can be obtained only with meticulous precautions, *Euglena* cells stopped growing. The addition of 1 mg/liter cadmium at the time of zinc removal led to an almost complete cessation of cell division, but the cells still grew in size. Cellular dry weight increased almost 100-fold. Many multinucleated cells were found, and in 5–7 days the cells became nonmotile (Falchuk *et al.*, 1975b). In very similar experiments, Nakano and colleagues (1978) reported that motility was rapidly lost initially, but that it gradually recovered with continued incubation.

It is likely that the effects of cadmium on *Euglena* are predominantly those of heavy metals rather than due to its chemical relationship to zinc. Mercury, which clearly acts as a heavy metal, can be tolerated by *Euglena* only at low concentrations. Three mercuric compounds did not effect Euglena growth at the indicated concentrations: mercuric chloride, 10^{-7} M; methylmercuric chloride, 5×10^{-9} M; and phenylmercuric acetate, 10^{-8} M. Five- to tenfold higher concentrations of each compound resulted in a lag phase proportional to concentration, after which the growth rate was close to normal and the final cell yield was independent of the presence of mercury (Simeray *et al.*, 1977). These effects could be explained either by selection of a subpopulation of resistant cells or the acquisition of resistance by all.

Much higher concentrations of mercury have profound effects on *Euglena* galactolipids. 10 mM mercury almost completely abolishes the synthesis of mono- and digalactosyldiglycerides. Since this synthesis depends on galactosyl transferase, Matson *et al.* (1972) also looked at the mercury sensitivity of the enzyme. Mercury ion (0.5 mM) and methyl mercury inhibited enzyme activity 40 and 90%, respectively.

2. Chromium and Molybdenum

Only the hexavalent form of chromium has been used in studies with *Euglena*. Cells were not killed in 10 minutes, even at concentrations of 1000 mg/liter chromium. Compared to other protozoa, *Euglena* is very resistant to chromium effects. The tolerated concentration at which at least some organisms were still alive after 3-hour exposure was 180 mg/liter chromium (Ruthven and Cairns, 1973). All *Euglena* cells survived 3-hour exposure to 1 mg/liter chromium (Yongue *et al.*, 1979). However, combinations of heat and chromium exposure led to more complicated results greater than the sum of the two independently. Nothing has been reported about the mechanism of the chromium effects on *Euglena* growth.

Many of the culture media commonly used for *Euglena* contain molybdenum at concentrations from 0.08 to 8 mg/liter. However, it has been shown not to be a required component (Kempner and Miller, 1972a). Growth is inhibited at concentrations above 1000 mg/liter molybdenum. In the range of 100–1000 mg/liter molybdenum, abnormal cell division occurs. Two to five percent of the cells were found as "clusters of three to nine actively moving cells attached to each other at one end" (Colmano, 1973).

In an early report on the stratification of *E. gracilis* by centrifugation (Kempner and Miller, 1968), it was stated that the stratified cells would reorganize, recover, and divide normally. When such cells were removed from the ultracentrifuge, washed, and placed in a high-salt medium (Kempner and Miller, 1965a), the number of cells per milliliter remained constant for approximately 10 hours, presumably a recovery period, after which the rate of increase in cell population

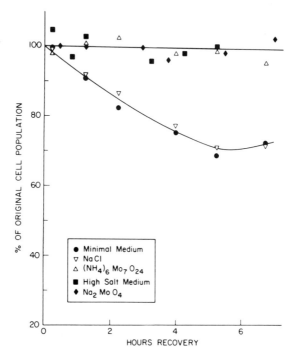

Fig. 8. Effects of culture medium on recovery of *Euglena* populations after centrifugal stratification. Cells centrifuged, washed, and resuspended in a high-salt medium (Kempner and Miller, 1965a), in minimal medium (Kempner and Miller, 1972a), or in minimal medium supplemented with the indicated salt.

became the same as the uncentrifuged controls. After the same degree of centrifugation but with cells in a minimal growth medium (Kempner and Miller, 1972a), there was a 30–50% decrease in cell population during the first 6 hours of recovery (Fig. 8). The loss of cell population could be prevented if the minimal medium was supplemented with sodium molybdate at the same concentration as in the high-salt medium. No other component showed any protective action. Substitution of either sodium chloride or ammonium molybdate revealed that molybdate ion was the specific protector. Addition of sodium molybdate to control cultures had no observed effect on cell growth or division. The mechanism of this unique effect is unknown.

3. *Cobalt*

In the past there was extensive interest in the role of cobalt in *Euglena* metabolism because of its occurrence in vitamin B_{12}. In recent years, however, the studies have principally concerned the vitamin itself (see Section IV,B) and only two studies of the metal have appeared. Coleman *et al.* (1971) varied the concentration of cobalt nitrate and observed the growth of *E. viridis*. When the cobalt concentration was changed from 0.04 to 2.88 mg/liter, the total dry weight of cells obtained during a 3-week growth period was reduced to one-third. They also showed that cobalt was concentrated by these cells increasingly as the external level went up; the cellular concentration was 2400 times greater than the medium at the highest concentration tested. Growth rate and final cell concentration of *E. gracilis* during exposure to the same cobalt salt [$Co(NO_3)_2$] were reported by Simeray and Delcourt (1979). No effects were observed below 4 mg/liter cobalt. As the external concentration was raised to 59 mg/liter cobalt, there was a progressive decrease in exponential growth rate and diminution of final cell population.

4. *Other Elements*

a. Arsenic. Cells growing in media containing arsenic showed progressively slower growth rates with increasing arsenic concentration. Growth was completely stopped when the amount of arsenate equaled the amount of phosphate (0.002 *M*) in the medium. One millimole arsenate caused a 50% inhibition of the phophorylation rate of isolated *Euglena* mitochondria (Kahn, 1974).

b. Copper. Copper sulfate was immediately lethal to *E. gracilis* at high concentrations of 500 mg/liter as copper (Ruthven and Cairns, 1973), whereas no effect on growth was seen at levels below 32 mg/liter (Simeray and Delcourt, 1979). In the range of 100–300 mg/liter copper, Simeray and Delcourt observed very slow growth for 4 days followed by a partial return to higher growth rate, which they suggested may be due to the development of resistance to copper by *Euglena* cells.

c. Iron. Knezek and Maier (1971) supplied up to 10 mg/liter iron either as the sulfate or in a chelated form. The influence on *Euglena* cell size and growth in a very complex medium was observed; inorganic iron was found to be a more efficient source of iron than the chelated metal. The effect of increased iron on the nitrogen, copper, and manganese content of cells was also studied.

d. Boron. *Euglena* sp. growing in natural ponds in Alabama and Mississippi where the boron content rarely exceeds 0.2 mg/liter were found to contain 3.8 mg boron per gram dry cells (Boyd, 1970).

e. Sodium Chloride and Sodium Phosphate. Benson (1972) examined the growth of *E. gracilis* in undefined media to which various amounts of the sodium salts were added. In the presence of vitamin B_{12}, increased growth was observed up to added concentrations of 212 m*M* NaCl and 125 m*M* NaH_2PO_4.

IV. Organic Compounds

A. GENERAL CARBON SOURCES

A general carbon source is any organic compound which supplies carbon to all biosynthetic pathways. It is usually identified by its ability to support an exponential increase of *Euglena* cell populations through successive subcultures. Although most often these chemicals are studied in the absence of other organic compounds (except vitamins), this is not a requirement. One of the popular growth media for *Euglena* (Table I) contains both glutamic and malic acids, and all aerated cultures contain CO_2 unless specifically excluded (Brody and White, 1973).

Continual exponential increases in *Euglena* populations implies that all the cellular contents increase at exactly the same rate. If this were not true, there would be a condition of unbalanced growth in which certain components increase more rapidly than others. This leads eventually to cells engorged with rapidly produced materials and/or deficient in other substances, which are synthesized more slowly. This has been discussed in detail by Wilson and Levedahl (1968), who pointed out that except for glutamic acid, several general carbon sources (ethanol, acetate, succinate) commonly used for *Euglena* did not support balanced growth. Freyssinet *et al.* (1972) showed that butyrate also failed to support balanced growth. These chemicals must have had a differential effect on one or more aspects of cellular biosynthesis. In experiments utilizing cells grown in these conditions, therefore, this additional variable must be considered. Furthermore, the ability of such compounds to support *Euglena* growth implies a stimulatory effect on cellular metabolism in the most general sense.

In discussing specific metabolic changes due to general carbon sources, it is necessary to compare biochemical or ultrastructural observations with those from cultures grown in other conditions. Thus it must be realized that what appears to be stimulation by one general carbon source can be viewed as inhibition by another.

It has been clearly established that *Euglena* will preferentially use ethanol, glucose, or glutamic acid (but not succinate) rather than photosynthetic CO_2 fixation. Comparisons among the organic compounds can be attempted in a very general way by considering the doubling times of similar cultures or the preferential utilization of one labeled carbon source over another. From these data, the following rough guide to cellular preferences is indicated:

fumarate $\sim$ pyruvate $>$ ethanol $\sim$ acetate $\sim$ glutamic acid $\sim$ lactate $>$ glucose
glucose $>$ CO_2 $>$ succinate $>$ butyrate $>$ butanol
glucose $>$ glycolate $>$ serine $>$ glycine   (in the light only)

1. *Carbon Dioxide*

Carbon dioxide can be utilized by *Euglena* via two independent mechanisms. The photosynthetic mechanism occurs only in the light and can supply the total carbon requirements of the cells. Discussion of many aspects of this process may be found in other sections of these volumes. The second process, sometimes called *dark* CO_2 *fixation* is more correctly independent of light. It is incapable of supporting the growth of *Euglena* since only a restricted group of compounds is supplied with carbon from CO_2. Carbon dioxide, as utilized in this manner, is therefore not a general carbon source *in sensu stricto*. However, in the presence of other general carbon sources, appreciable amounts of cellular carbon can be derived from CO_2, and its presence can significantly alter the flow of carbon from other organic compounds (Levedahl, 1968). Variation in atmospheric CO_2 concentration altered the metabolic fate of labeled glycine and formate in both light and dark (Foo and Cossins, 1978) and drastically changed enzyme activities associated with glycollate utilization (Lor and Cossins, 1978; Yokota *et al.*, 1978a). Furthermore, removal of CO_2 from the gas phase led to a disappearance of catalase activity in acetate-supplemented cultures (Brody and White, 1973).

The fixation of CO_2 in the dark was reported to vary with the growth phase of the culture and with other carbon sources by Peak and Peak (1976), who interpreted their results in terms of the anaplerotic replenishment of Krebs cycle intermediates depleted by growth requirements. These authors subsequently found (Peak and Peak, 1977) that this heterotrophic CO_2 fixation was regulated by ammonium ion (see further discussion in Section III,A,2). Reitz and Moore (1971) also reported variations in fatty acid distributions in cells grown in light and dark with ethanol, glucose, or carbon dioxide. Unfortunately, their cells

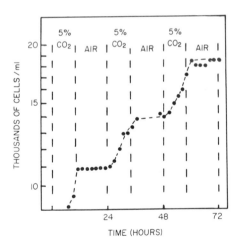

Fig. 9. Cell division in *Euglena* cultures during cyclic variation in CO_2 concentration. Time cycles: 14 hours of air (0.03% CO_2); 10 hours of 5% CO_2–95% air. Cells grown in autotrophic medium at 29°C with continuous light. (From Jones and Cook, 1978.)

were always incubated in the presence of large amounts of glutamic and malic acids, so that interpretation of their results is futile.

Light-dependent CO_2 fixation was shown to be controlled by a repression-derepression regulatory system that is independent of the synthesis of chlorophyll (Wolfovitch and Perl, 1972). Raising the CO_2 level from that of the air to 5% enhanced autotrophic growth of *Euglena*, especially at a culture pH greater than 5.5 (Jones and Cook, 1978). Cycling of the gas phase between these two concentrations enabled the entrainment of cellular division (Fig. 9), although only one-third of the cells divided at each cycle. These authors believe that many CO_2 effects in *Euglena* are due to a lowering of the internal pH. This fascinating hypothesis is worthy of further investigation.

2. Ethanol and Acetate

These two closely related compounds are the most commonly used carbon sources for routine growth of *E. gracilis* and *A. longa*, although neither supports balanced growth. Acetate cannot be used in acid media. At pH 4.5 and below, acetic, formic, propionic, and butyric acids are all lethal to *Euglena*, whereas malic, pyruvic, fumaric, and succininic acids are not so, even at pH 3.0 (Bates and Hurlbert, 1970). Toxicity occurs when the undissociated forms can enter the cells; the ionic forms are excluded. At pH 6, the transport of acetate into log phase *Euglena* showed a 15–30-minute delay, which Lux and Petzold (1976) interpreted as activation of a transport system for this substrate. Ethanol has also been reported (Cook, 1971b; Cook and Kaiser, 1973) to show a pH dependence of growth-supporting ability. The phenomenon was primarily observed at pH 4 in the light. Ethanol, butanol, and especially fumarate each inhibited *Euglena* growth in the light.

Pakhamova *et al.* (1970) reported that *A. longa* grew faster on ethanol than acetate. Ethanol-grown cells were found to have more carbohydrate and less protein and lipids than normal, but cells grown on acetate had the same nucleic acid content.

Mitochondria from ethanol-grown cells were compared to those from glutamate–malate cultures (Sharpless and Butow, 1970b). A succinoxidase activity, insensitive to antimycin or cyanide, was observed in ethanol-grown *Euglena*. The enzyme from the glutamate–malate culture was sensitive to both compounds. The mitochondria differed, not only biochemically, but also physically as seen in slightly different buoyant densities.

Dark-grown cells (with glutamate + malate + CO_2 or glucose + CO_2) placed in the light with ethanol had delayed greening. Ammonia (as a nitrogen source) relieved the inhibition due to ethanol (Schwelitz *et al.*, 1978a). In the dark, ethanol inhibited glycolysis, thereby making more carbon available for enhanced gluconeogenesis and paramylum synthesis (Garlaschi *et al.*, 1974).

The circadian rhythm of cellular motility showed increased period length in the presence of ethanol. Methanol and isopropanol inhibited motility but did not lengthen the period. Isopropanol stopped all motility; butanol acted similarly, but after 2–4 days motility recovered (Brinkmann, 1976). It has also been found (Feldman and Bruce, 1972) that acetate changed the period length of the circadian rhythm of *Euglena* phototaxis.

3. *Glucose*

As a general carbon source for *Euglena*, glucose has been more actively studied than perhaps any compound except CO_2. It is now clear that glucose utilization is directly dependent upon culture pH (Hurlbert and Bates, 1971). However, even under optimum conditions glucose was not an efficient source, perhaps due to the limited transport system controlling cellular entry (Lux and Petzold, 1976). Support for this hypothesis is found in the study by Hurlbert and Bates (1971). Their *bacillaris* strain grew on glucose at pH 4.5; when transferred to the same medium at pH 7, growth resumed only after a lag of more than 250 hours (a lag shortened by small amounts of glycine). These effects were due to a change in cellular permeability at neural pH. The authors indicated that the long lag was due to selection of a mutant. The rate of glucose incorporation was the same in glucose- and ethanol-grown cells, but very low in acetate-grown cells (Graves, 1971).

Compared to acetate-grown cells, *Euglena* grown on glucose had very few microbodies and very low levels of catalase (Brody and White, 1973). The acetate cells had low pyruvate kinase activity, but addition of glucose to the culture resulted in a large increase in the enzyme. Both the localization and specificity of fructokinase and glucokinase were the same in glucose- and CO_2-grown cells (Lucchini, 1971).

The presence of glucose during the greening process was inhibitory to chlorophyll formation (Schwelitz *et al.*, 1978b), and inhibition was not relieved by addition of a nitrogen source. The activity of photosystems I and II was low, and no pyrenoid region was seen in the plastids. Most of the glucose was removed from the medium in the light within 48 hours. Chloroplast development then proceeded rapidly and was completed within the next 24 hours.

4. *Glutamic and Malic Acids*

Frequently these two compounds are used together as carbon sources in an acidic medium, but glutamic acid alone will suffice to support *Euglena* growth. Glutamic acid was rapidly taken up during exponential growth phase (Lux and Petzold, 1976); in stationary phase cultures a 15-minute lag in uptake was observed, which was eliminated by preincubation with alanine.

Glutamic acid can be used as a sole source of carbon, nitrogen, and energy by

Euglena (Table I). In log phase the cells showed a constant cell volume (Kempner and Marr, 1970). In the simplest heterotrophic growth medium (Kempner and Miller, 1972a) *Euglena* actively metablized glutamic acid; the cells displayed rapid, balanced growth, which has been maintained invariant for years.

Glutamic acid induces the synthesis of the mitochondrial marker enzyme, succinic dehydrogenase, and also an NADP-requiring succinic semialdehyde dehydrogenase. The NAD-linked isozyme was unaffected by glutamate (Tokunaga *et al.*, 1976). It was suggested that the NADP enzyme was induced in order to use γ-aminobutyric acid as an alternate pathway from glutamic acid into the Krebs cycle. Oxidative phosphorylation in mitochondria from glutamate-malate-grown *Euglena* showed the same three sites (sites I, II, and III, sensitive to rotenone, antimycin, and cyanide, respectively) found in the mammalian respiratory chain (Sharpless and Butow, 1970a).

5. Succinic Acid

Succinic acid can be utilized as a sole carbon source by *Euglena*, but in an unbalanced growth since both protein and dry weight decrease during the exponential growth phase (Wilson and Levedahl, 1968; Votta *et al.*, 1971). The uptake of succinate by *Euglena* during this rapid growth was found to be delayed, suggesting an induction process for a specific transport mechanism (Lux and Petzold, 1976). It was not incorporated during the stationary phase.

Euglena growth on succinate was compared at pH 3.5 and 6.9. At the acid pH, 83% of the succinate is not ionized, whereas only 0.2% is not ionized at pH 6.9. The cells continued to grow as long as an adequate supply of the nonionized succinate was available; if it was too low, the cells entered the stationary phase (Votta *et al.*, 1971).

6. Lactate

Lactate-grown *Euglena* showed the same doubling time as those grown in glutamic–malate, but had double the respiration rate. A giant mitochondrion was observed only in lactate cultures; upon subculture into glutamate–malate medium, mitochondria returned to normal, establishing the reversibility of the phenomenon (Calvayrac, 1970). Autotrophic *Euglena* also contain giant mitochondria, but at the same time there was a decrease in respiration (Osafune *et al.*, 1975). Calvayrac *et al.* (1974) reported that *Euglena* grown synchronously on lactate or autotrophically displayed a "mitochondrial cycle." During the division phase of the cells they found small, discrete mitochondira. In the nondividing phase, there was a mitochondrial network, usually near the cell surface. They indicated that separate mitochondria could coalesce as part of this cycle. These observations are very similar to those of Leedale and Buetow (1970, 1976) in carbon-starved cells.

7. *Glycolate*

Glycolate cannot be used as a sole carbon source by *Euglena* unless there is a separate energy supply. This can be satisfied either by light or, in the dark, by glucose. Similar energy demands were made by glycine and serine (Murray *et al.*, 1970). In *Euglena*, these compounds were utilized by means of a glycolate pathway, which is very similar to that found in higher plants (Murray *et al.*, 1971).

8. *Peptone*

In a complex organic growth medium, peptone was reported to stimulate production by *Euglena* of an extracellular protease. It was found to be a single protein, an endopeptidase of molecular weight 41,000 (Nakano *et al.*, 1979).

9. *Carbon Starvation*

The withdrawal of available carbon sources from growing *Euglena* led to cessation of cell division and dramatic decrease in cell volume (Kempner and Marr, 1970). Both processes were reversible on readdition of a carbon source. Cytological changes have also been observed. Remarkable optical microscope pictures (Leedale and Buetow, 1970) showed a mitochondrial system in normal cells consisting of a "labile reticulum of threads." Upon carbon starvation the threads fragmented; some became digested by lysosomes and the remaining pieces thickened. Within a few hours after carbon replenishment, the threadlike reticulum was observed again. The reduction in cell size in a *Euglena* mutant during carbon starvation was observed (Leedale and Buetow, 1976), along with a reduction in paramylum grains, accumulation of lysosomes, and appearance of an endoplasmic reticulum around the nucleus. A large increase in cellular autolytic activity was shown by the large number of identifiable hydrolases (Baker and Buetow, 1976). During carbon starvation, carbohydrates are used first. With extended deprivation, *Euglena* digest protein and RNA in order to satisfy basic metabolic demands for utilizable organic material.

B. Vitamins B_1 and B_{12} and Their Derivatives

E. gracilis shows an absolute growth requirement for both vitamin B_1 (thiamine) and vitamin B_{12} (cyanocobalamin). Very few experiments concerning thiamine requirements of *Euglena* have ever been published (Cook, 1968), and none in recent years. Vitamin B_{12} has been a subject of great interest for a long time [see Valencia's (1974) excellent review]. Cook (1968) has reviewed the earlier literature about vitamin B_{12} in *Euglena* metabolism and growth, and the past decade has seen a proliferation of reports on this subject. Considerable advances in our knowledge of B_{12} effects in *Euglena* have come especially from extensive work in the laboratories of Carell and Valencia.

Euglena growth requirements are as well satisfied by hydroxocobalamin and sulphitocobalamin as by cyanocobalamin. Methyl-, nitrito-, or 5'-deoxyadenosylcobalamin (coenzyme B_{12}) are less effective. Inactive forms include the mono- and dicarboxylic acids of cyanocoblamin, monocarboxylic acid of hydroxocobalamin, and etiocobalamin, factor B (Adams and McEwan, 1971).

The most obvious effect of B_{12} deprivation* is a dramatic increase in cell size; with extended periods of starvation, Bertaux and Valencia (1973) have found numerous multinucleated "polybranched" forms. These observations clearly indicate primary blockage of the cell division process while cell growth continues. In general, organelle structure was not obviously affected by B_{12} deprivation, but there was an increase in the number of paramylum granules (Bertaux and Valencia, 1973). The nucleus increased in size, roughly in proportion to the increase in cell size (Bertaux *et al.*, 1978). The chromosomes became smaller, stained less intensely, and with prolonged deprivation, disappeared. However, the nucleolar DNA retained its normal condensed state. The nucleolus itself split into many fragments (Fig. 10). As the cell volume increased, the ridges of the pellicle became larger (Bre and Lefort-Tran, 1978).

Several earlier conflicting reports of macromolecular synthesis during B_{12} starvation have now been resolved. Cultures of *E. gracilis* grown in the absence of B_{12} showed progressively diminished synthetic rates as the endogenous reserve of the vitamin was exhausted. Cellular stasis was achieved in 5–7 days, independent of growth conditions, carbon source, or previous growth rate (Carell, 1969; Bertaux and Valencia, 1973; Bre *et al.*, 1975; Shehata and Kempner, 1978). There is the possible exception of *Euglena* grown in standing culture on lactate in the dark (Bre *et al.*, 1975), which do not appear to show a B_{12} requirement! During starvation in all other growth conditions, there was an increase in the cellular concentrations of chlorophyll and chloroplasts (Carell, 1969), a three-fold increase in protein (Carell, 1969; Bertaux and Valencia, 1972; Bertaux *et al.*, 1978; Shehata and Kempner, 1978), a twofold increase in RNA, and almost a doubling of cellular DNA (Carell *et al.*, 1970; Bertaux and

---

Fig. 10. (a) Normal nucleus of *Euglena* in G_1 phase. Cell volume, 1900 μm^3; 2% glutaraldehyde prefixation, 15 minutes; 4% glutaraldehyde–phosphate buffer (pH 7.3) fixation, 1 hour. Uranyl–lead staining. Nuclear envelope, ergastoplasm, and other membrane systems are not stained. 11,000 × (R. Valencia, unpublished). (b) Nucleus of cell deprived of vitamin B_{12} for 8 days. Cell volume, 6500 μm^3. Fixation as above, but with additional osmium fixation to show nuclear envelope. Nucleolar fragmentation occurs around the annular structure; chromatin condensation has largely disappeared. Light zone indicates the former chromosome. 11,000 ×. (From Bertaux *et al.*, 1978).

*This topic is also considered in detail in Volume IV, Chapter 2.

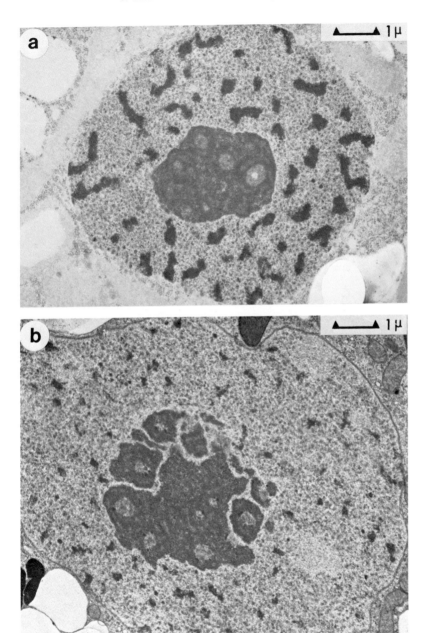

Figure 10

Valencia, 1972; Bertaux et al., 1978; Shehata and Kempner, 1978). Cell division was blocked (Bertaux and Valencia, 1971; Bre and Lefort-Tran, 1978; Shehata and Kempner, 1978), but since cell growth continued, there was an increase in cell volume (Bertaux and Valencia, 1971, 1973; Bertaux et al., 1978; Shehata and Kempner, 1978). Measurements of thousands of Euglena revealed that the distribution of normal cell volumes changed during B_{12} starvation (Shehata and Kempner, 1978; see also Fig. 13). In addition to the shift to larger average size, a second peak appeared, due to the accumulation of larger-than-average cells. It was concluded (Shehata and Kempner, 1978) that B_{12} deprivation interfered with at least two separate cellular processes. This explains the earlier failures to satisfy the B_{12} requirement with individual metabolic intermediates (Milner and Weissbach, 1969).

Specific enzymes are affected by B_{12} starvation of Euglena. Ornithine decarboxylase activity normally appeared during the light phase of synchronized Euglena growth, and then fell off to 10% in the dark phase. In the absence of vitamin B_{12}, but while protein synthesis was still continuing, the same pattern was repeated, although the activity peak diminished by 70% (Lafarge-Frayssinet et al., 1978). The methylation of DNA, which is normally quite high in Euglena, was depressed during B_{12} starvation (Bertaux and Valencia, 1974). Deoxynucleoside triphosphate pools increased during B_{12} starvation only in proportion to the increase in cell volume (Goetz and Carell, 1978).

Cell growth resumed after addition of vitamin B_{12} to starved cultures. Within 2 hours of replenishment, DNA synthesis was observed (Johnston and Carell, 1973; Goetz et al., 1974; Goetz and Carell, 1978; Lafarge-Frayssinet et al., 1978; Shehata and Kempner, 1979) plus a large transient increase in the deoxynucleoside triphosphate pools (Carell and Goetz, 1976; Goetz and Carell, 1978). No variation in cell number, volume, RNA, protein, or ornithine decarboxylase activity was seen during the first 6 hours of B_{12} replenishment (Johnston and Carell, 1973; Goetz et al., 1974; Lafarge-Frayssinet et al., 1978; Shehata and Kempner, 1979). Resumption of cell growth was implicit in the development of the pellicle: after 2–3 hours recovery, new ridges* with microtubules appeared between the old ridges (Fig. 11) (Bre and Lefort-Tran, 1978). Ornithine decarboxylase appeared 8 hours after B_{12} replenishment, and vastly surpassed the levels observed in normal cells. Two separate peaks of enzyme activity were found between 8 and 24 hours of recovery (Lafarge-Frayssinet et al., 1978).

Nuclear division resumed 2–3 hours after B_{12} replenishment (Bre and Lefort-Tran, 1978). Cell division was induced and a doubling of cell number was seen between 6 and 10 hours (Johnston and Carell, 1973; Goetz et al., 1974; Lafarge-Frayssinet et al., 1978; Bre and Lefort-Tran, 1978; Shehata and Kempner, 1979). Synchronous division of Euglena for two (Johnston and Carell,

---

*For further discussion, see Chapter 2, this volume.

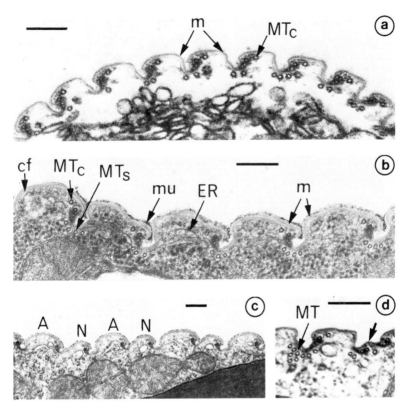

Fig. 11. Pellicle structure in normal, vitamin B_{12}-starved and recovering *Euglena*. (a) Pellicle of control cell emptied of internal contents. A three-layered, 120-Å thick membrane (m) covers the cell. The ridges are of uniform (2400 Å) width. Microtubules are found at the base of the groove and also near the ridge. The last of the ridge microtubules (MT_C) is located near the middle of the ridge, suggesting that pellicle growth proceeds from the back edge (the initiation point of a new ridge) towards the front edge. (b) Pellicle of a cell starved for 6 days. The ridges are two to three times wider than those of the control. A noticeable mucilagenous layer (mu) is seen on the outside of the cell. The three-layered, 120-Å membrane is unchanged, and below this is found a fibrous layer (cf). Three microtubules are located in the corner of the ridge, whereas four others (MT_S) are located near the base of the groove. ER, ergastoplasm. (c) Pellicle from a cell starved and then supplemented with vitamin B_{12}. Most cells enter mitosis after addition of the vitamin; they show a regular alternation of old ridges, A (which had enlarged during the starvation period) and the new ridges, N (induced by supplementation). (d) Detail of an early stage in the initiation of new ridges after the addition of B_{12}. The arrow indicates the bud of a future ridge, which already contains at least 5 to 6 microtubules. MT, group of at least eight microtubules at the bottom of a groove. Calibration bars represent 0.25 μm. (From Bre and Lefort-Tran, 1978.)

230   E. S. Kempner

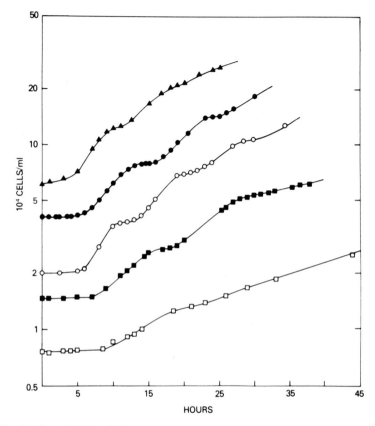

Fig. 12. Growth of vitamin B_{12}-starved *E. gracilis* after addition of vitamin. Increase in cell concentration shown for cultures previously starved for 25 (▲), 50 (●), 72 (○), 90 (■), or 140 (□) hours. For clarity, curves were displaced vertically. (Adapted from Shehata and Kempner, 1979.)

1973) or more (Shehata and Kempner, 1979) cell cycles occurred during B_{12} recovery, and the quality of the synchrony was dependent on the interval of the previous starvation (Shehata and Kempner, 1979; Fig. 12). Both volume and DNA content of individual cells was determined in normal, B_{12}-starved and recovering *Euglena* cultures. The distributions of these parameters in the cell populations are shown in Fig. 13. Although the volume distribution of starved cells was bimodal, the DNA distribution was not. The changes in these distributions during the first cycle of synchronized division indicate different cellular events. DNA duplication was completed within all cells within 8 hours. Since cell division decreases cellular volume by one-half, this process can be seen both in the shapes of the volume distributions and the sharp decrease in the means of the distributions between 6 and 10 hours. It was concluded that all cells were

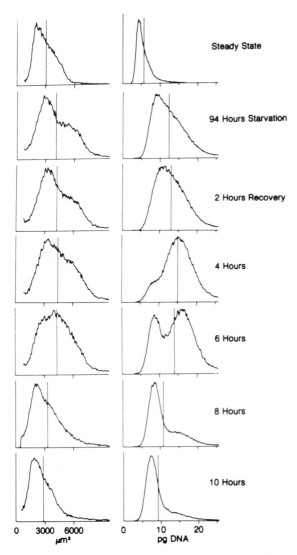

Fig. 13. Distribution of cellular volumes and cellular concentrations of DNA in cultures of *E. gracilis*. Steady-state growth distributions are modified during vitamin B_{12} starvation. After 94 hours, cultures are supplemented with the vitamin. Cellular recovery during the first division cycle is shown at 2, 4, 6, 8, and 10 hours. The mean of each distribution is indicated by the vertical line. (From Shehata and Kempner, 1979.)

blocked at a unique point in the cell cycle after the initiation of DNA synthesis and before nuclear division. Some cells were additionally blocked in reactions unrelated either to specific points in the cell cycle or to DNA synthesis. Recovery from this second process was completed within 4 to 6 hours after replenishment of the vitamin.

C. NORMAL METABOLITES AND THEIR DERIVATIVES

1. *Amino Acids, Analogs, and Derivatives*

Glutamic acid is the only amino acid that has been used as a sole carbon and nitrogen source for *E. gracilis,* although many other amino acids have been shown to be adequate sources of nitrogen (Cook, 1968). The addition of L-methionine to *Euglena* cultures utilizing glutamic acid resulted in severe inhibition of growth. Some inhibition was also obtained with homocysteine, cysteine, and threonine, all in the *levo* form. No deleterious effects of these amino acids were found if acetate or ethanol was used as the carbon source. Milner and Weissbach (1969) showed that this phenomenon was due to methionine interference with glutamic acid uptake. With acetate or ethanol as the carbon source, Owens and Blum (1967) reported that 9 mM cysteine inhibited growth of *Euglena,* but not of *Astasia*.

Effects of cystamine on photosynthetic growth in synchronous populations are revealed by a slight delay in division and also by a decrease in the size of the population steps; the failure to double cell number was progressively more severe with higher concentrations of cystamine up to 100 mM. Nissen and Eldjarn (1969) also studied cystamine added to photosynthetic cultures, but in the presence of SH-Sephadex, which kept cystamine reduced. Depression of cell population increase was seen with as little as 5 mM cystamine. Both sulfur compounds resulted in enlarged cells, indicating effects on cell division rather than cell growth.

In two impressive studies, Owens and Blum (1967, 1969) examined the effects of amino acid esters (methyl, ethyl, and butyl) on *Astasia* and *Euglena*. Both showed a lag period that was lengthened by higher concentrations of the esters, followed by exponential growth at almost the control rate. They showed that cellular hydrolysis of the esters released an alcohol that supported growth (as does the amino acid). It was the amino acid ester per se which was inhibitory, probably by interference with aminoacyl-tRNA synthetases. When hydrolysis had reduced the concentration of the ester below an inhibitory level, growth resumed. One compound, glutamic diethyl ester, was found to have unique effects only on *Euglena*. In addition to the growth inhibition by the other amino acid esters, it also resulted in cell doublets and higher multiples of joined cells with a common cytoplasm. The lesion in cell division was associated with the unwinding and resealing of the pellicle along the division plane.

The most widely studied amino acid analog in *Euglena* is that of methionine–ethionine. During greening, 3 μg/ml DL-ethionine inhibited pigment synthesis but did not disturb cell multiplication (Aaronson *et al.*, 1967). At 10 μg/ml, ethionine inhibited chlorophyll synthesis (Lyman, 1967), with a slight reduction in cell growth (Aaronson and Ardois, 1971), which was completely blocked by 100 μg/ml ethionine.

Acetate-stimulated induction of malate synthase and isocitrate lyase in light–dark synchronized cultures was blocked by 1 m*M* parafluorophenylalinine, although after 6 hours of incubation with analog there were indications that the enzymes had begun to be synthesized (Woodward and Merrett, 1975).

2. *Purines, Pyrimidines, and Analogs*

The growth of *Astasia*, but not of *Euglena*, was inhibited by 8-azaguanine. In an extensive comparison of biochemical processes and reactions, no simple explanation for this difference could be shown. Kahn and Blum (1965) suggested that in *Astasia* the analog may inhibit growth by mechanisms that do not involve RNA. Other analogs do affect the photoorganotrophic growth of *Euglena*, as shown by Krauss *et al.* (1977).

In light–dark (14:10) synchronized *Euglena* cultures, Davis and Merrett (1974) reported that two mitochondrial enzymes, fumarase and succinate dehydrogenase, doubled during the early dark phase. Addition of 6-methylpurine to the culture at the beginning of the light period blocked any further increase in enzyme levels, but if addition of the analog was delayed for 8 hours, the normal doubling occurred without interruption. Since this analog interferes with RNA synthesis, it was concluded that transcription must have occurred before the eighth hour of light, even though the synthesis did not occur until after the dark period began 6 hours later. On the other hand, during random heterotrophic growth, the addition of this analog or of 5-fluorouracil completely inhibited cell division, but had no effect on the induction (by acetate) of malate synthase or isocitrate lyase (Woodward and Merrett, 1975). This laboratory also studied the chloroplast enzyme ribulose-1,5-diphosphate carboxylase which is synthesized in light (Lord *et al.*, 1975). Continued enzyme synthesis was observed on return to darkness. Exposure to 5-fluorouracil during the light period blocked the synthesis of ribulose-1,5-diphosphate carboxylase, and this inhibition was not relieved by incubation without inhibitor in the dark. As with the mitochondrial enzymes, it appears that synthesis of this chloroplast enzyme required a light-dependent transcriptional step.

Evans and Smillie (1971) also reported effects of 5-fluorouracil on chlorophyll synthesis and cytoplasmic enzyme activity in dark-grown cells exposed to light. They found that exposure to the analog 48 hours before the transfer to light had no effect on chlorophyll synthesis—a finding subsequently confirmed by Theiss-Seuberling (1973). Treatment with 5-fluorouracil in the light did inhibit

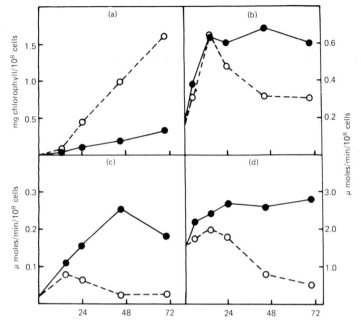

Fig. 14. Effects of 5-fluorouracil on cellular synthesis. Dark-grown heterotrophic cells were harvested and resuspended in autotrophic medium. After 2 days incubation in the dark, the cultures were illuminated and 5-fluorouracil was added. The synthesis of chlorophyll (a), glucose-6-phosphate dehydrogenase (b), NADP-isocitrate dehydrogenase (c), and NADPH-glutathione reductase (d) were observed in the presence (●) or absence (○) of 5-fluorouracil. (From Evans and Smillie, 1971.)

synthesis of chlorophyll and cytochrome *c*. Cytoplasmic enzymes normally continued to increase during the first 12 hours of light exposure, but then decreased as chloroplast synthesis accelerated. The presence of the analog during this period had no effect on the 12-hour enzyme increase, but the reduction in enzymatic activity was not observed (Fig. 14). It was suggested that the decrease might be due to hydrolysis of cytoplasmic proteins to provide amino acids for chloroplast protein synthesis.

Ishida *et al.* (1973b) found that *Euglena* incorporated more [^{14}C]uracil into RNA in the dark than in the light.

3. *Other Naturally Occurring Compounds*

Mannose did not support the growth of *Euglena*, and its presence inhibited growth on almost all other carbon sources (Blum and Wittels, 1968). Such profound effects are unexpected since almost no mannose was found to enter the cells. Exogenous mannitol, the analogous glycitol, is also known to be excluded from *Euglena,* even though large quantities of the endogenously synthesized compound are found intracellularly (Kempner and Miller, 1972b).

D. Antibiotics and Related Inhibitors

An excellent review of the effects of a variety of compounds on *Euglena* has been presented by Schmidt and Lyman (1976). The specificity of these drugs for DNA synthesis and protein synthesis in cytoplasm, mitochondria, and chloroplasts was described in detail. This report should be consulted for concise descriptions of modes of action. In a series of papers, Ebringer (1971, 1972a,b) has reported screening large numbers of antibiotics. Chlorophyll formation, cell bleaching, and the killing of *Euglena* were determined, and relationships between the inhibitory compounds were observed.

Three drugs (streptomycin, chloramphenicol, and cycloheximide) have been studied so extensively and their mechanism of action has been so well established that they have developed into research tools for *Euglena*. Streptomycin permanently bleaches *Euglena* and has given rise to a variety of strains now widely used. Ebringer *et al.* (1970) and Freyssinet (1977) have given recent results of streptomycin effects on *Euglena* that are consistent with the known mechanism. There is some confusion, however, as to whether streptomycin affects synthesis of α-tocopherol, which is normally found in *Euglena* chloroplasts (Baszynski *et al.,* 1969; Pigretti, 1973).

1. *Chloramphenicol*

Chloramphenicol inhibited the greening of dark-grown *Euglena* cells (Aaronson *et al.,* 1967) much more than it depressed cell division (Ben-Shaul and Markus, 1969). The antibiotic affected mitochondria as well as chloroplasts (Stewart and Gregory, 1969; see Table IV). The synthesis of organellar enzymes, ribulose-diphosphate carboxylase and NADP-glyceraldehyde-3-phosphate dehydrogenase, was arrested (Hovenkamp-Obbema and Stegwee, 1974).

2. *Cycloheximide*

Phenylalanine-tRNA synthetases in *Euglena* occur in three independent forms: cytoplasmic, mitochondrial, and chloroplastic. The effects of cycloheximide on the activity of these enzymes are complex, and it was concluded that the control of the synthesis of these enzymes is not simple (Spare *et al.,* 1978). RNA polymerases specific for the synthesis of cytoplasmic ribosomal RNA were totally inhibited by cycloheximide, suggesting a cytoplasmic origin for these enzymes, which then express their activity in the nucleolus where this RNA is synthesized (Ledoigt and Calvayrac, 1974).

Table IV
SUMMARY OF THE EFFECTS OF SEVERAL ANTIBIOTICS ON ORGANELLE DEVELOPMENT IN *Euglena*[a,b]

Antibiotic	Inhibitor of the development of cyanide-sensitive respiration	Inhibitor of mitochondrial cytochrome synthesis	Effector of bleached-cell formation
Chloramphenicol	+ +	+ + +	+ + +
Tetracycline	+ + +	+ + +	+
Lincomycin	±	—	+ + +
Kanamycin	±	—	+ + +
Neomycin	+	+ + +	+
Euflavine	+ + +	+ + +	—
Ethidium bromide	+ +	?	—

[a] (+) Indicates an observable effect on the parameters indicated at concentrations of the antibiotic that cause little or no effect on growth of the cells; (−) indicates no significant effect under the same conditions; (?) indicates effect not established (see text).
[b] From Stewart and Gregory (1969).

Ben-Shaul and Ophir (1970) found that cell division was more sensitive to cycloheximide than was chlorophyll synthesis, and that the drug reduced the number of thylakoids per chloroplast. In greening cells, cycloheximide arrested cell division, but the development of chloroplasts was only partially affected. They concluded that under these conditions the main effect of the drug was via the cytoplasmic origin of the chloroplast membrane system, any effects on chlorophyll synthesis being secondary.

In *Euglena*, Evans (1971) discovered that cycloheximide had an additional effect, independent of protein synthesis. The uptake of glucose and 2,4-dinitrophenol was inhibited by the drug due to effects on membrane transport, either by a change in the pH gradient across the cell membrane or effects on the membrane itself.

Chloramphenicol and cycloheximide block protein synthesis on 70 S (chloroplast) and 80 S (cytoplasmic) ribosomes, respectively. Because of this distinction in specificity, both inhibitors are frequently used to establish the site of synthesis of a specific protein. This approach has been utilized extensively by Merrett and his co-workers in an impressive series of excellent reports (Codd and Merrett, 1970; Lord and Merrett, 1971; Davis and Merrett, 1973, 1974, 1975; Lord *et al.,* 1975; Woodward and Merrett, 1975) concerning the glycolate pathway and glyoxylate cycle. Other significant applications of this dual antibiotic technique are found in studies by Ohmann (1969), Bishop and Smillie (1970), Richards *et al.* (1971), Wolfovitch and Perl (1972), Neumann and Parthier (1973), Marcenko (1974), and Matson and Kimura (1976).

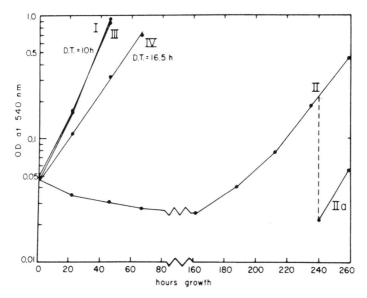

Fig. 15. Adaptation of *Euglena* cells to growth in the presence of dinitrophenol (DNP). (I) Normal cell growth; (II) growth in the presence of DNP; (IIa) growth of cells from II inoculated in fresh medium with DNP; (III) growth of adapted cells in fresh medium; (IV) growth of adapted cells in fresh medium containing DNP. D.T., doubling time. (From Kahn, 1973.)

3. *2,4-Dinitrophenol*

An "uncoupler" of oxidative phosphorylation is 2,4-dinitrophenol: It permits oxidative metabolism to proceed without net ATP synthesis or ADP consumption. Brief exposure to low concentrations (2–5 μM DNP) stimulated acetate oxidation and inhibited acetate assimilation in *Euglena* (Collyard and Danforth, 1970). No effect was found on endogenous respiration or total ATP content of cells. The uptake of amino acids was inhibited at concentrations of DNP greater than 1 μM (Lux and Petzold, 1976). The inhibition of plastid development by DNP depended greatly on both the pH of the extracellular medium and the cellular concentration (Evans, 1971). DNP entered the cell within a few minutes. After 22 hours, almost no labeled DNP could be detected in the cells, but chlorophyll synthesis was almost completely blocked, and the cells lost both K^+ and Mg^{2+} to the medium.

DNP added to the culture medium at 10 μM caused severe inhibition and killing of many *Euglena* cells; after continued, long-term exposure (160 hours) to DNP, growth resumed at almost the control rate (Kahn, 1973) (Figure 15). The cells could adapt to growth (photosynthetic or on organic compounds) in the presence of the uncoupler; cell death rapidly followed in the presence of DNP if light and carbon sources were removed; this was attributed to very high energy requirement of adapted cells. Respiration was very loosely coupled, showing no respiratory control. However, washed mitochondria isolated from adapted cells showed the same rates of respiration and phosphorylation and the same sensitivity to cyanide as those from normal cells. Furthermore, *Euglena* that were resistant to DNP were simultaneously adapted to five other uncouplers or inhibitors of oxidative phosphorylation (Kahn, 1974). It was concluded that this adaptation was due either to modification of the energy-coupling pathway or else to "formation of an altered mitochondrial membrane impermeable to many uncouplers and inhibitors of oxidative phosphorylation" (Kahn, 1974). It was subsequently shown that the adaptation did not involve preferential synthesis of major new mitochondrial membrane proteins, but damage to cellular amino acid and sugar transport systems was observed (Kahn and McConnell, 1977).

4. *DCMU*

For several years DCMU [3-(3,4-dichlorophenyl)-1,1-dimethyl urea] has been an important tool in the study of photosynthesis in *Euglena*. It inhibits synthesis of chlorophyll but not the development of chloroplasts.

Euglena grown in the dark until the stationary phase were transferred to the light for 20 days with DCMU. No change in dry weight was observed and the cells remained colorless (Pohl and Wagner, 1972). Identifiable chloroplast structures were synthesized in the presence of 10^{-5} M DCMU as well as reduced quantities of chloroplast-specific compounds such as galactolipids, chlorophylls, and polyunsaturated C_{16} and C_{18} fatty acids (Göbel *et al.*, 1976). However, at 10^{-3} M DCMU, these fatty acids were not synthesized (Pohl and Wagner, 1972). DCMU added to exponential phase cells led to slower, but still exponential growth with diminished chlorophyll synthesis. The amount of chloroplast RNA was unaffected, but deranged chloroplast structures showing vesicles in the thylakoids were found. The progressive appearance of vesicles was correlated with loss of chlorophyll (Calvayrac and Ledoigt, 1976). A KCN-sensitive enzyme was transiently observed in the mitochondria during the greening process; DCMU blockage of photosynthesis also inhibited formation of this enzyme (Brody and White, 1973). Similarly, the DCMU effects on photosynthesis were responsible for depression in synthesis of δ-aminolevulinate, a major precursor for chlorophyll (Richard and Nigon, 1973). DCMU enhanced the light-dependent bleaching of chlorophyll and also the release of ethane from the oxidation of unsaturated fatty acids (Elstner and Pils, 1979).

Successive subculture in the presence of DCMU led to the development, over a 10-week period, of DCMU-resistance in all *Euglena* cells. The adapted cells had near-normal appearing chloroplasts, increased cellular protein and decreased paramylum (Calvayrac *et al.*, 1979a). Chlorophyll content and oxygen evolution were no longer depressed by DCMU. This resistance was maintained indefinitely in the absence of DCMU without evidence of reversion (Calvayrac *et al.*, 1979b).

5. *Nalidixic Acid and Furylfuramide*

In a remarkable study, Lyman (1967) showed that *Euglena* growth in both light and dark in an acidic heterotrophic medium was unaffected by up to 200 μg/ml nalidixic acid. However, ability to form green colonies was completely abolished by 20 hours exposure in the light to only 50 μg/ml, and the effect persisted after removal of the inhibitor. Using ethionine to block chloroplast protein and chlorophyll synthesis, it was shown that nalidixic acid effects required protein synthesis, and it was suggested that the mode of action was via specific inhibition of chloroplast DNA synthesis. Ebringer (1970) confirmed these results at 500 μg/ml in a neutral (pH 7.8) growth medium, and found that, at pH 6, higher concentrations of nalidixic acid and longer exposure times were required for the same degree of effect. He subsequently showed (1971) that many antibiotics which inhibit DNA synthesis permanently bleached *Euglena*.

The nalidixic acid blockage occurred at an early stage of chloroplast formation, but did not change the size and shape of mitochondria (Neumann and Parthier, 1973). Further elucidation of the mechanism of action was shown concisely by Lyman *et al.* (1975) by using the mutant P_4ZUL that can make chloroplasts but has no photosynthetic electron transport system, and also by using DCMU that specifically blocks that system in the wild type. They concluded that nalidixic acid action required a functioning photosynthetic electron transport system, and that light enhancement of the inhibition was associated with the transport system. Ultraviolet light inactivation of green colony formation showed that nalidixic acid caused a loss in the number of chloroplast genomes. Chloroplast DNA replication was completely inhibited by nalidixic acid, and this DNA was undetectable by density-gradient analyses after only 1.5 generations exposure to the inhibitor (Pienkos *et al.*, 1974). This disappearance was much too rapid to be accounted for by dilution (Lyman *et al.*, 1975), implying degradation of preexisting molecules.

One of the nitrofuran derivatives, furylfuramide, has been found to bleach *E. gracilis* cells, possibly by a similar mechanism to that of nalidixic acid: specific inhibition of chloroplast DNA replication, followed by degradation of the inhibitor (Ikushima, 1975).

6. *Actinomycin D*

Isolated *Euglena* chloroplasts incorporated tritiated nucleotides into chloroplast DNA, but this process was significantly inhibited by high concentrations of actinomycin D (10 μg/ml) (Scott *et al.*, 1968), suggesting a DNA polymerase system. The drug had a variety of effects on *E. gracilis* when added directly to the culture medium at concentrations up to 2 μg/ml (O'Donnell-Alvelda, 1968). After 6 hours exposure, cellular metaboly was reduced, generation time was almost doubled, while cell division was completely stopped in 30% of the cells exposed to the highest concentration. Within 2 to 6 hours nuclear division was arrested, primarily at prophase or early anaphase. Nucleoli were much less electron-dense in treated cells, although there was no effect on pre-existing nucleolar material and chromatin. Actinomycin, which blocks RNA synthesis, prevents deposition of new nucleolar material and chromosomal matrix. Similar effects on the nucleolus of *A. longa* have been seen, but only at much higher concentrations (Chaly *et al.*, 1979), and it was claimed that morphological effects could be seen within 1 hour of exposure to actinomycin.

Chlorophyll synthesis in *Euglena* was blocked by high concentrations (45 μg/ml) of actinomycin D (Bovarnick *et al.*, 1969), whereas colony formation was much more sensitive. Colonies that did form were all green, but fewer colonies were formed with progressively longer incubations with the antibiotic. Ebringer (1971) and Theiss-Seuberling (1973) confirmed these observations and also found only a slight effect of low concentrations of actinomycin on NADP glyceraldehyde-3-phosphate dehydrogenase synthesis.

7. *Antimycin A*

Isolated *Euglena* mitochondria have been treated with antimycin A specifically to reduce the concentration of *b*-type cytochromes. Respiratory rates using succinate, lactate (Sharpless and Butow, 1970a), or glycollate (Collins *et al.*, 1975) as substrates in the presence of antimycin have been reported. Sharpless and Butow (1970a) concluded that *Euglena* does have several *b*-type cytochromes at phosphorylation site II, but they also found alternate reactions which could bypass the antimycin-sensitive step.

The respiration rate of intact *Euglena* cells was depressed within a few minutes of addition of 0.5 μg/ml antimycin to a culture; the exponential increase in cell number is stopped, but after a lag of 20 hours growth resumed due to an adaptation process (Sharpless and Butow, 1970b). Growth rate in the presence of the inhibitor depended on the carbon source; with ethanol it could be as high as the control rate. These adapted cells showed an enhanced respiration rate. These authors also found a new type of succinoxidase activity within 10 hours exposure to antimycin. This new activity was stimulated by AMP and appeared in mitochondria with buoyant density slightly greater than normal. Giant mitochondria were seen within a few hours exposure to antimycin, but subsequent cultivation in the absence of the inhibitor led to rapid recovery of cellular respiration and reappearance of normal mitochondria (Calvayrac and Butow, 1971). A new cytochome, cytochrome 556, was identified by Calvayrac and Claisse (1973) from low temperature spectra of whole *Euglena* cells. Synthesis of this cytochrome ceased shortly after addition of antimycin, but did not resume when adaptive growth commenced. They drew a parallel between the amount of cytochrome 556 and formation of giant mitochondria in *Euglena*.

8. *Myomycin*

Myomycin had very little effect on heterotrophic growth of green cells in the light, although chlorophyll synthesis was inhibited somewhat. However, bleached colonies appeared very quickly and within 5 days exposure almost all cells formed bleached colonies. Fluorescence microscopy indicated a degeneration of plastids until they resembled those of dark-grown cells. Electron microscopic examination indicated normal nuclei and mitochondria and confirmed the disintegration of plastids (Fasulo *et al.*, 1976). Etiolated cells from dark-grown cultures showed very similar patterns when grown in the presence of myomycin; Vannini *et al.* (1978) concluded that the antibiotic acted as an inhibitor of protein synthesis by chloroplast ribosomes. Cells that were bleached by continuous growth in the presence of myomycin showed no detectable chlorophyll. Indefinite growth in the pH 6.3 culture medium yielded no revertants. If the pH of the medium was raised to 8.0 (which is outside the stability range of myomycin) the cells slowly regained ability to green. For 3 weeks at pH 8.0 there was a slow accumulation of carotenoids that made the cells appear orange. Eventually, chlorophyll synthesis became detectable and oxygen evolution was observed during the subsequent 14 days (Fasulo *et al.*, 1979b). Unfortunately, no data on the growth of normal cells at pH 8 or that of cells washed free of myomycin at pH 8 were given; these would permit distinction between slow myomycin degradation and cell recovery.

9. *Lincomycin*

Heterotrophic growth of *Euglena* in the light was unaffected by lincomycin (2 mg/ml) with only a very slight effect on respiratory activity or cytochrome spectra. Almost all of the cells were bleached, however (Table IV). The bleaching was not reversible and revertants were not seen (Stewart and Gregory, 1969). Lower concentrations of lincomycin and several 7-halo and 7-oxy derivatives were tested (Ebringer and Foltinova, 1976). Although no bleached colonies were found after incubation in 500 μg/ml lincomycin, all colonies were bleached at 1500 μg/ml. The 7-halo derivatives of lincomycin were more effective in bleaching and also more toxic to *Euglena* than the 7-oxy derivatives. At 2 mg/ml, Freyssinet (1977) reported little effect on uptake or incorporation of carbonate, leucine, or sulfate. However, in cell-free preparations with 20-fold less inhibitor, lincomycin specifically inhibited chloroplasts and chloroplast ribosomes, with little effect on the activity of cytoplasmic ribosomes. Utilizing this specificity, Freyssinet (1978) then studied the synthesis of cytoplasmic and chloroplast ribosomal proteins using the double antibiotic approach with cycloheximide and lincomycin. In a formidable experiment, he showed that all cytoplasmic ribosomal proteins were synthesized in the cytoplasm; of 39 chloroplast ribosomal proteins, 9 were synthesized in the chloroplast, 12 in the cytoplasm, and 6 were inhibited by both antibiotics.

10. *Rifampicin*

Rifampicin had no effect on the growth of *Euglena* in neutral pH media either in the light or dark (Ishida *et al.*, 1973a,b), but did inhibit uracil incorporation equivalently in both light conditions (Ishida *et al.*, 1973b). Leucine incorporation in the light was stimulated by rifampicin (Ishida *et al.*, 1973a). However, in acidic or neutral media, rifampicin had no effect on chlorophyll synthesis (Brown *et al.*, 1969; Theiss-Seuberling, 1973), cell growth, or RNA synthesis unless it was added together with 1% dimethyl sulfoxide, DMSO (Brown *et al.*, 1969). Rifampicin, 20 μg/ml, with 1% DMSO had profound effects on *Euglena* structure: disordered chloroplasts, segmented mitochondria with a vacuolar appearance, but no observable differences in nucleus and cytoplasm between control and treated cells. The distribution of ^{32}P in RNA synthesized in the dark was the same with and without rifampicin and DMSO. In the light, however, the combination completely inhibited incorporation into chloroplast 17 S RNA and decreased the ratio of 23 S to 28 S RNA. Heizmann (1974) did not observe complete inhibition of transcription in dark-grown cells, which were starved for 4 days before exposure to rifampicin and DMSO.

11. *Erythromycin and Spectinomycin*

Erythromycin and spectinomycin treatment of intact *Euglena* cells had little effect on uptake and incorporation of carbonate, sulfate, or leucine, but amino acid incorporation by isolated chloroplasts was significantly diminished (Freyssinet, 1977). Chlorophyll synthesis and cell bleaching were more sensitive to spectinomycin than erythromycin (Ebringer, 1972a).

12. *Anisomycin*

Anisomycin, like cycloheximide, is known to block protein synthesis in 80 S cytoplasmic ribosomes. In a very careful study, Neumann and Parthier (1973) found that mixotrophic cell division in *Euglena* was temporarily inhibited by the antibiotics. After a lag period, the amount of chlorophyll synthesis, CO_2 fixation, and chloroplast leucyl–tRNA synthetase increased and vastly exceeded that of untreated cells (Fig. 16). They suggested that the effects on cytoplasmic ribo-

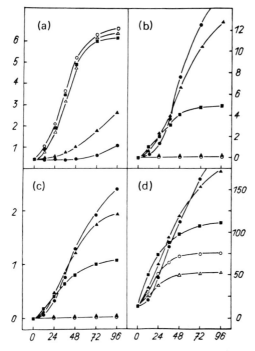

Fig. 16. Effects of several drugs on *E. gracilis.* Dark-grown cells illuminated at time zero. (a) Cell multiplication (millions of cells/ml); (b) chlorophyll synthesis (pg chlorophyll/cell); (c) RUDP carboxylase activity, (μmole CO_2) fixed/mg protein/minute; (d) leucyl-tRNA synthetase activity (pmoles [^{14}C]leucine bound/mg tRNA/minute). Untreated controls (■); chloramphenicol added 12 hours prior to illumination (○); nalidixic acid added 12 hours before illumination (△); cycloheximide added at time zero (●); anisomycin added at time zero (▲). (From Neumann and Parthier, 1973.)

somes limited the supply of proteins needed for the structure or regulation of chloroplasts, leading to a lag in development. There may also have been a slow internal inactivation of anisomycin (and also cycloheximide) during metabolism.

13. *Nigericin*

Nigericin is an ionophore whose primary action is stimulation of transmembrane exchange of protons and alkali metals. In *Euglena,* however, the effects seem to differ. Evans reported (1971) that 1.4 μM nigericin (and also a 3.6 μM cycloheximide) led to a decrease in polyribosomes, with concomitant increase in the number of free ribosomes, but there was no effect on chloroplast development or respiration. At these low concentrations, both drugs inhibited the uptake of glucose and the uncoupler of oxidative phosphorylation, 2,4-dinitrophenol (DNP). The inhibition of chloroplast formation by DNP was relieved by nigericin by excluding DNP from the cells.

14. *Olivomycin*

Astasia longa cells are sensitive to olivomycin, an antibiotic structurally similar to chromomycin. Cell growth is stimulated by 22 μg/ml (three times greater than controls in 72 hours), but continued exposure leads to cell lysis and decrease in cell population. After 240 hours exposure, the cells adapt to olivomycin and cell growth resumes (Sukhareva-Nemakova *et al.*, 1975). The lipid fraction of cell dry weight increases, but no drastic changes occur among the classes of lipids. Among the fatty acids, the C_{20} unsaturated fatty acids almost disappear. As the adaptation process takes place, fatty acid distribution and cellular lipid content return to normal.

15. *Mitomycin and Its Derivatives*

Mitomycins interact with DNA in cells. They act as bifunctional alkylating agents and cross-link the DNA. Some mitomycins (A and C) have a high affinity for nuclear DNA in *Euglena* and are toxic at low concentrations; bleaching of colonies is therefore not seen. Other derivatives (B, N-methyl or porfiromycin) are not as toxic and there are many surviving cells that show bleaching effects (Ebringer *et al.*, 1969). It would be interesting to know if there were stoichiometric relationships between the killing/bleaching ratio of these antibiotics and the ratio of nuclear to chloroplast DNA present.

16. *Virginiamycin*

Virginiamycin can be separated into two components, M and S. Individually or together they have no effect on the heterotrophic growth rate of *E. gracilis*. Chlorophyll synthesis is unaffected by S, inhibited by M, and abolished by the combination of the two together (Van Pel *et al.*, 1973). Cells grown in the presence of the M fraction and plated without antibiotic gave rise to green colonies, indicating a phenotypic effect; chloroplasts showed severe derangement of structure. Cells grown in virginiamycin M together with virginiamycin S plated out to give only white colonies, suggesting a genetic change; chloroplasts were not present in these cells, but plastid-like structures were observed in which lamellae were replaced by tubules. The mitochondria appeared normal in all virginiamycin-exposed cells. The blockage of chloroplast synthesis was shown (Van Pel and Cocito, 1973) to be due to a blockage of chloroplast ribosomal RNA synthesis, an effect unique to *Euglena*.

17. *Ethidium Bromide, Acriflavine, and Acridines*

The DNA-intercalating dye, ethidium bromide, inhibited growth of *Euglena* in the dark (Stewart and Gregory, 1969) and in the light (Stewart and Gregory, 1969; Nass and Ben-Shaul, 1973). Chlorophyll synthesis was somewhat diminished by the drug, but the synthesis of new chloroplast structure was normal (Nass and Ben-Saul, 1973). Bleached colonies were not induced by ethidium bromide (Stewart and Gregory, 1969). Mitochondrial structure was aberrant in light or dark, and mitochondrial DNA synthesis was much more sensitive to the drug than was nuclear DNA (Nass and Ben-Shaul, 1973); cell respiration (Stewart and Gregory, 1969) and cytochrome oxidase activity (Nass and Ben-Shaul, 1973) were significantly diminished. These results indicate a distinct specificity of ethidium bromide for *Euglena* mitochondrial synthesis (Table IV).

Like ethidium bromide, acriflavin intercalates in DNA. The effects on *Astasia* were clearly shown by Gulikova *et al.* (1974). The electron microscope revealed disordered mitochondrial structure. Acriflavin inhibited thymidine incorporation into mitochondrial (but not nuclear) DNA. Furthermore, protein synthesis in mitochondria was much more sensitive to the drug than that on cytoplasmic ribosomes. The mechanism of action of acriflavin, like ethidium bromide, is due to disturbance of transcription of mitochondrial DNA.

Closely related to acriflavine is the group of acridine compounds. The 1-nitroacridines are potent metabolic inhibitors. One of these, ledakrin, was found to rapidly inhibit both DNA and RNA synthesis in synchronized *Euglena* cultures; protein synthesis inhibition was observed only in the following division cycle (Chotkowska and Konopa, 1973). The drug was covalently bound to DNA *in vivo* and cross-links the DNA strands, suggesting a precisely defined mechanism of action (Konopa *et al.*, 1976).

18. *Other Compounds*

A variety of other organic compounds have been added to *Euglena* cultures. One or two reports of each have appeared. Cell growth, photosynthesis, respiration, or the formation of bleached colonies were most commonly measured, but the mechanisms of these effects have not usually been established. For the most part, these studies indicated that *E. gracilis* is relatively resistant to a wide spectrum of chemicals. The conclusion seems inescapable that *Euglena* cells must possess a permeability barrier considerably greater than do most other microorganisms.

Formaldehyde, propionaldehyde, and glutaraldehyde had only slight effects on *Euglena* when added at low concentrations in culture fluid (Gittleson, 1975) or entering as atmospheric pollutants (de Koning and Jegier, 1970b). Some inhibition of growth was observed on exposure to trienoic anacardic acid (Gellerman *et al.*, 1969) and to the pesticides 2,4 D and 2,4,5 T (Poorman, 1973). Cell growth was unaffected by 10% diesel or lubricating oils (Dennington *et al.*, 1975), amitrole and atrazine (Valentine and Bingham, 1976), and several anesthetics (Ogli, 1977), although chloroform did show some effect; the mechanism of action of this compound is apparently not by DNA or RNA blockage, but more likely on mitotic arrest by inhibition of microtubular protein polymerization.

Euglena cells were also resistant to Panacide (Saxena *et al.*, 1979); to curvularin, cyanein, and cytochalasins A, B, and D (Betina and Micekova, 1972); to isoniazide (Pigretti, 1973); to gramicidin, chlorpromazine, and atractyloside (Kahn, 1974); to polychlorinated biphenyls and to DDT (Mosser *et al.*, 1972).

Structural damage to *Euglena* membranes and organelles was reported by Silverman and Hikida (1976) for colchicine, β-mercaptoethanol, Triton X-100, and tannic acid and by Mikołajczyk and Diehn (1978) for Triton X-100 and cetyltrimethylammonium bromide. Protection from freezing injury by methanol was observed by Morris and Canning (1978).

Euglena growth was inhibited by monorden (Betina and Micekova, 1972), oligomycin, and tri-*n*-butylchlorotin (Kahn, 1974), colicin E2 (Smarda *et al.*, 1975), euflavine, tetracycline, and neomycin (Stewart and Gregory, 1969).

The toxicity of the polychlorinated biphenyls to *Euglena* decreased with increasing amounts of chlorine per molecule; the mechanism of action is by inhibition of photosynthesis or chlorophyll product (Ewald *et al.*, 1976), but later studies suggested that decreased permeability of the cell membrane to bicarbonate may be the principal action of these compounds in *Euglena* (Bryan and Olafsson, 1978).

Lonergan and Sargent (1978) reported that photoelectron flow in *Euglena* was inhibited by ethoxzolamide and acetazolamide; the latter also decreased carbonic anhydrase activity and oxygen evolution. Very specific damage was caused by the herbicide SAN 9789 (Vaisberg and Schiff, 1976); in *Euglena,* the drug uniquely blocks carotenoid synthesis.

Coumarin inhibited cell division of *Euglena,* but not cell growth. Some lobed multinucleated giant cells were found (Vannini *et al.*, 1977) similar to those reported after exposure to cadmium or molybdenum. There were increases in the number of plastids per cell, in chlorophyll content, and in oxygen evolution (Fasulo *et al.*, 1979a). It was suggested that coumarin prevents formation of microtubules needed in mitosis, resulting in a lengthening of the G_2 phase of the cell cycle.

Stimulation of *Euglena* growth by gibberellic acid, abscisic acid, androstendiol, cholesterol, or vitamin D_3 was found by Bralczyk *et al.* (1978).

V. Conclusions

There is a wide chemical spectrum of compounds known to stimulate or inhibit *Euglena* and a diverse set of cellular reactions to them. The nature and degree of these reactions sometimes depends on other culture conditions: light, temperature, pH, etc. From the thicket of observations about these effects, certain general principles appear to have emerged.

The most obvious of these effects is on cellular permeability. Most algae, protozoa, and other microorganisms are affected by a greater variety or by lower concentrations of these compounds, especially the antibiotics, than is *Euglena*. It has been suggested repeatedly that *Euglena* is relatively impermeable to many of these agents. Several low-molecular-weight metabolites revealed, directly or indirectly, that metabolic consequences depended on cellular entry: ionized forms were excluded, but *Euglena* was permeable to the undissociated molecules. By extension, these phenomena may play an important intracellular role; the passage of metabolites between organelles is probably also controlled by the permeability of subcellular structures.

It is noteworthy that several chemical agents, both organic and metallic, interfere with cell division, resulting in the appearance of giant, branched forms. Do these chemicals have a common molecular basis for their action and will their use reveal more about the mechanisms of cell division?

The involvement of nitrogen ion in the control of cellular lipids has been illustrated dramatically in several *Euglena* studies. These suggest a novel biochemical explanation for a fundamental cellular process which has long been perplexing.

The active interest in both phosphate and vitamin B_{12} in *Euglena* metabolism has been described. Each appears to have manifold roles, which are not yet completely understood and which offer rewarding areas for future studies.

I shall not extol the virtues of *Euglena:* I shall not dwell at length on its unique phylogenetic position, its wide latitude of growth conditions, its well-established composition, structure, and biochemistry. Rather I suggest that there has now been revealed several fascinating prospects for unravelling more of the mysterious and wondrous ways displayed by this protozoan chameleon.

References

Aaronson, S., and Ardois, G. (1971). *J. Phycol.* **7**, 18–20.
Aaronson, S., Ellenbogen, B. B., Yellen, L. K., and Hutner, S. H. (1967). *Biochem. Biophys. Res. Commun.* **27**, 535–538.
Adams, J. F., and McEwan, F. (1971). *J. Clin. Pathol.* **24**, 15–17.
Baker, W. B., and Buetow, D. E. (1976). *J. Protozool.* **23**, 167–176.
Baszynski, T., Dudziak, B., and Arnold, D. (1969). *Ann. Univ. Marie Curie-Sklodowska,* **24**, 1–8.
Bates, R. C., and Hurlbert, R. E. (1970). *J. Protozool.* **17**, 134–138.
Begin-Heick, N. (1970). *Can. J. Biochem.* **48**, 251–258.
Begin-Heick, N., and Blum, J. J. (1967). *Biochem. J.* **105**, 813–819.
Ben-Shaul, Y., and Markus, Y. (1969). *J. Cell Sci.* **4**, 627–644.
Ben-Shaul, Y., and Ophir, I. (1970). *Can. J. Bot.* **48**, 929–934.
Benson, R. E. (1972). *Proc. Soc. Exp. Biol. Med.* **139**, 1096–1099.
Bertaux, O., and Valencia, R. (1971). *J. Physiol. (Paris)* **63**, 167A.
Bertaux, O., and Valencia, R. (1972). *J. Physiol. (Paris)* **65**, 349A–350A.

Bertaux, O., and Valencia, R. (1973). *C. R. Hebd. Seances Acad. Sci.* **276**, 753-756.

Bertaux, O., and Valencia, R. (1974). *Colloq. Int. C.N.R.S.* **240**, 331-343.

Bertaux, O., Moyne, G., Lafarge-Frayssinet, C., and Valencia, R. (1978). *J. Ultrastruct. Res.* **62**, 251-269.

Betina, V., and Micekova, D. (1972). *Z. Allg. Mikrobiol.* **12**, 355-364.

Bishop, D. G., and Smillie, R. M. (1970). *Arch. Biochem. Biophys.* **137**, 179-189.

Blum, J. J. (1965). *J. Cell Biol.* **24**, 223-234.

Blum, J. J., and Wittels, B. (1968). *J. Biol. Chem.* **243**, 200-210.

Blum, J. J., Sommer, J. R., and Kahn, V. (1965). *J. Protozool.* **12**, 202-209.

Bonaly, J., Bariaud, A., Delcourt, A., and Mestre, J.-C. (1978). *C. R. Hebd. Seances Acad. Sci.* **287**, 463-466.

Bovarnick, J. G., Zeldin, M. H., and Schiff, J. A. (1969). *Dev. Biol.* **19**, 321-340.

Boyd, C. E. (1970). *Am. Midl. Nat.* **84**, 565-567.

Bralczyk, J., Weilgat, B., Wasilewska-Dabrowska, L. D., and Kleczkowski, K. (1978). *Plant Sci. Lett.* **12**, 265-271.

Bre, M.-H., and Lefort-Tran, M. (1978). *J. Ultrastruct. Res.* **64**, 362-376.

Bre, M.-H., Diamond, J., and Jacques, R. (1975). *J. Protozool.* **22**, 432-434.

Brinkman, K. (1976). *J. Interdiscip. Cycle Res.* **7**, 149-170.

Brody, M., and White, J. E. (1973). *Dev. Biol.* **31**, 348-361.

Brown, R. D., Bastia, D., and Haselkorn, R. (1969). *In* "RNA-Polymerase and Transcription" (L. Silvestri, ed.), pp. 309-328. North-Holland Publ., Amsterdam.

Bryan, A. M., and Olafsson, P. G. (1978). *Bull. Environ. Contam. Toxicol.* **20**, 374-381.

Buetow, D. E. (1965). *J. Cell. Comp. Physiol.* **66**, 235-242.

Buetow, D. E., and Schuit, K. E. (1968). *J. Protozool.* **15**, 770-773.

Calvayrac, R. (1970). *Arch. Mikrobiol.* **73**, 308-314.

Calvayrac, R., and Butow, R. A. (1971). *Arch. Mikrobiol.* **80**, 62-69.

Calvayrac, R., and Ledoigt, G. (1976). *Plant Sci. Lett.* **7**, 249-263.

Calvayrac, R., Bertaux, O., Lefort-Tran, M., and Valencia, R. (1974). *Protoplasma* **80**, 355-370.

Calvayrac, R., Bomsel, J.-L., and Laval-Martin, D. (1979a). *Plant Physiol.* **63**, 857-865.

Calvayrac, R., Laval-Martin, D., Dubertret, G., and Bomsel, J. L. (1979b). *Plant Physiol.* **63**, 866-872.

Calvayrac, R. M., and Claisse, M. L. (1973). *Planta* **112**, 17-24.

Carell, E. F. (1969). *J. Cell Biol.* **41**, 431-440.

Carell, E. F., and Goetz, G. H. (1976). *Biochem. J.* **156**, 473-475.

Carell, E. F., Johnston, P. L., and Christopher, A. R. (1970). *J. Cell Biol.* **47**, 525-530.

Chaly, N., Lord, A., and Lafontaine, J. G. (1979). *Can. J. Bot.* **57**, 2031-2043.

Chisolm, S. W., and Stross, R. G. (1976a). *J. Phycol.* **12**, 210-217.

Chisolm, S. W., and Stross, R. G. (1976b). *J. Phycol.* **12**, 217-22.

Chotkowska, E., and Konopa, J. (1973). *Arch. Immunol. Ther. Exp.* **21**, 767-774.

Codd, G. A., and Merrett, M. J. (1970). *Planta* **95**, 127-132.

Coleman, R. D., Coleman, R. L., and Rice, E. L. (1971). *Bot. Gaz. (Chicago)* **132**, 102-109.

Collins, N., Brown, R. H., and Merrett, M. J. (1975). Biochem. J. **150**, 373-377.

Collyard, K. J., and Danforth, W. F. (1970). *J. Protozool.* **17**, 334-340.

Colmano, G. (1973). *Bull. Environ. Contam. Toxicol.* **9**, 361-364.

Colombetti, G., and Diehn, B. (1978). *J. Protozool.* **25**, 211-217.

Constantopoulos, G. (1970). *Plant Physiol.* **45**, 76-80.

Cook, J. R. (1968). *In* "The Biology of *Euglena*" (D. E. Buetow, ed.), Vol. 1, pp. 243-314. Academic Press, New York.

Cook, J. R. (1971a). *Exp. Cell Res.* **69**, 207-211.

Cook, J. R. (1971b). *J. Cell Physiol.* **78**, 273-276.

Cook, J. R., and Kaiser, H., Jr. (1973). *J. Cell Physiol.* **82**, 489-496.

Cramer, M., and Myers, J. (1952). *Arch. Mikrobiol.* **17**, 384-402.

Davis, B., and Merrett, M. J. (1973). *Plant Physiol.* **51**, 1127-1132.

Davis, B., and Merrett, M. J. (1974). *Plant Physiol.* **53**, 575-580.

Davis, B., and Merrett, M. J. (1975). *Plant Physiol.* **55**, 30-34.

Davis, E. A., and Epstein, H. T. (1971). *Exp. Cell Res.* **65**, 273-280.

de Koning, H. W., and Jegier, Z. (1969). *Arch. Environ. Health* **18**, 913-916.

de Koning, H. W., and Jegier, Z. (1970a). *Atmos. Environ.* **4**, 357-361.

de Koning, H. W., and Jegier, Z. (1970b). *Arch. Environ. Health* **20**, 720-722.

Dennington, V. N., George, J. J., and Wyborn, C. H. E. (1975). *Environ. Pollut.* **8**, 233-237.

Doemel, W. N., and Brooks, A. E. (1975). *Water Res.* **9**, 713-719.

Ebringer, L. (1970). *J. Gen. Microbiol.* **61**, 141-144.

Ebringer, L. (1971). *Experientia* **27**, 586-587.

Ebringer, L. (1972a). *J. Gen. Microbiol.* **71**, 35-52.

Ebringer, L. (1972b). *Neoplasma* **19**, 579-589.

Ebringer, L., and Foltinova, P. (1976). *Acta Fac. Rerum Nat. Univ. Comenianae, Microbiol.* **5**, 113-125.

Ebringer, L., Mego, J. L., and Jurasek, A. (1969). *Arch. Mikrobiol.* **64**, 229-234.

Ebringer, L., Nemec, P., Santova, H., and Foltinova, P. (1970). *Arch. Microbiol.* **73**, 268-280.

Edmunds, L. N., Jr. (1965). *J. Cell. Comp. Physiol.* **66**, 159-182.

Edmunds, L. N., Jr., Jay, M. E., Kohlmann, A., Liu, S. C., Merriam, V. H., and Sternberg, H. (1976). *Arch. Microbiol.* **108**, 1-8.

Elstner, E. F., and Pils, I. (1979). *Z. Naturforsch., C. Biosci.* **34C**, 1040-1043.

Epstein, H. T., and Allaway, E. (1967). *Biochim. Biophys. Acta.* **142**, 195-207.

Evans, W. R. (1971). *J. Biol. Chem.* **246**, 6144-6151.

Evans, W. R., and Smillie, R. M. (1971). *J. Exp. Bot.* **22**, 371-381.

Ewald, W. G., French, J. E., and Champ, M. A. (1976). *Bull. Environ. Contam. Toxicol.* **16**, 71-80.

Falchuk, K. H., Fawcett, D. W., and Vallee, B. L. (1975a). *J. Cell Science* **17**, 57-78.

Falchuk, K. H., Fawcett, D. W., and Vallee, B. L. (1975b). *J. Submicrosc. Cytol.* **7**, 139-152.

Falchuk, K. H., Krishan, A., and Vallee, B. L. (1975c). *Biochemistry* **14**, 3439-3444.

Falchuk, K. H., Mazus, B., Ulpino, L., and Vallee, B. L. (1976). *Biochemistry* **15**, 4468-4475.

Falchuk, K. H., Hardy, C., Ulpino, L., and Vallee, B. L. (1978). *Proc. Natl. Acad. Sci. U.S.A.* **75**, 4175-4179.

Fasulo, M. P., Vannini, G. L., Bruni, A., and Mares, D. (1976). *Z. Pflanzenphysiol.* **80**, 407-416.

Fasulo, M. P., Vannini, G. L., Bruni, A., and Dall'Olio, G. (1979a). *Z. Pflanzenphysiol.* **93**, 117-127.

Fasulo, M. P., Vannini, G. L., and Mares, D. (1979b). *Protoplasma* **101**, 301-315.

Feldman, J. F., and Bruce, V. G. (1972). *J. Protozool.* **19**, 370-373.

Fennikoh, K. B., Hirshfield, H. I., and Kneip, T. J. (1978). *Environ. Res.* **15**, 357-367.

Folkman, Y., and Wachs, A. M. (1973). *Water Res.* **7**, 419-435.

Foo, S. S. K., and Cossins, E. A. (1978). *Phytochemistry* **17**, 1711-1715.

Freyssinet, G. (1977). *Biol. Cell.* **30**, 17-26.

Freyssinet, G. (1978). *Exp. Cell Res.* **115**, 207-219.

Freyssinet, G., Heizmann, P., Verdier, G., Trabuchet, G., and Nigon, V. (1972). *Physiol. Veg.* **10**, 421-442.

Garlaschi, F. M., Garlaschi, A. M., Lombardi, A., and Fiorti, G. (1974). *Plant Sci. Lett.* **2**, 29-39.

Gellerman, J. L., Walsh, N. J., Werner, N. K., and Schlenk, H. (1969). *Can. J. Microbiol.* **15**, 1219-1223.

Gittleson, S. M. (1975). *Acta Protozool.* **14**, 371-377.

250　　　　　　　　　　　　　　　　E. S. Kempner

Göbel, E., Riessner, R., and Pohl, P. (1976). Z. Naturforsch., C. Biosci. **31C**, 687-692.
Goetz, G. H., and Carell, E. F. (1978). Biochem. J. **170**, 631-636.
Goetz, G. H., Johnston, P. L., Dobrosielski-Vergona, K., and Carell, E. F. (1974). J. Cell Biol. **62**, 672-678.
Graves, L. B., Jr. (1971). J. Protozool. **18**, 543-546.
Greenblatt, C. L., and Schiff, J. A. (1959). J. Protozool. **6**, 23-28.
Gulikova, O. M., Zaitseva, G. N., and Pakhomova, M. V. (1974). Mikrobiologiya **43**, 1058-1063.
Heizmann, P. (1974). Biochem. Biophys. Res. Comm. **56**, 112-118.
Hovenkamp-Obbema, R., and Stegwee, D. (1974). Z. Pflanzenphysiol. **73**, 430-438.
Hurlbert, R. E., and Bates, R. C. (1971). J. Protozool. **18**, 298-306.
Hutner, S. H., Bach, M. K., and Ross, G. I. M. (1956). J. Protozool. **3**, 101-112.
Hutner, S. H., Zahalsky, A. C., Aaronson, S., Baker, H., and Frank, O. (1966). Methods Cell Physiol. **2**, 217-228.
Hutner, S. H., Zahalsky, A. C., and Aaronson, S. (1968). In "The Biology of Euglena" (D. E. Buetow, ed.), Vol. 2, pp. 193-214. Academic Press, New York.
Ikushima, T. (1975). Annu. Rep. Res. React. Inst., Kyoto Univ. **8**, 83-85.
Ishida, M. R., Kikuchi, T., Matsubara, T., Tsushimoto, G., and Mizuma, N. (1973a). Annu. Rep. Res. React. Inst., Kyoto Univ. **6**, 24-28.
Ishida, M. R., Kikuchi, T., Matsubara, T., Tsushimoto, G., and Mizuma, N. (1973b). Annu. Rep. Res. React. Inst., Kyoto Univ. **6**, 78-81.
Johnston, P. L., and Carell, E. F. (1973). J. Cell Biol. **57**, 668-674.
Jones, C. R., and Cook, J. R. (1978). J. Cell. Physiol. **96**, 253-260.
Kahn, J. S. (1973). Arch. Biochem. Biophys. **159**, 646-650.
Kahn, J. S. (1974). Arch. Biochem. Biophys. **164**, 266-274.
Kahn, J. S., and McConnell, R. T. (1977). J. Bioenerg. Biomembr. **9**, 363-372.
Kahn, V., and Blum, J. J. (1965). J. Biol. Chem. **240**, 4435-4443.
Kempner, E. S., and Marr, A. G. (1970). J. Bacteriol. **101**, 561-567.
Kempner, E. S., and Miller, J. H. (1965a). Biochim. Biophys. Acta **104**, 11-17.
Kempner, E. S., and Miller, J. H. (1965b). Biochim. Biophys. Acta **104**, 18-24.
Kempner, E. S., and Miller, J. H. (1968). Exp. Cell Res. **51**, 141-149.
Kempner, E. S., and Miller, J. H. (1972a). J. Protozool. **19**, 343-346.
Kempner, E. S., and Miller, J. H. (1972b). J. Protozool. **19**, 678-681.
Kirk, J. T. O., and Keylock, M. J. (1967). Biochem. Biophys. Res. Commun. **28**, 927-931.
Knezek, B. D., and Maier, R. H. (1971). Soil Sci. Plant Anal. **2**, 37-44.
Konopa, J., Chotkowska, E., Koldej, K., Matuszkiewicz, A., Pawlak, J. W., and Woynarowski, J. M. (1976). Mater. Med. Pol. (Engl. Ed.) **8**, 258-265.
Krauss, G.-J., Maneva, L., Golovinsky, E., and Reinbothe, H. (1977). C. R. Acad. Bulg. Sci. **30**, 563-566.
Lafarge-Frayssinet, C., Bertaux, O., Valencia, R., and Frayssinet, C. (1978). Biochim. Biophys. Acta **539**, 435-444.
Ledoigt, G., and Calvayrac, R. (1974). Planta **121**, 181-191.
Leedale, G. F., and Buetow, D. E. (1970). Cytobiologie **1**, 195-202.
Leedale, G. F., and Buetow, D. E. (1976). J. Microsc. Biol. Cell. **25**, 149-154.
Levedahl, B. H. (1968). In "The Biology of Euglena" (D. E. Buetow, ed.), Vol. 2, pp. 85-96. Academic Press, New York.
Liedtke, M. P., and Ohmann, E. (1969). Eur. J. Biochem. **10**, 539-548.
Lonergan, T. A., and Sargent, M. L. (1978). Physiol. Plant. **43**, 55-61.
Lor, K.-L., and Cossins, E. A. (1978). Phytochemistry **17**, 659-665.
Lord, J. M., and Merrett, M. J. (1971). Biochem. J. **124**, 275-281.
Lord, J. M., Armitage, T. L., and Merrett, M. J. (1975). Plant Physiol. **56**, 600-604.
Lucchini, G. (1971). Biochim. Biophys. Acta **242**, 365-370.

Lux, H., and Petzold, U. (1976). *Biochem. Physiol. Pflanz.* **170**, 397-404.
Lyman, H. (1967). *J. Cell Biol.* **35**, 726-730.
Lyman, H., and Siegelman, H. W. (1967). *J. Protozool* **14**, 297-299.
Lyman, H., Jupp, A. S., and Larrinua, I. (1975). *Plant Physiol.* **55**, 390-392.
Marcenko, E. (1974). *Cytobiologie* **9**, 280-289.
Matson, R. S., and Kimura, T. (1976). *Biochim. Biophys. Acta* **442**, 76-87.
Matson, R. S., Mustoe, G. E., and Chang, S. B. (1972). *Environ. Sci. Technol.* **6**, 158-160.
Mikolajaczyk, E., and Diehn, B. (1978). *J. Protozool* **25**, 461-470.
Miller, S., and Diehn, B. (1978). *Science* **200**, 548-549.
Mills, W. L. (1976). *J. Environ. Sci. Health, Part A* **A11**, 567-572.
Milner, L., and Weissbach, H. (1969). *Arch. Biochem. Biophys.* **132**, 170-174.
Morimoto, H., and James, T. W. (1969a). *Exp. Cell Res.* **58**, 55-61.
Morimoto, H., and James, T. W. (1969b). *Exp. Cell Res.* **58**, 195-200.
Morris, G. J., and Canning, C. E. (1978). *J. Gen. Microbiol.* **108**, 27-31.
Morsoli, R., and Begin-Heick, N. (1973). *Can. J. Biochem.* **51**, 1402-1411.
Mosser, J. L., Fisher, N. S., Teng, T.-C., and Wurster, C. F. (1972). *Science* **175**, 191-192.
Murray, D. R., Giovanelli, J., and Smillie, R. M. (1970). *J. Protozool.* **17**, 99-104.
Murray, D. R., Giovanelli, J., and Smillie, R. M. (1971). *Aust. J. Biol. Sci.* **24**, 23-33.
Nakano, Y., Okamoto, K., Toda, S., and Fuwa, K. (1978). *Agric. Biol. Chem.* **42**, 901-907.
Nakano, Y., Sudate, Y., and Kitaoka, S. (1979). *Agric. Biol. Chem.* **43**, 223-229.
Nass, M. M. K., and Ben-Shaul, Y. (1973). *J. Cell Sci.* **13**, 567-590.
Neumann, D., and Parthier, B. (1973). *Exp. Cell Res.* **81**, 255-268.
Nichols, K. M., and Rikmenspoel, R. (1978). *Exp. Cell Res.* **116**, 333-340.
Nissen, P., and Eldjarn, L. (1969). *Physiol. Plant.* **22**, 364-370.
O'Donnell-Alvelda, E. (1968). *Cytologia* **32**, 568-581.
Ogli, K. (1977). *Med. J. Osaka Univ.* **28**, 15-21.
Ohmann, E. (1969). *Arch. Mikrobiol.* **67**, 273-292.
Osafune, T., Mihara, S., Hase, E., and Ohkuro, I. (1975). *Plant Cell Physiol.* **16**, 313-326.
Owens, I., and Blum, J. J. (1967). *J. Biol. Chem.* **242**, 2893-2902.
Owens, I., and Blum, J. J. (1969). *J. Protozool.* **16**, 211-215.
Pakhomova, M. V., Shanina, N. A., and Zaitseva, G. N. (1970). *Mikrobiologiya* **39**, 953-957.
Parenti, F., Dell'Aquila, A., and Parenti-Rosina, R. (1971). *Exp. Cell Res.* **65**, 117-122.
Parenti, F., DiPierro, S., and Perrone, C. (1972). *J. Protozool.* **19**, 524-527.
Peak, J. G., and Peak, M. J. (1976). *J. Protozool.* **23**, 165-167.
Peak, J. G., and Peak, M. J. (1977). *J. Protozool.* **24**, 441-444.
Piccinni, E., and Coppellotti, O. (1977). *Comp. Biochem. Physiol. B.* **57B**, 281-284.
Pienkos, P., Walfield, A., and Hershberger, C. L. (1974). *Arch. Biochem. Biophys.* **165**, 548-553.
Pigretti, M. M. (1973). *Rev. Latinoam. Microbiol.* **15**, 99-106.
Pohl, P., and Wagner, H. (1972). *Z. Naturforsch., B: Anorg. Chem., Org. Chem., Biochem., Biophys., Biol.* **27B**, 53-61.
Poorman, A. E. (1973). *Bull. Environ. Contam. Toxicol.* **10**, 25-28.
Prask, J. A., and Plocke, D. J. (1971). *Plant Physiol.* **48**, 150-155.
Price, C. A., and Vallee, B. L. (1962). *Plant Physiol.* **37**, 428-433.
Raison, J. K., and Smillie, R. M. (1969). *Biochim. Biophys. Acta* **180**, 500-508.
Reitz, R. C., and Moore, G. S. (1972). *Lipids* **7**, 217-219.
Richard, F., and Nigon, V. (1973). *Biochim. Biophys. Acta* **313**, 130-149.
Richards, O. C., Ryan, R. S., and Manning, J. E. (1971). *Biochim. Biophys. Acta* **238**, 190-201.
Ruthven, J. A., and Cairns, J., Jr. (1973). *J. Protozool.* **20**, 127-135.
Saxena, P. N., Amila, D. V., and Ahmad, M. R. (1979). *Indian J. Exp. Biol.* **17**, 223-224.
Schmidt, G. W., and Lyman, H. (1976). *Bot. Monogr. (Oxford)* **12**, 257-299.

252 E. S. Kempner

Schwelitz, F. D., Cisneros, P. L., Jagielo, J. A., Comer, J. L., and Butterfield, K. A. (1978a). *J. Protozool.* **25**, 257-261.

Schwelitz, F. D., Cisneros, P. L., and Jagielo, J. A. (1978b). *J. Protozool.* **25**, 398-403.

Scott, N. S., Shah, V. C., and Smillie, R. M. (1968). *J. Cell Biol.* **38**, 151-157.

Sharpless, T. K., and Butow, R. A. (1970a). *J. Biol. Chem.* **245**, 50-57.

Sharpless, T. K., and Butow, R. A. (1970b). *J. Biol. Chem.* **245**, 58-70.

Shehata, T. E., and Kempner, E. S. (1977). *Appl. Environ. Microbiol.* **33**, 874-877.

Shehata, T. E., and Kempner, E. S. (1978). *J. Bacteriol.* **133**, 396-398.

Shehata, T. E., and Kempner, E. S. (1979). *J. Protozool.* **26**, 626-630.

Silverman, H., and Hikida, R. S. (1976). *Protoplasma* **87**, 237-252.

Simeray, J., and Delcourt, A. (1979). *C. R. Seances Soc. Biol. Ses. Fil.* **173**, 33-35.

Simeray, J., Delcourt, A., and Mestre, J. C. (1977). *C. R. Seances Soc. Biol. Ses. Fil.* **171**, 901-906.

Smarda, J., Ebringer, L., and Mach, J. (1975). *J. Gen. Microbiol.* **86**, 363-366.

Spare, W., Lesiewicz, J. L., and Herson, D. S. (1978). *Arch. Microbiol.* **118**, 289-292.

Speiss, E., and Richter, G. (1970). *Arch. Mikrobiol.* **75**, 37-58.

Speiss, E., and Richter, G. (1971). *Arch. Mikrobiol.* **78**, 118-127.

Stewart, P. R., and Gregory, P. (1969). *Microbios* **3**, 253-266.

Sukhareva-Nemakova, N. N., Kalenik, N. M., and Silaev, A. B. (1975). *Izv. Akad. Nauk SSSR, Ser. Biol.* **3**, 362-370.

Theiss-Seuberling, H.-B. (1973). *Arch. Mikrobiol.* **92**, 331-344.

Tokunaga, M., Nakano, Y., and Kitaoka, S. (1976). *Biochim. Biophys. Acta* **429**, 55-62.

Vaisberg, A. J., and Schiff, J. A. (1976). *Plant Physiol.* **57**, 260-269.

Valencia, R. (1974). *J. Physiol. (Paris)* **69**, 5A-76A.

Valentine, J. P., and Bingham, S. W. (1976). *Can. J. Bot.* **54**, 2100-2107.

Vannini, G. L., Fasulo, M. P., and Bruni, A. (1977). *Z. Pflanzenphysiol.* **84**, 183-187.

Vannini, G. L., Fasulo, M. P., Bruni, A., and Dall'Olio, G. (1978). *Protoplasma* **96**, 335-349.

Van Pel, B., Bronchart, R., Kebers, F., and Cocito, C. (1973). *Exp. Cell Res.* **78**, 103-110.

Van Pel, B., and Cocito, C. (1973). *Exp. Cell Res.* **78**, 111-117.

Votta, J. J., Jahn, T. L., and Levedahl, B. H. (1971). *J. Protozool.* **18**, 166-170.

Wilson, B. W., and Levedahl, B. H. (1968). *In* "The Biology of *Euglena*" (D. E. Buetow, ed.), Vol. 1, pp. 315-332. Academic Press, New York.

Wodzinski, R. S., Labeda, D. P., and Alexander, M. (1977). *J. Air Pollut. Control Assoc.* **27**, 891-893.

Wolfovitch, R., and Perl, M. (1972). *Biochem. J.* **130**, 819-823.

Woodward, J., and Merrett, M. J. (1975). *Eur. J. Biochem.* **55**, 555-559.

Yokota, A., Nakano, Y., and Kitaoka, S. (1978a). *Agric. Biol. Chem.* **42**, 115-120.

Yokota, A., Nakano, Y., and Kitaoka, S. (1978b). *Agric. Biol. Chem.* **42**, 121-129.

Yongue, W. H., Jr., Berrent, B. L., and Cairns, J., Jr. (1979). *J. Protozool.* **26**, 122-125.

Zielinski, R. E., and Price, C. A. (1978). *Plant Physiol.* **61**, 624-625.

CHAPTER 7

CHLOROPLAST MOLECULAR STRUCTURE WITH PARTICULAR REFERENCE TO THYLAKOIDS AND ENVELOPES

Guy Dubertret and Marcelle Lefort-Tran

I. Introduction

Research into the biology of chloroplasts and mitochondria started with identification of their structures through the use of adapted techniques. Conventional thin-sectioning methods and, later, freeze–fracture techniques revealed the structural organization of the organelle membranes. As soon as the necessary biochemical and biophysical knowledge developed, the processes of photosynthesis and respiration began to be better understood and appeared to be complementary to and indissociable from subcellular structures, thus emphasizing structure–function relationships at the cellular level.

THE BIOLOGY OF *EUGLENA*, VOL. 3

Concerning the chloroplasts, the questions raised are: What do the mature chloroplasts do for the cell and how do they do it (Bogorad, 1975)? The specialized function of chloroplasts in providing and organizing the machinery for primary conversion of light energy into useful chemical energy is associated with thykaloid membranes (Haan *et. al.,* 1973; Heber and Krause, 1971). By means of two photochemical reactions in Photosystem I and Photosystem II, light quanta absorbed by light-harvesting chlorophylls allow charge separations in reaction centers. Electrons are then extracted from water and transported to $NADP^+$, and this light-induced electron flow is coupled to ATP synthesis. Finally, energy (ATP) and reducing power ($NADPH + H^+$) generated by these membrane-associated "light" reactions allow the "dark" reactions of CO_2 fixation and reduction as carbohydrates in the nonstructured soluble phase of the stroma.

These light reactions could occur *a priori* in free multimolecular complexes, including light-harvesting systems, reaction centers, and components of an electron transport chain. The photosynthetic membranes should then be considered as a passive element in energy transduction mechanisms, whose sole function would be to act as an inert support for such complexes. However, the discovery that such complexes work in a cooperative manner, and the requirements set by the chemiosmotic hypothesis of Mitchell (1966), tend to give to photosynthetic membranes (discovered by electron microscopy) a dynamic function (a) in ensuring a relative structural organization of these complexes, thus allowing cooperation, (b) in ensuring a polarized orientation of these complexes suitable for proton translocation across the membrane, and (c) in delimiting a compartment in which translocated protons can accumulate.

The problem has been to superimpose functionally defined structural requirements with morphological pictures of the ultrastructure of chloroplast and photosynthetic membranes. The origin of the development of the "topochemical" analyses (Lavorel, 1976) corresponds to the discovery of the photosynthetic unit by Emerson and Arnold (1932). After this date, the relationship between structure and function of chloroplast membranes has been strengthened with the purpose of explaining the second by the first.

Besides the thylakoids, a second category of chloroplastic membranes is the envelope, the importance of which appeared very recently with the development of research on the interaction between the cell compartments. Photosynthetic membranes, i.e., thylakoids (associated with light-dependent reactions) and the soluble stroma (associated with the dark biochemical reactions) are enclosed by a system of limiting membranes: two in most higher plants and three in *Euglena* chloroplasts. Unfortunately, the envelope has received little attention despite its importance in the functional and structural integrity of the chloroplast.

This chapter discusses the structure of *Euglena* chloroplast membranes, including envelopes and thylakoids, in relation to their functions and to those of higher plants.

Most of the experimental studies have been concerned with two strains of *Euglena gracilis: E. gracilis* strain Z Pringsheim and *E. gracilis* var. *bacillaris* Pringsheim. Some mutants from Brandeis University (Prof. J. A. Schiff) have been used as *"fixed pictures"* to study the course of chloroplast development, for example, W_3BUL, but any intensive research correlating function, biochemistry, and structure (as in *Chlamydomonas*, for example, Ohad *et al.*, 1967; Goodenough and Staehelin, 1971; Bogorad, 1975; and many others) has been lacking. In fact, even though *Euglena* is one of the cells most often used in biochemical and physiological studies, its photosynthetic characteristics were, until now, neglected. With *Euglena* it is possible to use a large range of nutritional conditions that allow vegetative growth without the induction of cell differentiation (cyst, gametes, etc.) and to obtain very homogeneous cell populations. On the other hand, it is possible to induce chloroplast differentiation in the absence of cell growth and division when *Euglena* is maintained on "resting medium" (Stern *et al.*, 1964). Lastly, but as importantly, mild methods are available for the isolation of its chloroplasts in an intact state. These interesting biological features have been recently utilized in our laboratory along with some advanced technology to study photosynthetic membranes while bearing in mind the necessary relationship between structure and function.

II. *Euglena* Chloroplast Envelope

A. GENERAL REMARKS

Being the bridge and barrier between the cytoplasm and the chloroplast (Heber and Walker, 1979), the envelope is the most permanent structure of chloroplasts. Even when the thylakoids are not differentiated (as in etioplasts in the dark or in chloroplast mutants induced by genetic defects), the envelope is present.

In higher plants and green and red algae, chloroplasts are surrounded by two membranes that constitute the chloroplastic envelope (Figs. 1a,b). In other organisms, except dinoflagellates and euglenoids, which have an envelope consisting of three membranes (Fig. 1c), the chloroplasts are enclosed by four membranes. In this last case, the two external membranes are connected with the cellular endoplasmic reticulum and are called chloroplast ER (Dodge, 1974; Gibbs, 1978) (Fig. 1d).

The passive function of the chloroplast envelope is to separate from the cytoplasm the soluble enzymes of the Calvin–Benson cycle. Active functions are concerned with (a) the free transport of low-molecular-weight metabolites; (b) the selective permeability (for a few species) of anions in relation with three specific transporters (Heldt, 1969; Strotmann and Berger, 1969), probably mainly associated with the inner membrane of the envelope (Heldt and Rappley, 1970; Heldt and Sauer, 1971); (c) the synthesis of chloroplastic galactolipids (Joyard and Douce, 1976); (d) the origin of the photosynthetic membranes from the inner

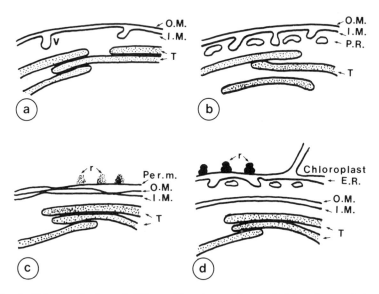

Fig. 1. Main organizations of different chloroplast envelopes of higher plants and algae are shown. (a) In C₃ plants the chloroplast envelope is composed of two membranes: outer membrane (O.M.) and inner membrane (I.M.) which in early stages forms vesicles (V); and thylakoids (T). (b) Mesophyll and bundle sheet chloroplasts of C₄ plants have an I.M. of the envelope that infolds to form a reticulum of anastomosing tubules called the peripheral reticulum (P.R.). (c) In *Euglena* the chloroplast envelope has a third external membrane, the perichloroplastic membrane (Per. m.), in close connection with the other two membranes (O.M. and I.M.); cytoribosomes (r). (d) In most groups of algae the double-membrane envelope of the chloroplast is surrounded by cisternae of endoplasmic reticulum (E.R.) that bears cytoribosomes (r). (*Ochromonas danica* is taken as an example, after Gibbs, 1978.) Vesicles and tubules are present between the chloroplast E.R. and the O.M. of the envelope.

sheet of the envelope (Lefort, 1959; Mühlethaler and Frey-Wyssling, 1959; Menke, 1962, Virgin *et al.*, 1963); and (e) the identification and the transfer of chloroplast proteins synthesized in the cytoplasm (Ellis, 1977; Dobberstein *et al.*, 1977; Priestley, 1977; Chua and Schmidt, 1978,1979) at specialized areas in the outer and inner membranes of the envelope.

In comparison with the chloroplast envelope of higher plants, we will discuss the structure and functions of the different membranes of the chloroplast envelope of *Euglena*, and we will be concerned especially with the ontogenic significance of the third perichloroplastic membrane and the relations between the three layers (Lefort-Tran *et al.*, 1980).

B. ORGANIZATION AND PRESUMED FUNCTIONS

The triple membranous structure of the chloroplast envelope of *Euglena* was observed in electron micrographs of sectioned cells (Gibbs, 1962,1970; Leedale,

1967,1968, Dodge, 1974; Taylor, 1976). An interpretation of this puzzling organization has been the subject of debate. The external sheet or perichloroplastic membrane (Fig. 2), which very closely surrounds the young proplast, appears often in juxtaposition with the intermediary membrane. Contacts are observed that become more numerous in fully developed chloroplasts (Figs. 3

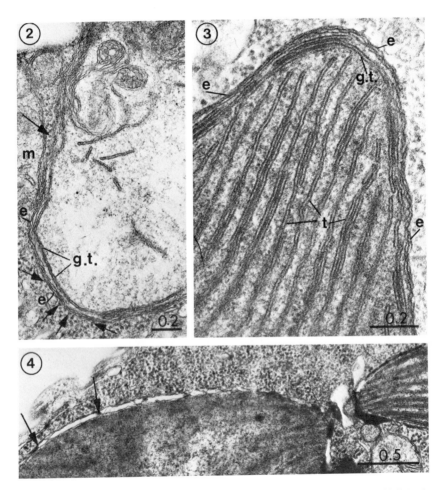

Fig. 2. Euglena proplast with a triple layered envelope (e) and the "girdle thylakoids" (g.t.). Cytoribosomes are seen close to the envelope (arrows). (m) Mitochondria. × 42,000.

Fig. 3. Greening chloroplast under intermittent light. The three membranes of the envelope (e) are imbricated together. Independent thylakoids (t) and "girdle thylakoids" (g.t.) are shown. × 67,500.

Fig. 4. Fully developed chloroplasts in which the three layers of the envelope appear to be loosely attached together and a dense material appears to be infiltrated between them (arrows). × 33,700.

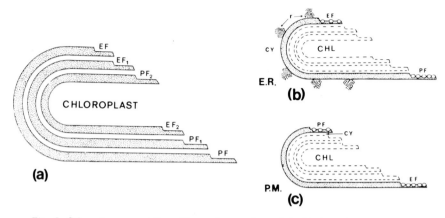

Fig. 5. Schematic representation of freeze-fracture faces of the *Euglena* chloroplast (CHL) envelope according to the nomenclature of Branton *et al.* (1975). (a) The inner membrane is depicted
with the convex protoplasmic face PF_2 and the exoplasmic concave face EF_2; the intermediary and
perichloroplastic membranes with concave protoplasmic faces PF_1 and PF and convex exoplasmic
faces EF_1 and EF. (b) Interpretation of the perichloroplastic membrane with an E.R. of ergastoplasmic origin. The PF face with the higher density of particles is close to the cytoplasm (CY) and
eventually cytoribosomes (r) could be associated with it. (c) Interpretation of the perichloroplastic
membrane (P.M.) as a plasma membrane: No ribosomes are attached and the convex fracture face
would be a PF with few particles.

and 4). By the freeze-fracture technique, instead of four fracture faces as with
"normal" chloroplasts, six complementary fracture faces are obtained (Fig. 5).
These faces are labeled according to Branton's nomenclature (Branton *et al.*,
1975), in which the face of the inner membrane close to the stroma is considered
as a protoplasmic face (labeled PF_2) and the complementary concave face as EF_2.
The two outer membranes of the *Euglena* envelope have the same polarity: The
convex membrane faces are exoplasmic EF_1 and EF; the concave faces are
protoplasmic PF_1 and PF.

1. The Inner Membrane

a. Structure. The inner membrane displays a significantly high density of
particles (Σ EF_2 + PF_2: Table I) that are unequally distributed on the two
complementary fracture faces. Their size does not exceed 75Å. Whatever the
treatment, that is, (1) *in situ* chloroplasts with rapid cryofixation (Gulik-
Krywicki and Costello, 1977; Plattner *et al.*, 1973) or (2) conventional technique
on isolated chloroplasts, the EF_2 fracture face has fewer particles than the PF_2
(Figs. 6 and 7). The partition coefficient $K_p = PF/EF$ keeps a positive value that
is close to the partition coefficient found in spinach chloroplast-envelope inner
membrane, 1820/979 = 2 (Sprey and Laetsch, 1976b). This result is opposite

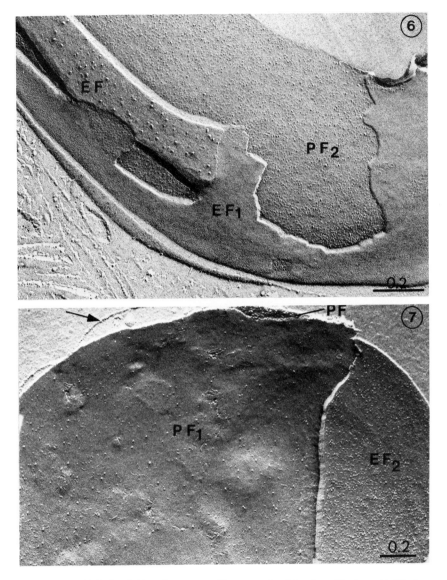

Figs. 6. and 7. Isolated chloroplasts prepared by conventional freezing with glycerol infiltration without glutaraldehyde prefixation.

Fig. 6. Fracture showing successively from the outer surface of the chloroplast: (a) EF of the perichloroplastic membrane with few large particles; (b) smooth EF_1 of the intermediary membrane; (c) PF_2 of the inner membrane of the envelope. × 69,300.

Fig. 7. Complementary concave fractures. PF appears as a small surface and under a transverse section (arrow). On the PF_1 face of the intermediary membrane a few particles are observed; the EF_2 surface of the exoplasmic face of the inner membrane has many particles. × 48,600.

Table I

PARTICLE DENSITY ON FRACTURE FACES OF ENVELOPE CHLOROPLAST MEMBRANES

Genus	Perichloroplastic membrane			Intermediary membrane (particles/μm^2 ± SE)			Inner membrane		
	EF	PF	Σ EF + PF	EF_1	PF_1	Σ EF_1 + PF_1	EF_2	PF_2	Σ EF_2 + PF_2
$Euglena$[a]									
In situ chloroplasts (rapid cryofixation)	455 ± 40	1480 ± 340	1935 ± 380	15	136 ± 36	151 ± 36	229 ± 90	2246 ± 89	2475 ± 180
Isolated chloroplasts (conventional technique)[b]	410	1818 ± 110	2228 ± 110	106	117 ± 43	223 ± 43	1034 ± 270	1718 ± 118	2752 ± 388
$Euglena$[c]	?	?	?	—[g]	—[g]	—[g]	1675	1160	2835
$Bangia$[d]				170	200	370	140	240	380
$Tetraseluris$[e]				1041	2023	3064	—		
$Spinacia$[f]				133 ± 23	150 ± 16	283 ± 39	979 ± 93	1820 ± 167	2800 ± 260
$Glaucocystis$ $nostochinearum$[h]				273 ± 13	446 ± 42	774	1130	2895	4025

[a] M. Lefort-Tran et al. (1980).
[b] Glycerol infiltration without glutaraldehyde prefixation.
[c] Miller and Staehelin (1973).
[d] Bisalputra and Bailey (1973).
[e] Melkonian and Robenek (1979).
[f] Sprey and Laestch (1976).
[g] Fracture faces observed, not recorded.
[h] Lefort-Tran (unpublished results).

the value previously given by Miller and Staehelin (1973) in which the polarity of the inner membrane appears reversed (Table I).

Because of the higher density and the smaller size of the particles, the inner membrane of the chloroplast envelope appears in freeze–fracture clearly different from the outer membrane in *Euglena*. This is also the case for other organisms studied so far.

b. Functions. A specific role in the transport of proteins synthesized in the cytoplasm has been postulated for this membrane (Strotmann and Berger, 1969; Heldt and Rappley, 1970; Chua and Schmidt, 1979). As if in correlation with the capacity of the inner membrane of spinach chloroplasts to synthesize galactolipids and carotenoids (Joyard and Douce, 1976), a dense material is present, interestingly, between the two internal sheets of the *Euglena* envelope (Fig. 4).

The inner envelope membrane and the internal membrane system in greening tissues are frequently continuous with each other. Consequently, it has been suggested that thylakoid membranes are formed by continued invagination from the inner sheet of the plastid envelope (Mühlethaler and Frey-Wyssling, 1959; von Wettstein, 1959; Menke, 1960,1962; Kirk and Tilney-Basset, 1967).

During illumination the induced alkalinization of the stroma does not modify the pH of the suspension medium of chloroplasts (Heber and Krause, 1971). Therefore, it seems that no communication between the perichloroplastic space and the stroma exists. If this were true, it would imply that there is no continuity between the most internal sheet of the envelope and the thylakoids (Sprey and Laetsch, 1978; Douce and Joyard, 1978,1979). An interesting variation of the capacity of invagination of the inner membrane is the well-known peripheral reticulum in chloroplasts of C_4 plants (Shummay and Weier, 1967; Laetsch, 1968; Rosado-Alberio *et al.*, 1968) (Fig. 1b). In this case, a study of the size and the particle distribution of the peripheral reticulum (P.R.) corresponds to a study of the inner membrane of the envelope (Sprey and Laetsch, 1968). Nevertheless, no relation has been observed between the thylakoids and the anastomozing tubules of the P.R. (Laetsch and Rice, 1969; Laetsch, 1971). The functional role of the P.R. is a matter of speculation.

In *Euglena* proplastids, at least two layers of thylakoids follow the outlines of the triple-layered envelope (Fig. 2). These form "girdle thylakoids" that appear connected with the vesicules of the prolamellar body when such a structure exists in dark-grown cells (Schiff, 1970; Salvador *et al.*, 1971; also see this volume, Chapter 8). During greening in *Euglena*, direct connections between inner membrane invaginations and the photosynthetic membranes have never been found before in electron microscopic sections (see also Fig. 3). Nevertheless, Fig. 8 shows a cryofracture of an isolated young chloroplast greened under intermittent light and suggests that such a connection does exist. The inner membrane of the envelope is fractured along the EF_2 face. Two points of contact (arrows, Fig. 8)

Fig. 8. Chloroplast of *Euglena* greening under intermittent light (400 flashes, see text). Arrows indicate morphological relations between the first thylakoids (t) and the inner sheet (EF₂) of the envelope (compare with Figs. 31 and 33a); (s) stroma. × 34,700.

appear between the thylakoids and the inner membrane. This suggests a degree of continuity between the two types of membranes in the early stages of greening. Because such examples are scarce, however, the role of the inner sheet of the plastid envelope in thylakoid biogenesis in chloroplasts, and especially in *Euglena,* must remain indeterminate.

2. The Intermediary Membrane

a. Structure. Convex (Fig. 6) or concave (Figs. 7, 9, and 10) fracture faces of an intermediary membrane appeared with very smooth surfaces as previously

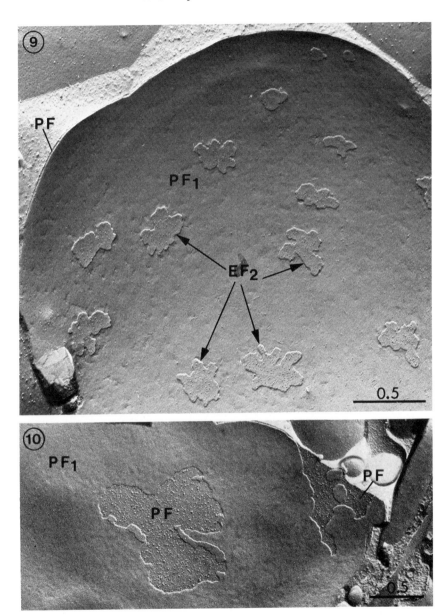

Figs. 9 and 10. In situ chloroplast prepared by spray–freezing without pretreatment. On the concave fracture face of the intermediary membrane PF_1 are attached isolated fragments of the particulated EF_2 fracture face. The perichloroplastic membrane is seen only as a thin edge PF. The surface of PF is visible in Fig. 10. ×40,500; ×30,000.

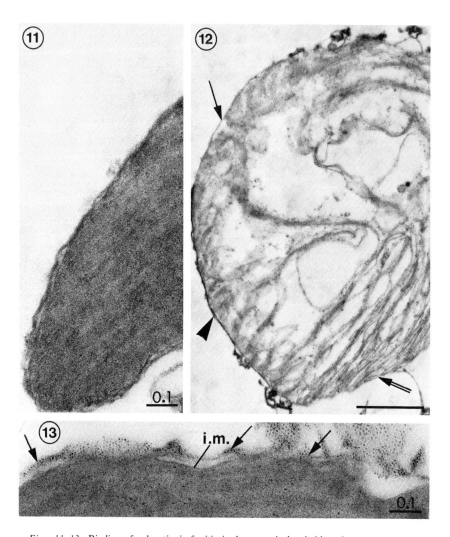

Figs. 11–13. Binding of polycationic ferritin is shown on isolated chloroplasts.

Fig. 11. Part of an isolated intact chloroplast of *Euglena*. The three membranes of the envelope are present. No binding sites are observed with polycationic ferritin. ×83,200.

Fig. 12. Isolated, partially disrupted chloroplast treated with cationized ferritin. The ligand is not observed bound to the inner membrane of the chloroplast envelope (single arrow) nor to the thylakoids (double-tailed arrow), but it is present where the envelope is better preserved (arrowhead). ×18,500.

Fig. 13. Details of binding sites of cationized ferritin on the intermediary membrane of the isolated *Euglena* chloroplast (arrows). The inner membrane (i.m.) is free of the ligand. ×83,100.

reported by Holt and Stern (1970) but were not clearly observed in the study of Miller and Staehelin (1973). Compared to the internal membrane, the intermediary membrane is very poor in particles not exceeding 151 ± 36 particles/ μm^2. This value is close to that given for the external envelope membrane in chloroplasts of *Spinacia* by Sprey and Laetsch (1976b). This result appears to be a strong argument for correlating the most external membrane in higher plant chloroplasts with the intermediary sheet in *Euglena* chloroplasts. The protein nature of the intramembrane particles is generally accepted. The low-particle density of the intermediary membrane may be correlated with a high lipid/protein ratio as demonstrated in spinach chloroplast envelope (Douce and Joyard, 1978). In higher plants and green algae, the external sheet is mainly composed of galactodiglycerides and sulfolipids. These appear to be uniformly distributed as shown by means of immunocytological studies with specific antibodies against the envelope lipids (Billecocq, 1975). As demonstrated by polycationic ferritin binding (Hackenbrock and Miller, 1975) on isolated *Euglena* chloroplasts (Lefort-Tran *et al.,* unpublished results), the equivalent intermediary membrane of the chloroplast envelope in *Euglena* carries strong negative charges (Fig. 13). The binding is not observed on intact chloroplasts (Fig. 11) but never occurs on thylakoids, even if the chloroplasts are open or on the inner membrane of the envelope when the outer ones are broken (Fig. 12). The strong electronegativity of the intermediary membrane of the *Euglena* chloroplast envelope is comparable to that obtained with outer membrane in spinach chloroplast by ferricationic cytochrome *c* adsorption (Neuburger *et al.,* 1977). This provides evidence that the polar and ionic heads of the lipid molecules, together with all the charged side chains of the amphipathic globular proteins, are exposed at the external surfaces of the envelope membranes. This membrane sheet is considered to be nonspecifically permeable to low-molecular-weight metabolites.

   b. Functions. Connections between the inner and the outer membranes of the envelope have not been excluded (Douce and Joyard, 1979; Priestley, 1977). In the *Euglena* chloroplast, a degree of imbrication is observed in thin sections between these two membranes (Figs. 2 and 3). It was shown that when the inner membrane is cleaved by the freeze–fracture procedure, isolated fragments are left (Fig. 9), as if adhesion domains within the intermediary membrane induce changes in the fracture properties of the membrane. With the rapid cryofixation technique, on the two complementary faces of the intermediary membrane, organized areas of particles are observed (Figs. 14 and 15) whose sizes are less than 60 Å. Such preferential distribution of particles should correlate either with a specialized function of the associated membrane areas or with the transfer of protein through the envelope at the contact sites between the two inner and outer membranes, as postulated by Chua and Schmidt (1979).

   Studies on the synthesis and the transfer of the small subunit of the ribulose-

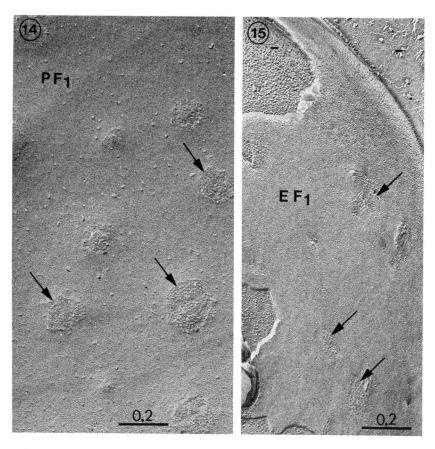

Figs. 14 and 15. Complementary fracture faces of the intermediary membrane concave PF_1 and convex EF_1, respectively. On the PF_1, a few particles are dispersed on the smooth background and well organized areas of pit particles (arrows) are observed. On the EF_1, corresponding patches of particles are found (arrows). $\times 67,500$; $\times 69,300$.

1,5-bisphosphate carboxylase (RubPcase) relate to the role played by the envelope in the transport into the chloroplast of polypeptides synthesized in the cytoplasm. It is well known that small subunits (SSU) of RubPcase are synthesized on 80-S cytoplasmic ribosomes (Kawashima and Wildman, 1971, 1972; Gray and Kekwick, 1974a,b; Gray *et al.*, 1978; Roy *et al.*, 1976), whereas the large subunits (LSU) are made within the chloroplast (Chan and Wildman, 1972; Blair and Ellis, 1973; Bottomley *et al.*, 1974; Cahen *et al.*, 1977; Malnoë *et al.*, 1979). In order to be assembled with the LSU in forming the holoenzyme molecules, the SSU must cross the envelope membranes at a rate of 10^4 molecules/plastid/hour (Bradbeer and Borner, 1978; Bradbeer *et al.*, 1979). In the case of spinach chloroplast envelopes, protein electrophoretic patterns revealed a major polypeptide that had the same mobility as the SSU of the RubPcase ($\sim$13K daltons) (Pineau and Douce, 1974). Through immunoelectrophoresis with immune serum against the enzyme RubPcase, the 13K-dalton polypeptide was characterized as the SSU. Since the electronegativity of the outer membrane envelope excludes the illegitimate adsorption of the molecule to the envelope, the biological significance of the SSU of RubPcase being found among the polypeptides of the envelope must be considered (Pineau et al., 1979).

The association of the SSU with the envelope may be a reflection of the transport of the peptide under the mature form (12K daltons). This would imply that the additional sequence which contains the necessary information for specific binding to a receptor of the chloroplast envelope (Chua and Schmidt, 1978) is cleaved early in the transfer, or that the life of the precursor is very short as compared to that of the mature SSU (Pineau et al., 1979). It is interesting to consider here the electrophoretic results obtained by Vasconcelos (1976) and Vasconcelos et al. (1976). Table II compares the polypeptides obtained by electrophoretic analyses of Euglena and spinach thylakoids, stroma, and envelope fractions.

Independently of some variations in molecular weights resulting from the different biological materials and methods used, it does appear that the two major polypeptides resolved from the soluble fraction were about 52K and 12K daltons, values which correspond to the LSU and the SSU of RubPcase. These components are also found in the polypeptide profile of the envelope but not in that of the thylakoid fraction. It seems reasonable to expect the Euglena envelope to have the same role in transfer of the SSU of RubPcase, as postulated for Spinacia

Table II

MOLECULAR WEIGHTS OF THE MAIN POLYPEPTIDES[a] IN THYLAKOIDS (T), SOLUBLE FRACTION (S), AND ENVELOPE MEMBRANE (E) OF Euglena AND SPINACH[b]

Chloroplast fraction	Euglena (Vasconcelos, 1976)	Spinach (Pineau and Douce, 1974)
T	42, 31. 90, 77, 60, 28, 25, 21, 14	54, 33, 26, 24. 70, 29, 15
S	52, 12 39, 31	52, 14
E	48 35, 31, 25, 12	27, 33, 52. 12, 12.5

[a] Molecular weights of polypeptides in kilodaltons; separated by SDS polyacrylamide gel electrophoresis.

[b] Major bands are underlined.

and *Avena* chloroplasts (Cobb and Wellburn, 1974, 1976). The intramembranous differentiations found in the intermediary membrane of the *Euglena* envelope (Figs. 14 and 15) as well as the specific sticking of patches of the inner membrane (Fig. 9) may be the first morphological evidences of the postulated mechanism for the transfer of polypeptides through the envelope (Chua and Schmidt, 1979).

3. The Perichloroplastic Membrane

The presence of three membranes* surrounding chloroplasts in all euglenoid species is generally accepted. However, the significance of this organization has been misinterpreted for a long time (Leedale, 1967,1968; Gibbs, 1970; Dodge, 1974; Kivik and Vesk, 1974; Taylor, 1976).

4. Structure and Significance

Gibbs proposed previously (1970) that the three membranes of the euglenoid chloroplast originated from a fusion of two of the four membranes of the chloroplast envelope, i.e., the chloroplast endoplasmic reticulum complex (ER) (Fig. 1c,d). It is now accepted that this is not the case in dinoflagellates. In *Euglena* the nature and the origin of the third outer membrane envelope have been much debated. Some authors (Bisalputra and Bailey, 1973; Bisalputra, 1974) have advanced the idea that the outer sheet is ergastoplasmic in nature and bears cytoplasmic ribosomes. According to Leedale (1968), each *Euglena* chloroplast is surrounded with an E.R. sheath that is associated with tubular extensions of the outer nuclear membrane. In a recent contribution Gibbs (1978) suggests the similarity of this supplementary membrane to the plasmalemma of an hypothetic symbiont. Though attractive, this hypothesis may be questioned.

It has been shown (Lefort-Tran *et al.*, 1980) that the molecular internal organization of the perichloroplastic membrane differs from the two other sheets by the density of intramembranous particles (Table I). According to the hypothesis that the membrane is equivalent to the plasmalemma of a symbiont, the cryofracture faces PF and EF would correspond, respectively, to the convex and the concave faces of the membrane (Fig. 5c) (see Branton *et al.*, 1975). The convex PF face should have a low particle density ($430/\mu m^2$), whereas the concave EF face should have a higher density ($1480/\mu m^2$), according to Gibb's hypothesis. Therefore, the K_p = PF/EF would be < 1, which would be very exceptional for a plasma membrane (Dempsey *et al.*, 1973; Lefort-Tran *et al.*, 1978).

On the contrary, if the perichloroplastic membrane is derived from the E.R. (Fig. 5b), and if the convex fracture face, poor in particles, is the EF face (Fig. 6) according to Branton's nomenclature, then the concave PF (Figs. 7 and 10) face should be enriched in particles, giving a partition coefficient K_p = (1480 $\pm$

---

*This topic is also discussed in Chapter 1 of this volume.

340)/(455 ± 40) = ~3.2. This value matches well with most of the known K_p's (Satir and Satir, 1974).

The polarity of this extrachloroplastic membrane is consistent with the vacuolar nature of the perialgal membrane around the *Chlorella* which is symbiotic to *Paramecium bursuria* and in which the *Chlorella* retains its cell wall and plasma membrane (Karakashian *et al.*, 1968; Vivier *et al.*, 1967). In this example, Meier *et al.* (1980) found, respectively, 3178 and 833 particles per μm^2 for the P concave and the E convex faces, values which give a partition coefficient K_p = 3.8. In this last case the identification of fracture faces is unambiguous because the perialgal vacuole is directly homologous to the phagocytic vacuole (Karakashian *et al.*, 1968).

Independently, cytoribosomes have been observed on thin sections of *Euglena* chloroplasts, and these cytoribosomes seem to be associated with the perichloroplastic membrane (Figs. 2 and 3). When chloroplasts are gently isolated, the third perichloroplastic sheet is often retained as observed on freeze–fracture micrographs (Figs. 6, 9, and 10). Particulated structures are stabilized on this third sheet. Their contrast and structure are comparable to those of cytoribosomes. In immunofluorescence studies, antibodies against ribosomal proteins of the large subunit give a positive reaction (Fig. 16a,b) (Lefort-Tran *et al.*, 1980). These latter cytological observations agree well with Freyssinet's findings

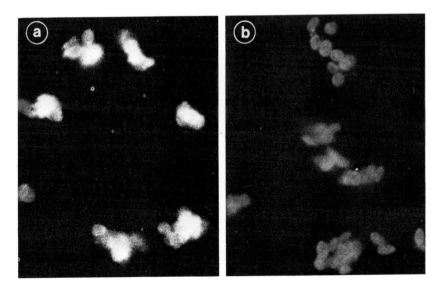

Fig. 16. Cytoimmunofluorescence on isolated *Euglena* chloroplast. (Sheep globulin and antiglobulin of rabbit, labeled with isothiocyanate of fluorescein were used.) (a) Immunserum against *Euglena* ribosomal 80 S proteins with a positive reaction. (b) Control without immune serum with a negative reaction. The weak light is due to the red fluorescence of chlorophyll. ×800.

(1977a,1978) indicating that, in isolated fractions of *Euglena* chloroplasts, cytoribosomes are still found even after numerous washings that eliminate free ribosomes. Deoxycholate and Triton X-100 did not eliminate the large subunit of cytoribosomes as shown by the observed ratio of RNA 25 S/20 S (G. Freyssinet, personal communication). Nevertheless, as in the work of Laulhere and Dorme (1977), it can not be decided whether the cytoribosomes are functionally attached or if they are contaminants.

Even so, and whether or not the perichloroplastic membrane is covered by cytoribosomes, which may or may not be functional in the synthesis of chloroplastic proteins, the ergastoplasmic nature of this membrane does not eliminate the endosymbiotic hypothesis as an explanation for the origin of chloroplasts.

According to the latter hypothesis, postulated long ago by Schimper (1885) and Mereschkowsky (1905), the inner membrane of the plastid envelope is derived from the plasma membrane of a blue-green algal ancestor. The polarity of the inner membrane as observed in freeze–fracture and the identical polar lipid composition of blue green algae and of algal or higher plant chloroplasts tend to favor the endosymbiotic theory. Nevertheless, in this hypothesis, the origin of the outer membrane of the plastid envelope is not explained. It has been assumed that this membrane arose from the endoplasmic reticulum or that it represents a boundary membrane of an endocytotic vacuole formed by a plastid-free proto-eukaryote (Schnepf, 1966). Joyard and Douce (1976) have pointed out in *Spinacia* that the chemical composition and enzymatic activities of this membrane are quite different from those of the microsomal fraction. On the one hand, it is not very surprising that in the course of evolution the properties of the initial endocytotic vacuole may have been modified or lost and on the other hand, in an actual endosymbiotic association as in *Glaucocystis nostochinearum* (Lefort, 1965), the original cell-wall-less blue-green alga remains surrounded by its own plasma membrane. The inner membrane of *Glaucocystis* has a high density of particles $\sim4025/\mu m^2$ (see Table I), which is similar to the inner sheet of the chloroplast envelope (I.M.). The perisymbiont membrane around the cyanelle has only ~700 particles/μm^2 and its concave fracture face PF_1 is the most particulated (Table I); therefore, the perisymbiont membrane may be comparable to the outer membrane (O.M.) of the chloroplasts.

The relation between the outer nuclear membrane and the third sheath surrounding *Euglena* chloroplasts, the similar polarity of the two outer chloroplast membranes, and the reverse polarity of the inner one are, in our opinion, strong arguments for an ergastoplasmic nature of the perichloroplastic membrane in *Euglena*.

C. CONCLUSIONS

Euglena chloroplasts have four compartments, i.e., the perichloroplastic space, the intermediary space of the envelope, the matrix of the chloroplast, and the intrathylakoid lumen. When compared to green algal and higher plant chloroplasts that have only three compartments, the four-compartment structure of the *Euglena* chloroplast raises the problem of the significance of this unique organization.

Despite the special nature of such an architecture, the two inner sheets of the *Euglena* chloroplast are identical to the general chloroplast model represented by an inner protein-rich membrane with specific properties for synthesis and transport of macromolecules. The outer membrane is lipid-rich, electronegatively charged, with nonspecific permeability only to low-molecular-weight substances.

If the general model holds for the *Euglena* chloroplast, then the recognition sites (Neville and Chang, 1978) for chloroplast protein transfer would be restricted to the contact areas between the inner and outer membranes that are formed by organized areas of pits and particles on the intermediary membrane in *Euglena*.

III. Thylakoids

The primary photosynthetic reactions, including the absorption of light quanta by light-harvesting chlorophylls, the charge separation and the generation of reducing power in reaction centers, and the electron transport from water to NADP have long been associated with photosynthetic membranes. The very low time constants of these processes (for a review, see Junge, 1977) imply that molecular components involved in these reactions are closely associated in multimolecular complexes whose relative organization would be ensured by photosynthetic membranes. Thornber *et al.* (1967) calculated that about 70–80% of chloroplast membrane proteins are associated with chlorophylls as chlorophyll–protein complexes I and II, thereby demonstrating that the majority of membrane proteins is involved in the organization of chlorophyll in light-harvesting antennae of system I and system II photosynthetic units. The preeminent place that these light-harvesting complexes should therefore occupy in the molecular architecture of thylakoids explains the extensive studies during the last 10 years on the structure, organization, and possible morphological expression of photosynthetic units in photosynthetic membranes. Most analyses have been performed in higher plants and green algae, and the results are compared in this chapter with the few data available for *Euglena*.

A. STRUCTURE OF PHOTOSYNTHETIC UNITS

1. *Definition of Photosynthetic Units*

Emerson and Arnold (1932) calculated that *Chlorella* cells exposed to repetitive, saturating short flashes evolve only one O_2 molecule per 2500 chlorophyll molecules, thereby demonstrating that most of the chlorophylls are not involved in the photochemical reaction. On the basis of this finding Gaffron and Wohl (1936) proposed the concept of a photosynthetic unit (PSU) formed by the close association of a light-harvesting antenna containing about 2500 chlorophylls and a reaction center where the photochemical reaction leading to O_2 evolution would occur. However, it was later shown that the evolution of one O_2 molecule requires eight photoreactions in two different photosystems, four in Photosystem II (PSII), and four in Photosystem I (PSI). This leads to a correction of the size of this overall photosynthetic unit, which would be formed by two reaction centers (PSI and PSII) associated to a light-harvesting antenna containing about 400–600 (2400 ÷ 4) chlorophyll molecules. This also leads to the concept that such overall units could be formed by two basic photosynthetic units (system I and system II), in which each reaction center would be associated with its own light-harvesting antenna.

These two definitions, which correspond to the overall ratio of the total number of chlorophyll molecules to the number of evolved O_2 molecules or some component of system I or II reaction centers, do not take into account the possible existence of unorganized chlorophylls and suppose that chlorophylls are equally distributed among the two photosystems.

The functional size of the photosynthetic unit of system I or system II then would correspond to the amount of light-harvesting chlorophyll functionally associated with system I or system II reaction centers. This size can be estimated in a relative manner for system II units from kinetic parameters of fluorescence induction in the presence of 3-(3,4-dichlorophenyl) 1,1-dimethylurea (DCMU). The kinetic parameters also provide information on the characteristics of energy excitation migration among chlorophylls and between system II units (Fig. 17).

Several questions arise concerning the structure of PSU: (a) Do they correspond to individual, structured entities in the photosynthetic membrane ("puddle" model) or are they organized as a chlorophyll continuum in which reaction centers would be dispersed with a statistically defined ratio ("lake" model), or (b) do they consist of several building blocks and, if so, what is their chemical composition and their relative organization, and (c) how does the structure of these building blocks allow light-harvesting and trapping properties of the photosynthetic unit? Several approaches have been used to answer these questions.

2. Biochemical Approach

Progress in methods for solubilizing photosynthetic membranes with detergent and in separating polypeptides by hydroxyapatite chromatography or by sodium dodecyl sulfate polyacrylamide gel electrophoresis (SDS-PAGE) have shown that at least 60%, if not all, of the chlorophylls of photosynthetic membranes are associated with proteins in chlorophyll–protein complexes, and that about 75% of membrane proteins are involved in such complexes (for a review, see Thorn-

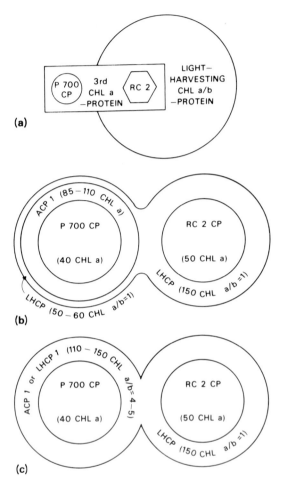

Fig. 17. Three models (a, b, and c) for the relative organization of four chlorophyll–protein complexes in a photosynthetic unit including 400 chlorophyll molecules, one PSI and one PSII reaction center. The number of chlorophylls would be 40 Chl *a* in P-700 Chl *a* protein, 20 Chl *a* in reaction center of photosystem II, 120 Chl *a* in a third Chl *a*-protein, and 110 Chl *a* + 110 Chl *b* in the light harvesting chlorophyll *a/b*-protein. RC2, reaction center of system 2; P700, reaction center of system 1. CP, chlorophyll–protein complex; LHCP, light-harvesting chlorophyll *a/b*-protein complex; ACP, active chlorophyll–protein complex. [Model (a) is reproduced from Thornber *et al.,* 1977. Alternative models (b) and (c) are separate package models reproduced from Boardman *et al.,* 1978.]

ber, 1975; Anderson, 1975; Boardman *et al.*, 1978). The green extract obtained after solubilization of photosynthetic membranes may be separated by SDS-PAGE into three main chlorophyll-containing bands: two chlorophyll–protein complexes termed CP1 and CP2 and a free pigment band corresponding to SDS-solubilized chlorophyll.

CP1 is almost certainly ubiquitous in the photosynthetic plant kingdom (Brown *et al.*, 1975) and accounts for some 28% of the chloroplast membrane protein and for 10–18% of the total chlorophyll in higher plants (Thornber *et al.*, 1967). It contains P-700, the reaction centers of PSI, with a chlorophyll/P-700 molar ratio close to 40/1. Thornber (1975) termed this complex P-700-chlorophyll *a*-protein complex. It does not seem that its distribution along photosynthetic membranes is restricted to some regions, since it has been shown that both sub-chloroplast fractions enriched in PSI or in PSII activities derived, respectively, from stroma and grana membranes, contain CP1 (Remy,1971; Sane *et al.*, 1970).

CP2 is present in all chlorophyll *b*-containing organisms (Thornber, 1975). It contains chlorophyll *a* and chlorophyll *b* in equimolecular amounts and accounts for some 50% of chloroplast membrane proteins and for about 30–60% of the total chlorophyll in higher plants and green algae. These proportions vary widely, depending on environmental conditions, during the greening of higher plants and green algae or in mutants (for a review, see Boardman al., 1978) and seem to be related to the chlorophyll *b* content of these organisms (Brown *et al.*, 1975; Genge *et al.*, 1974). *Euglena*, which naturally contains little chlorophyll *b*, contains low amounts of CP2 (Brown *et al.*, 1975; Ortiz and Stutz, 1980). Genge *et al.* (1974) calculated that only 4% of *Euglena* chlorophyll is involved in this complex. The fact that a photosynthetically competent chlorophyll *b*-less barley mutant (Thornber and Highkin, 1974) totally lacks all constituents of CP2 suggests that this complex must function solely in harvesting light. It was renamed the light-harvesting chlorophyll *a/b*-protein complex (LHCP) by Thornber (1975). Since its distribution along photosynthetic membranes parallels that of PSII activity and seems to be restricted to stacked regions (Remy, 1971; Sane *et al.*, 1970), and since light absorbed by chlorophyll *b* is principally used for system II photochemistry, it is generally accepted that LHCP serves principally as light-harvesting antennae for PSII reaction centers.

The SDS-solubilized chlorophylls in the free pigment zone at the gel front represent about one-third of the total chlorophyll, and it is thought that they may derive from CP1, CP2, or other unstable chlorophyll–protein complexes rather than from the lipid phase of the membrane (Markwell *et al.*, 1979). Thornber *et al.* (1977) ascribed 10–20% of free chlorophylls to a third chlorophyll–protein complex corresponding to the reaction center of photosystem II, and the remaining 80–90% free chlorophylls to a fourth chlorophyll–protein complex serving as a light-harvesting complex for PSI.

Figure 17a presents a model proposed by Thornber *et al.* (1977) for the relative organization of these four chlorophyll–protein complexes in a photosynthetic unit containing 400 chlorophyll molecules, one PSI, and one PSII reaction center. This model resembles those of Seely (1973) and of Butler and Kitajima (1975), in which a continuous array of chlorophyll accounts for the migration of energy excitation in the photosynthetic unit. Figures 17b,c represent alternative organizations of these complexes in separate package models for photosynthetic units, as proposed by Boardman *et al.* (1978), taking into account fractionation of photosynthetic units into PSI and PSII by digitonin or Triton X-100.

These models are based on the assumption that these additional chlorophyll–protein complexes really exist. Recent results obtained using conservative procedures for solubilizing photosynthetic membranes and for separating chlorophyll–protein complexes during gel runs have led to the characterization of new chlorophyll-containing bands (for review and for comparison, see Machold *et al.*, 1979). Among these new chlorophyll proteins, one of apparent molecular weight of about 41K daltons has been described by several authors as being involved in photosystem II (Hayden and Hopkins, 1977; Remy and Hoarau, 1978, Anderson *et al.*, 1978; Henriques and Park, 1978; Delepelaire and Chua, 1979; Machold *et al.*, 1979). The assignment of other chlorophyll–proteins to system I or system II light-harvesting complexes is still unknown. The models of photosynthetic units proposed by Thornber (1975) and Boardman *et al.* (1978) thus remain valid until more precise determinations of the number of chlorophyll–protein complexes, of their three dimensional structure, and of their organization become available.

Thornber (1975) proposed that the rectangular portion of his model (including the P-700 chlorophyll *a*-protein, the third chlorophyll *a*-protein, and the reaction center of PSII; see Fig. 17a) is common to all photosynthetic plants, and that the nature of the major light-harvesting component has varied during the course of evolution. According to this model, the photosynthetic units of *Euglena* would principally differ from those of higher organisms by the amount of LHCP associated with this common core unit.

3. Functional Approaches

a. Higher Plants and Green Algae. Most analyses have been performed on higher plants and on the widely used alga *Chlorella,* principally using analyses of light saturation curves for O_2 evolution and measurement of O_2 evolution under repetitive flashes. Unit size appeared to be approximately constant in most of the algae analyzed (Schmid and Gaffron, 1971), but seemed to be rather variable in higher plants, depending on the leaves analyzed. However, mutation can change unit size by reducing chlorophyll concentration of cells without altering that of the reaction centers (Wild, 1969). This suggests that the structure of light-harvesting antennae is controlled by several genes and therefore consists of several different building blocks. Most experiments on greening organisms demonstrate an increase in the size of photosynthetic units during chloroplast differentiation, resulting from a step-wise insertion, first of reaction centers and then of light-harvesting chlorophylls, into the photosynthetic membrane (Herron and Mauzerall, 1972; Egneus *et al.*, 1972; Henningsen and Boardman, 1973; Baker and Leech, 1977).

However, and particularly during chloroplast differentiation, these useful estimations of the overall chlorophyll/reaction center ratios must be combined with other methods of investigation that provide information on the distribution chlorophylls among the two photosystems, on the eventual existence of unorganized chlorophylls, and on functional interrelationships between light-harvesting antennae and reaction centers. These other methods are mainly based on kinetics analysis of PSII transients (for a review, see Williams, 1977). Joliot and Joliot (1964) demonstrated a nonlinear relation between the redox state of PSII reaction centers and either fluorescence yield or O_2 evolution in mature chloroplasts, thereby demonstrating that energy transfers occur between system II units. These results are confirmed by the linear increase of fluorescence lifetime, with an increasing proportion of closed system-II reaction centers during fluorescence induction. System II units in mature chloroplasts thus are organized either as individual entities, interconnected to each other to allow free energy migration among several light-harvesting antennae (puddle model), or as a nonindividually structured chlorophyll continuum containing a definite number of reaction centers.

Fluorescence induction measurements in the presence of DCMU has appeared to be a potent tool in analyzing system II light-harvesting and -trapping properties of photosystem II in modified situations (for explanations, see Fig. 18). It has been used by Bennoun and Chua (1976) for analyzing mutations affecting *Chlamydomonas* in PSII.

This approach to investigating the greening process was primarily introduced by Butler (1965) and by Diner and Mauzerall (1973). The results obtained on the formation of system II units during greening of a *Chlorella* mutant (Dubertret and Joliot, 1974), of the *y-1 Chlamydomonas* mutant (Cahen *et al.*, 1976) and of bean (Melis and Akoyunoglou, 1977; Akoyunoglou, 1977) are consistent with conclusions obtained with conventional methods that PSII reaction centers are formed first and later completed by increasing amounts of functional light-harvesting chlorophylls. Moreover, the absence of energy transfers between the first-formed PSII units, as deduced from the exponential rise of variable fluorescence during the first stages of greening of the *Chlorella* mutant (Dubertret and Joliot, 1974) and of bean (Akoyunoglou, 1977) suggests that these first-formed system II units might be physically isolated from the others. In this case, photosynthetic units would correspond to individually structured entities, interconnected or not in the photosynthetic membrane according to the ''puddle'' model.

b. Euglena. Few analyses of light-harvesting and light-trapping properties of photosynthetic units have been performed on mature chloroplasts of *Euglena*. From measurements of O_2 evolution under repetitive, saturating short flashes on fully green *Euglena* cells, Dubertret and Lefort-Tran (1978a) calculated an overall

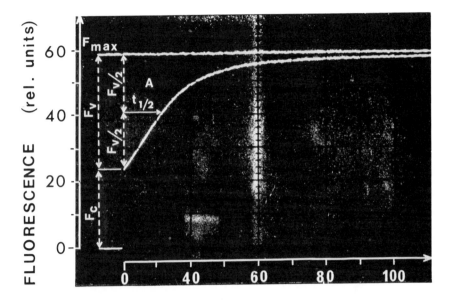

T I M E   (milliseconds)

Fig. 18. Fluorescence induction curve recorded at room temperature in the presence of 2×10^{-5} M DCMU. F_{max}, F_c, F_v = maximal, constant, and variable fluorescence, respectively; $t_{\frac{1}{2}}$ = half-rise time of variable fluorescence; A = area above the induction curve. [Principles of interpretation of the measured parameters are detailed in Dubertret and Lefort-Tran (1978a)]. The fluorescence induction area was utilized for the relative estimation of active system II reaction center concentration (Malkin and Kok, 1966). The reciprocal of half-rise time of variable fluorescence was taken as a relative measure of the optical cross section of the light-harvesting antennae of system II units. The exponential or sigmoidal shape of induction curves was interpreted as reflecting the absence or the occurrence of energy transfers between system II units (Joliot and Joliot, 1964). Variation in the F_v/F_{max} ratio can be interpreted as reflecting change either in the proportion of unorganized chlorophylls or in the concentration of active system II reaction centers on well-organized chlorophylls or also as a change in the concentration of permanent quenching traps, depending on the rise time and on the shape of the induction curve.

photosynthetic unit of about 1600 chlorophyll molecules per O_2 molecule evolved instead of the 2400 chlorophyll molecules in *Chlorella* (Myers and Graham, 1971) and in some higher plants (Schmid and Gaffron, 1971). The size of the optical cross section of the PSII photosynthetic unit, as measured with a half-rise time of variable fluorescence, appeared to be reduced by two-thirds as compared to *Chlorella*, *Chlamydomonas*, or spinach (G. Dubertret, unpublished data). The small size of both types of photosynthetic units probably results from the low amount of chlorophyll *b* and of light-harvesting chlorophyll *a/b*–protein complex in *Euglena* (Brown *et al.*, 1975; Genge *et al.*, 1974).

Formation of PSU during chloroplast development. In an analysis of the greening process of etiolated nondividing *gracilis* Z cells, Dubertret and Lefort-Tran (1978a) observed a decrease of photosynthetic capacity expressed on per chlorophyll basis, and a concomitant decrease in the values of light intensities required for saturating photosynthesis. This evolution, which appears to be similar to that observed in higher plants, could be interpreted in the same way as that resulting from a decreasing concentration of reaction centers on a chlorophyll basis, and therefore as resulting from an increase in the size of the photosynthetic unit during greening. However, direct measurement of the size of photosynthetic units, either with the O_2 method (Emerson and Arnold, 1932) or with the fluorescence method (fluorescence induction area, see Fig. 18) showed, on the contrary, a marked decrease of the size of PSU during chloroplast differentiation (Fig. 19).

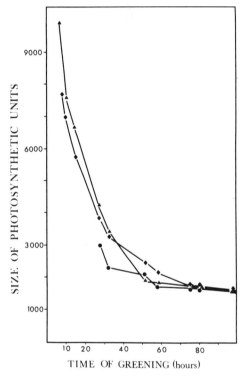

Fig. 19. Changes in photosynthetic unit size during greening of *Euglena* under 1200 lux white light. (●) Number of chlorophyll molecules involved in the evolution of one O_2 molecule under a regime of repetitive saturating short flashes; (▲) ratio of the total chlorophylls to PSII reaction centers as estimated from fluorescence induction area (relative units); (◆) optical cross section of system II light-harvesting antennae as estimated from the reciprocal of half-rise time of variable fluorescence ($t_{\frac{1}{2}}$). Curves are normalized to the same final value.

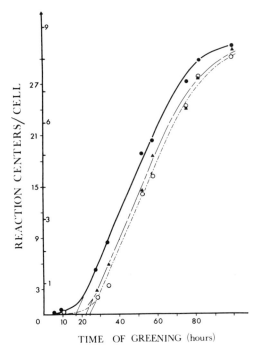

Fig. 20. Time course of chlorophyll synthesis and PSII reaction center formation during greening of *Euglena*. Concentration of PSII reaction centers is calculated from measurements of O_2 evolution under a regime of repetitive saturating short flashes (quantitative values) or from fluorescence induction area (relative values). Curves are normalized to the same final value. The formation of reaction centers begins with a delay of about 10 hours compared to chlorophyll synthesis. Key: (●) Fluorescence measurements; (▲) O_2 measurements; (◆) μg chlorophyll/10^6 cells.

This resulted from a delay in reaction center formation compared to chlorophyll synthesis throughout the greening process (Fig. 20). This apparent contradiction results from a marked decrease (from 30 milliseconds to 200 milliseconds) of the turnover rate of the overall photosynthetic reaction during greening of *Euglena*, as shown by saturation curves for O_2 flash yield as a function of dark interval between each flash (Fig. 21). In other words, the high regeneration rate of the few reaction centers present during the early stages of greening entails high values for O_2 evolution on a chlorophyll basis and requires high light intensities for saturation of o_2 evolution, in spite of the large size of photosynthetic units.

These results are consistent with those obtained with the analysis of kinetics of trapping of the absorbed energy during fluorescence induction in the presence of DCMU. The concomitant changes of fluorescence induction parameters during greening, as the concentration of active system II reaction centers on a

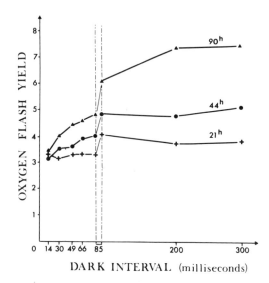

Fig. 21. Relation between O_2 flash yield and dark interval between each flash in experiments on O_2 evolution under a regime of repetitive saturating short flashes with cells greened for 21 (+), 44 (●), and 90 (▲) hours under 120-lux white light. Flashes are saturating in the 85–300 millisecond range. The gap at 85 milliseconds is due to the specifications of the stroboscope, which does not provide saturating flashes in the 14–85 millisecond range. Note the rapid turnover rate of the photosynthetic reaction during early stages of greening since 14 milliseconds are sufficient to regenerate reaction centers in 21-hour greened cells instead of about 300 milliseconds in fully greened cells.

chlorophyll basis increases, are shown in Fig. 22: Variable fluorescence increases at the expense of constant fluorescence (increase of F_v/F_{max}, half-rise time ($t_{\frac{1}{2}}$) of variable fluorescence increases, and the shape of induction curves progressively changes from exponential to sigmoidal. These changes are similar to those observed in green cells containing well organized light-harvesting chlorophylls, in which increased concentration of active, ''open'' system II reaction centers are experimentally obtained after inactivation by light. Early synthesized chlorophylls thus behave like functional system II light-harvesting antennae, lacking active system II reaction centers and thus appear to be rapidly organized in a way allowing for free energy migration according to the ''lake'' or to the ''connected puddles'' models. During the first stages of greening, these chlorophylls are in large excess compared to the few active system II reaction centers present and will therefore increase the optical cross section of their light-harvesting antennae (short $t_{\frac{1}{2}}$). Moreover, this very low number of active reaction centers compared to light-harvesting chlorophylls will be responsible for a low efficiency for trapping the absorbed energy, which then will be largely dissipated as heat or constant fluorescence (high F_c level and low F_v/F_{max} ratio).

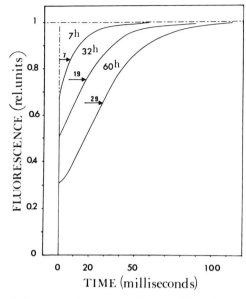

Fig. 22. Changes in fluorescence induction in cells greened for 7, 32, and 60 hours. Curves are normalized to the same F_{max} level. As greening proceeds, variable fluorescence increases at the expense of constant fluorescence, the $t_{\frac{1}{2}}$ of variable fluorescence increases, the shape of induction curves progressively changes from exponential to sigmoidal, and the area above the curve increases. This evolution is similar to that observed in fully greened cells, in which increasing proportions of system II reaction centers are allowed to regenerate in the dark after having been closed by light.

Finally, the dispersion of active system II reaction centers (low reaction center/chlorophyll ratio) prevent efficient energy transfers between system II units (exponential shape of induction curve). As greening proceeds, the addition of active system II reaction centers (increase of PSII centers/chlorophyll ratio) allows most of the absorbed energy to be trapped (decrease of F_c to the benefit of F_v, and increase of F_v/F_{max} ratio), causes a decrease of the optical cross section of each resulting unit (increase of $t_{\frac{1}{2}}$) and allows interunit energy transfers to occur (sigmoidal shape of induction curves).

This sequential process for system II unit formation, during which newly synthesized chlorophylls appear to be first organized as functional light-harvesting antennae before being completed by the late formation of active system II reaction centers, is unusual compared to that observed in most greening organisms. It results from the previously described unusual delay in their formation compared to chlorophyll synthesis (Fig. 20).

The question then arises as to why the lag for PSII reaction center formation is longer than that for chlorophyll synthesis in *Euglena*. Experiments of lag elimination in chlorophyll synthesis which used preillumination prior to exposure of

the etiolated cells to continuous light led Holowinsky and Schiff (1970) to propose that this lag corresponds to the time necessary for the light production of
components required for maximal rate in chlorophyll synthesis. These components also seem to be required for the synthesis of other components involved in
chloroplast development, which display a lag very close to that measured for
chlorophyll synthesis under the same conditions (for a review, see Freyssinet,
1976). Attempts to identify these components were performed by analyzing the
chloroplastic or cytoplasmic localization of light-induced RNA (Heizmann
et al., 1972; Heizmann, 1974; Cohen and Schiff, 1976) or protein syntheses
(Schwartzbach et al., 1976), and by determining the relaxation times of different
light-induced events at the onset of greening (Freyssinet, 1976). The results
obtained were consistent with a model proposed by Freyssinet (1977b) whereby
the fairly constant length of the lag phase for processes involved in chloroplast
development would be related to the time required for the light-induced formation of chloroplast rRNA necessary for chloroplast protein synthesis.

One can suppose that the lag in reaction center formation corresponds to a
similar photoinduced process. In this case, its longer duration compared to
chlorophyll synthesis implies that reaction centers are synthesized by a specific
cellular machinery, the formation of which would require a longer time after
induction by light. Preilluminations should then eliminate the lag in their accumulation in greening cells, on the condition that the subsequent dark period be
long enough to allow completion of the light-induced formation of the synthesizing machinery before exposure of cells to continuous light.

Experiments presented in Fig. 23 show that preillumination exerts a quantitatively equal shortening effect on lag duration of both chlorophyll and PSII reaction center accumulation, but has no effect on the delay of active system II
reaction center formation compared to chlorophyll synthesis (Dubertret, 1981a).
A lag in reaction center formation can thus be shortened to a minimum corresponding to this delay but cannot be totally eliminated, even after dark periods
much longer than the normal duration. The hypothesis of a light- induced formation of a synthesizing machinery for reaction centers thus does not account for
their late formation, which could then result from a late triggering of their
synthesis, when a definite amount of chlorophyll is attained in early-greened
cells. Figure 24 shows the kinetics of chlorophyll and PSII reaction center accumulation in cells greened under different light intensities (Dubertret and
Lefort-Tran, 1978b; Dubertret, 1981a). The light intensities used appear to determine the accumulation rates of these components during the greening process
but do not significantly affect the duration of their respective lag phases and,
therefore, the duration of the delay in the formation of PSII reaction centers
compared to chlorophyll synthesis. The amounts of chlorophyll already accumulated in early-greened cells when reaction center formation begins (arrow on
abscissa scale) widely varies (Fig. 24) depending on the light intensities used,

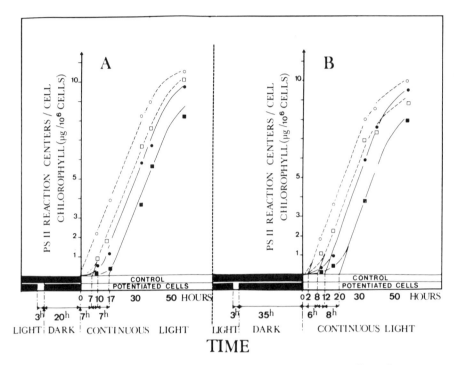

Fig. 23. Time courses of accumulation of chlorophyll (○, ●) and system II reaction centers (□, ■) during greening of non-preilluminated control cells (——) and of cells preilluminated for 3 hours (---) and then submitted to 20 (A) or 35 (B) hours of darkness before exposure to continuous 1200-lux white light. PSII reaction center concentration is estimated from fluorescence induction areas. The lag phase for each of these curves is measured by extrapolating the linear phase of the maximal accumulation rate on the time scale. The duration of the dark period subsequent to the preillumination remains without effect on the delay in reaction center formation compared to chlorophyll synthesis.

thereby demonstrating that the formation of PSII reaction centers is not triggered when a definite concentration of chlorophyll is attained in early greened cells.

However, these experiments on greening after preillumination or under different light intensities clearly show that these experimental conditions do not exert any effect on the duration of the delay in the appearance of competent system II reaction centers compared to chlorophyll synthesis. The constancy of this delay suggests that it results from a time-dependent insertion of a reaction center component in the developing thylakoids, either from already accumulated inactive pools, or after delayed synthesis.

This hypothesis has been tested by using inhibitors of protein synthesis (Dubertret, 1981a). It has been demonstrated in *Euglena* that PSII activity requires polypeptides of chloroplastic origin (Gurevitz *et al.*, 1977; Cahen *et al.*,

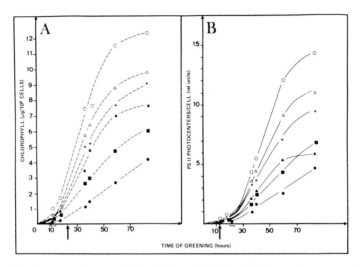

Fig. 24. Time course of accumulation of chlorophyll (A) and of PSII reaction centers (B) in cells greened under different light intensities. PSII reaction center concentration is estimated from fluorescence induction areas. Lag phase of each of these processes is indicated on the time scale. The amount of chlorophyll already synthesized when reaction center formation begins [arrow on time scale of (A)] largely varies, depending on light intensities used for greening, and light intensities remain without effect on the delay in reaction center formation compared to chlorophyll synthesis. Key: (○) 800; (△) 400; (+) 1500; (■) 150; (▲) 2500; (●) 50 lux.

1978). Therefore, the addition of inhibitors of protein synthesis by 70 S choroplastic ribosomes, e.g., streptomycin (Sm), to greening cells, should not inhibit PSII accumulation for a time, if PSII is formed from preexisting pools. On the contrary, if directly formed after delayed synthesis of components, PSII accumulation should rapidly stop after addition of such inhibitors.

When Sm is added to cultures at the onset of greening, the synthesis of chlorophyll and the development of chloroplast structures and functions is blocked at a stage corresponding to that observed at the end of the lag phase in non-poisoned cells (Ben-Shaul *et al.*, 1972; Bovarnick *et al.*, 1974). However, difficulties in interpretation arise from uncertainties concerning the inhibitory effects of Sm on chloroplast development. As a matter of fact, Bovarnick *et al.* (1974) showed that Sm exerts a diminished inhibitory effect on chlorophyll synthesis when added to cultures in late stages of greening, thereby suggesting that the permeability of the chloroplast envelope to Sm could decrease throughout greening. However, Diamond (1976) found that this decreasing inhibitory effect of Sm was effective on some but not on all photosynthetic parameters. This latter observation rules out the hypothesis of a membrane permeability barrier, and Diamond (1976) proposed that the Sm-insensitive chloroplast development could occur either at the expense of plastid pools constituted prior to the late addition of Sm, or due to membrane-bound chloroplast ribosomes that could remain inaccessible (Philippovitch *et al.*, 1970) or resistant (Fehlman *et al.*, 1975) to specific antibiotics.

Figure 25 shows further analyses, including measurement of chlorophyll synthesis and PSII reaction center accumulation in the presence of Sm added to cells already greened for 24 hours and then transferred to high or low light intensities (Dubertret, 1981a). Under both experimental conditions, chlorophyll synthesis is no longer markedly inhibited by Sm addition. Under high light intensity (Fig. 25b), Sm does not immediately prevent the formation of active system II reaction centers: Their concentration continues to increase for about 30 hours and then dramatically decreases. This observation and the fact that an increased RubPcase activity (Diamond, 1976) or RubPcase synthesis (as analyzed with polyacrylamide gel electrophoresis or with immunoelectrophoresis, Pineau, 1980) immediately stops after Sm addition confirms that the antibiotic enters the cells and is effective in blocking syntheses by 70 S chloroplastic ribosomes at these stages of greening. However, Fig. 25a shows that although it is efficient in inhibiting protein synthesis by plastid ribosomes, Sm, when added to 24-hour greened

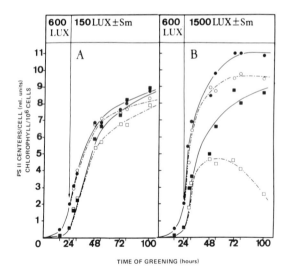

Fig. 25. Combined effects of streptomycin and light intensity on chlorophyll synthesis (●, ○) and on system II reaction center formation (□, ■) during greening. Cells were first greened under 600 lux, then exposed to 150 lux (A) or 1500 lux (B) in the presence (---) or in the absence (——) of streptomycin (Sm). Inhibitory effects of streptomycin are manifest only under high light intensity, and they are more important on system II reaction center formation than on chlorophyll synthesis.

cells, does not prevent further development of functional and mature chloroplasts and, more specifically, further development of system II reaction centers, if poisoned cells are kept under low light intensities.

PSII reaction centers thus are not synthesized on Sm-resistant, membrane-bound ribosomes. Their decreasing concentration in cells exposed to high light intensities seems to result from selective photodestruction that has been found to be more important on reaction centers than on chlorophylls (G. Dubertret, unpublished data).

These experiments lead to the conclusion that Sm-insensitive chloroplast development occurs at the expense of cytoplasmic syntheses and of a pool of plastid components accumulated early before Sm addition. This pool would be large enough to ensure for at least 80 hours the full development of mature chloroplasts, on the condition that the turnover rate of plastid components is limited. The high turnover rate of some system II reaction center components under high light intensity and their rapid consumption in this pool (since they are not renewed in the presence of streptomycin) would then be responsible for the progressive decrease in PSII reaction center formation and for their subsequent disappearance (Fig. 25b). The formation of such a pool of plastid components appears to be light-induced, since it has been shown that important chloroplast development occurs in the presence of Sm, when added at the onset of greening to preilluminated, potentiated cells (Dubertret, 1981a).

If such a pool really exists, the question arises as to why it is not incorporated directly after synthesis in the developing thylakoids. The delay in system II reaction center formation leads one to assume that some limiting step should occur between the light-induced synthesis and accumulation of some of their components and their assembly as active complexes in photosynthetic membranes. It is possible that this limiting step could correspond to a time-dependent process for insertion or reorganization in photosynthetic membranes of these early accumulated plastid components. This possibility has been tested with experiments similar to those described by Gurevitz et al. (1977). Changes in the functional interrelationships between light-harvesting chlorophylls and system II reaction centers were analyzed in cells transferred to darkness at different stages of greening in the presence or in the absence of specific protein inhibitors (Dubertret, 1981b).

Figure 26 shows that the formation of active system II reaction centers continues in cells returned to darkness after 32 hours of greening. Formation continues for a time (about 15 hours), which appears to be similar to the shift in time of the two accumulation curves, as if residual formation of system II reaction centers compensates for their initial delay compared to chlorophyll synthesis. Gurevitz et al. (1977) showed that this dark formation of active system II reaction centers is inhibited by chloramphenicol, when added to early greened cells. However, in our experimental conditions, later addition of Sm or cyc-

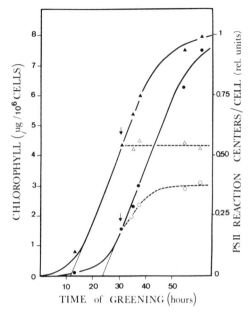

Fig. 26. Time course of chlorophyll (▲, △) and PSII reaction center (●, ○) accumulation in cells greened under continuous light (——) and in cells transferred back to darkness (---) after 30 hours of greening (arrow). Chlorophyll synthesis immediately stops in the dark, whereas continued formation of PSII reaction centers for about 15 hours is responsible for their concentration doubling in the dark.

loheximide to cells returned to darkness after the lag phase for chlorophyll synthesis no longer significantly prevents the continued formation of active PSII reaction centers in the dark. The absence of a requirement for *de novo* synthesis by chloroplastic or cytoplasmic ribosomes thus supports the hypothesis of early-formed light-induced plastid pools and also suggests that the limiting step in reaction center formation could correspond to some conformational changes allowing for the insertion or reorganization of system II reaction centers in the photosynthetic membrane.

Experiments on greening after late addition of Sm showed that the size of the proposed early-formed plastid pool is large enough to allow further development of fully mature chloroplasts. However, Figure 26 shows that in the dark, only a part of this pool is used for the formation of active system II reaction centers. Moreover, a quantitative analysis of reaction center formation in cells transferred to darkness at different stages of greening (Fig. 27) shows that the part of this plastid pool used for the formation of active PSII reaction centers in the dark is variable and appears to be just sufficient to decrease the chlorophyll/PSII ratio to a constant, lower limit corresponding to that found in fully greened cells. It

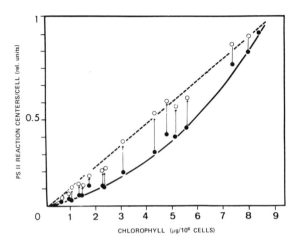

Fig. 27. Amounts of dark-formed reaction centers (●→○) in cells transferred back to darkness at different stages of greening and therefore containing increasing amounts of chlorophylls. The incurvation of the solid line (——) represents the relation between chlorophyll and reaction centers as greening proceeds in the control culture kept in the light. It is due to the initial delay in PSII reaction center formation compared to chlorophyll synthesis, i.e., to decreasing chlorophyll/PSII ratios until a lower limit value is attained at the end of greening. The dashed line (-) corresponds to the theoretical case where chlorophyll/reaction center ratios would remain constant during greening, with the same value as that found in fully greened cells. The amount of PSII reaction centers formed in cells transferred to darkness at different stages of greening is variable but appears to be just sufficient enough to lower the initial Chl/PSII ratio to the value found in fully greened cells.

thus seems that the low proportion of accumulated components used for the dark formation of active PSII reaction centers results from a limited number of potential insertion sites in the developing membrane. The number of such sites appears to be related to the amount of chlorophyll in excess responsible for chlorophyll/PSII ratio values higher than in fully greened cells, i.e., to the amount of light-harvesting chlorophylls not yet completed with active reaction centers. This suggests that there is only one insertion site for each developing, not yet functional system II unit.

These results lead us to propose a sequential model for the formation of system II units in *Euglena*. Light induces the synthesis of chlorophyll and system II reaction center components. The first chlorophylls synthesized after a lag phase of about 15 hours are rapidly incorporated in the developing thylakoids. During this time, synthesized reaction center components would accumulate as an inactive plastid pool and would not be incorporated in the developing thylakoids until reorganization of chlorophylls, lasting about 10 hours, would lead to the formation of functional light-harvesting antennae. The concomitant formation of one insertion site in each of these functional light-harvesting antennae then would allow the incorporation of accumulated reaction center components and the formation of active system II units.

We have already stated that the sequential process for photosynthetic unit formation in most greening organisms, as analyzed with similar fluorescence methods (Dubertret and Joliot, 1974; Cahen *et al.*, 1976; Melis and Akoyunoglou, 1977; Akoyunoglou, 1977) or other methods (Herron and Mauzerall, 1972; Egneus *et al.*, 1972; Henningsen and Boardman, 1973; Baker and Leech, 1977), consists in the early formation of active reaction centers, followed by a progressive increase of functional light-harvesting antennae due to reorganization of newly incorporated, temporarily inefficient chlorophylls. The sequence of synthesis events in *Euglena* does not seem to be basically different since PSII reaction center components appear to be synthesized and accumulated during the lag phase for chlorophyll synthesis. The main difference in PSII unit formation in *Euglena* then would lie in the late insertion of an active form of early synthesized PSII components in already constituted, mature, and functional light-harvesting antennae.

c. Structural Organization of Photosynthetic Units in Thylakoids. Concerning the structure of the PSII unit, it could be proposed that it corresponds to individually structured entities, if it were possible to demonstrate that they can be physically isolated from each other, thereby unable to exchange excitation energy. Exponential increases in variable fluorescence, which express functional disconnection of PSII units, have already been observed during greening of a *Chlorella* mutant (Dubertret and Joliot, 1974) and in bean greened under intermittent light (Akoyunoglou, 1977). However, the interpretation that this functional disconnection results from a physical isolation of PSII units requires us to take into account other parameters of fluorescence induction curves. As a matter of fact, exponential increases of variable fluorescence can be observed in some conditions without a physical disconnection of the PSII unit.

When *gracilis* Z cells are greened under intermittent light (15 seconds light/15 minutes darkness) (Dubertret and Lefort-Tran, 1978b; Dubertret and Lefort-Tran, 1981), chlorophyll *b* and probably the light-harvesting chlorophyll–protein *a/b* complex are not synthesized (chl *a*/chl *b* >> 18). As a consequence, the rate of chlorophyll accumulation and the final chlorophyll concentrations attained in cells greened under such light conditions are reduced compared to the control greened under continuous light (Fig. 28). In contrast, the formation of active system II reaction centers is much less affected by intermittent light, entailing low overall chlorophyll/PSII reaction center ratios (overall photosynthetic unit sizes 2 to 3 times lower than in control).

The exponential rise of variable fluorescence (Fig. 29) indicates that energy transfers do not occur between system II units formed under intermittent light. It is important to notice that this exponential shape appears in spite of a high level

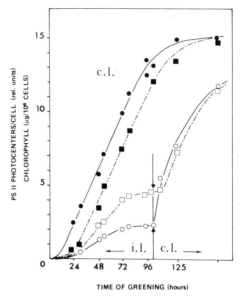

TIME OF GREENING (hours)

Fig. 28. Time course of chlorophyll (——) and PSII reaction center (- - -) accumulation in control cells greened under continuous 1200-lux white light (●, ■) and in cells first greened under intermittent light (○, □) (i.l., 15 seconds light/15 minutes of darkness), then transferred to continuous light (c.l., arrow). Intermittent light considerably reduces chlorophyll accumulation and particularly chlorophyll *b* synthesis (Chl *a*/Chl *b* > 18). Reaction center formation is much less affected by these light conditions. As a result, chlorophyll/reaction center ratios are about 3–4 times lower than in the control. Transfer to continuous light induces an immediate synthesis of chlorophylls *a* and *b*, and the Chl *a*/Chl *b* ratio decreases to normal values (5–7) in about 8 hours. PSII reaction center formation begins after a lag of about 12 hours, which is responsible for an increase of the Chl/PSII ratio until values close to those found in cells fully greened under continuous light are attained about 8 hours after the onset of continuous illumination.

of variable fluorescence ($F_v/F_{max} = 0.66$ in cells greened for 100 hours under intermittent light), of a high concentration of active system II reaction centers per chlorophyll (more than twice as in fully greened control), and of a small size of functional light-harvesting antennae ($t_{\frac{1}{2}}$ of F_v 3 to 4 times longer than in control). These characteristics exclude the possibility that the absence of energy transfers results from a too-low concentration of active system II reaction centers on chlorophyll allowing for free energy migration; this situation, which is found in *Euglena* during the first stages of greening under continuous light, entails high levels of constant fluorescence (low F_v/F_{max} ratio) and short $t_{\frac{1}{2}}$ of variable fluorescence (see formation of the photosynthetic unit in *Euglena* in Section III, A, 3). These characteristics are also not consistent with the possibility that active system II reaction centers are functionally isolated from each other by chlorophyll, not yet organized to allow for energy migration, because these

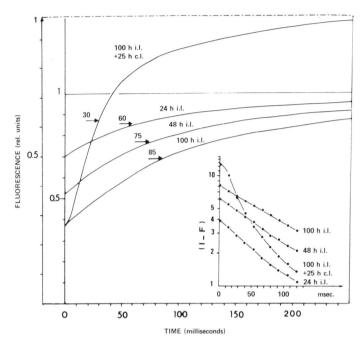

Fig. 29. Changes in fluorescence induction in cells first greened for 24, 48, and 100 hours under intermittent light (i.l.) and then transferred for 25 hours to continuous light (c.l.). Insert: log plot of $1-F_v$. Cells greened for 100 hours under intermittent light still display an exponential rise of variable fluorescense (see the log transformation of the curve) in spite of (a) a high concentration of PSII reaction centers per chlorophyll (large area above fluorescence induction curve), (b) a very small optical cross section of system II light-harvesting antennae (long $t_{\frac{1}{2}}$ of F_v), and (c) a good trapping efficiency of the absorbed energy (f_v/F_{max} as high as 0.63). Rapid chlorophyll synthesis after exposure of cells to continuous light is concomitant with an increase in the size of system II light-harvesting antennae (decrease of $t_{\frac{1}{2}}$), with an increase of variable fluorescence and with a change from exponential to sigmoidal of the shape of the induction curve, which acquires characteristics found in fully greened cells.

unorganized chlorophylls would be responsible as in other greening organisms (Cahen *et al.*, 1976; Akoyunoglou, 1977) for a high level of constant fluorescence (low F_v/F_{max} ratio). This functional disconnection could also result from a deviation of excitation energy towards traps other than active system II reaction centers, either toward system I by "spillover" or toward inactive system II reaction centers, including deficient electron donor sides, which behave as permanent quenching traps (Butler and Kitajima, 1975).

The simplest hypothesis to account for these characteristics of fluorescence induction curves then appears to be that in the special case of *Euglena* system II units developed under intermittent light are organized as small discrete entities (short $t_{\frac{1}{2}}$ of F_v), physically isolated from each other (exponential rise of F_v) and exhibiting a good trapping efficiency of the absorbed energy (high F_v/F_{max} ratio).

Functional analyses thus allow us to define system II units as corresponding to structured entities, individually organized in the photosynthetic membrane (puddle model), mostly interconnected in groups of at least five units (Paillotin and Swenberg, 1979), which may be physically disconnected from each other under some conditions. Functional and biochemical approaches thus converge in defining photosynthetic units as structured entities corresponding to multimolecular complexes whose basic structure, however, remains largely hypothetical. Indeed, the tridimensional, molecular structures of reaction center chlorophyll–protein complexes and of light-harvesting chlorophyll–protein complexes have to be made precise, as has been accomplished by Fenna and Matthews (1977) for a water-soluble bacteriochlorophyll a-protein complex. Moreover, their relative organization in the photosynthetic unit should be experimentally confirmed. Another question concerns the morphological expression of such entities in the photosynthetic membranes.

B. Structure of Thylakoids and Probable Morphological Expression of Photosynthetic Units

After the concept of the photosynthetic unit emerged, the development of electron microscopy showed the existence of particles associated with the outer surface of thylakoids (Park and Pon, 1963). Park and Biggins (1964) calculated their size ($\sim$150 Å) to be about 230 chlorophyll molecules, and they proposed that these particles correspond to the *quantasome*, i.e., to the morphological expression of photosynthetic units. However, Howell and Moudrianakis (1967a,b) succeeded in removing these particles of isolated thylakoids and demonstrated that they correspond to carboxydismutase and to a coupling factor. Park and Pfeifhofer (1968) and later Park and Sane (1971) proposed that photosynthetic units could correspond to the particles seen by Park and Branton (1966) inside freeze–fractured photosynthetic membranes.

1. Ultrastructure of Thylakoids in Higher Plants and Green Algae

As in other biological membranes, the above particles are asymmetrically distributed across the photosynthetic membrane. They appear small in size and densely packed on protoplasmic faces (PF, according to Branton *et al.,* 1975, formerly C faces) but larger and dispersed on exoplasmic faces (EF, formerly B faces). Moreover, Remy (1969) and later Phung Nhu Hung *et al.* (1970) pointed out that large freeze–fracture particles ($\sim$160 Å) could only be detected in stacked membrane regions. This heterogenous distribution along photosynthetic membranes was extended to the densely packed, small particles (80–100 Å) by

Goodenough and Staehelin (1971), who recognized four fracture faces in thylakoids of *Chlamydomonas:* the complementary EFs and PFs faces (formerly Bs and Cs) are characteristic of stacked membranes, whereas the complementary EFu and PFu faces (formerly Bu and Cu) can be found only in unstacked stroma membranes. This differentiation in particle size distribution along the photosynthetic membranes cannot be observed in mutants containing unstacked thylakoids (Goodenough and Staehelin, 1971). It disappears when thylakoids are experimentally unstacked by incubating isolated chloroplasts in a low-salt medium (Goodenough and Staehelin, 1971). Translational movements in the plane of the membrane are then responsible for intermixing of the different categories of particles, whose size and number remain conserved (Ojakian and Satir, 1974; Staehelin, 1976).

It thus had to be determined, within the framework of the hypothesis of Park and Pfeifhofer (1968) and of Park and Sane (1971), which kind of freeze–fracture particles could correspond to photosynthetic units (for a review, see Arntzen and Briantais, 1975). Branton and Park (1967) suggested that the large EFs particles (~160 Å) could best account for the space required by the 200–300 chlorophyll molecules of light-harvesting antennae. Experiments on mechanical or detergent fractionation of chloroplast membranes (Sane *et al.*, 1970; Goodchild and Park, 1971) demonstrated that the distribution of PSII activity along chloroplast membranes parallels the distribution of large particles in stacked and unstacked thylakoids, whereas PSI distribution appears to be approximatively homogenous. It thus was tempting to associate PSII activity with the large particles seen on outer fracture faces and PSI activity with the inner fracture faces containing small particles (Arntzen *et al.*, 1972). Direct evidence that large EFs particles are involved in harvesting light for PSII have recently been provided by Henriques and Park (1976), Miller *et al.* (1976) and Armond *et al.* (1977). They demonstrated a close correlation between the size of these particles and the amount of LHCP, respectively, in differentiating lettuce chloroplasts, in a mutant of barley lacking chlorophyll *b* and LHCP, and in pea chloroplasts developed under intermittent light.

2. Ultrastructure of Thylakoids in Euglena

a. Mature Chloroplasts. The particulate structure of *Euglena* thylakoids was first described by Holt and Stern (1970) in freeze–fractured cells. Later, Schwelitz *et al.* (1972) recognized in *Euglena* thylakoids the two principal, complementary types of fracture faces observed in higher plants and green algae, one termed Y, in which particles are widely dispersed (EF-type face) and the other, termed X, exhibiting closely packed particles (PF-type face). They did not find large, 160-Å EFs particles characteristic of stacked regions in higher plants and *Chlamydomonas.* Moreover, the uniformity of particle size on each fracture

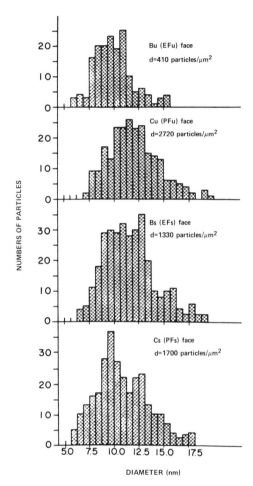

Fig. 30. Histograms of particle sizes and particle density found on the four fracture faces of *Euglena* thylakoids. Most of particles on Bs faces (EFs faces in the new nomenclature of Branton *et al.,* 1975) exhibit a diameter ranging from about 90–130 Å. Only few large 160–170 Å particles, typical of stacked, EFs faces of higher plants and green algae can be observed. (Reproduced from Miller and Staehelin, 1973.)

face (~110 Å) did not allow them to recognize lateral differentiation in particle size distribution corresponding to stacked or unstacked regions. However, careful analyses of freeze–fractured, isolated chloroplasts allowed Miller and Staehelin (1973) to distinguish the four complementary fracture faces characteristic of green algae and higher plants in which thylakoid stacking is observed. Typical particle size distribution and particle density per μm^2 on each of these faces, as measured by Miller and Staehelin (1973) in *Euglena,* are shown in Fig. 30. The main difference with particle size distribution in *Chlamydomonas* (Goodenough and Staehelin, 1971), spinach (Staehelin, 1976), or barley (Miller *et al.*, 1977) is that EFs faces display 90–130-Å particles and seem to be almost devoid of large, 160-Å particles. Moreover, the fact that the total number of particles per unit area in both complementary faces of freeze–fractured thylakoid membranes is constant, whether they are stacked or unstacked, and the observation that particle density in the inner, etched thylakoid surface (ES) does not vary from stacked (ESs) to unstacked (ESu) regions was taken by the authors as strong evidence that all these *Euglena* thylakoid membranes are similar, if not identical, in their internal composition.

The lack of lateral differentiation of thylakoid membranes in *Euglena*, since they are appressed by two or three along their entire length, does not allow us to separate stroma and grana lamellae, as in higher plants. It is therefore difficult to associate photochemical activities with some membrane ultrastructural characteristic in mature chloroplast of *Euglena*. Attention thus has focused on establishing structure–function relationships in *Euglena* chloroplasts modified by mutation or during the transitory steps of membrane differentiation during greening.

b. Modified Chloroplasts. A first analysis of this type, including freeze–fracturing techniques, was attempted by Schwelitz *et al.* (1972) by comparing the ultrastructural organization of the photosynthetic membrane of a wild-type *gracilis* Z and a photosynthetic mutant P_4, devoid of PSII primary electron acceptor and lacking plastoquinone A. Although thylakoids were not appressed in the mutant, no differences in size or distribution of particles within the membranes were observed when compared with wild-type membranes. Thus, these investigators did not succeed in associating freeze–fracture particles with any function or biochemical component of the photosynthetic membrane, and they suggested that the major ultrastructural change observed, i.e., the lack of pairing of thylakoids, could be related to changes in the lipid composition of the chloroplast membranes of the mutant.

Polack-Charcon *et al.* (1975) also failed to find any correlation between thylakoid ultrastructure and functions in two photosynthetic mutants of *Euglena*. Although thylakoids are unpaired in the mutant *E-52* that lacks PSII activity and cytochrome 558 or are concentrated in the center of the chloroplast of the *E-28* mutant, incapable of cyclic or noncyclic photophosphorylation particle size and density on the four fracture faces appeared to be similar to those of the wild-type.

Ophir and Ben Shaul (1974) analyzed changes in particle size distribution in freeze–fractured thylakoids of chloroplasts isolated from greening *Euglena* cells. They observed a significant increase in particle size on both the EF and PF of developing membranes (mainly after 24 hours of greening) from a mean value of ~80 Å in prolamellar bodies to ~120 Å in thylakoids of cells greened for 48 hours. A concomitant slow decrease in their density repartition per μm^2, from

~880–530 and ~3610–2660 on the EF and PF faces, respectively, was also observed. Parallel measurements of O_2 evolution did not provide strong evidence for any relationship between the observed changes of membrane ultrastructure and the development of the overall photosynthetic reaction.

We have already stated that when nondividing etiolated *Euglena* cells are greened under intermittent light, chlorophyll *b* and probably LHCP are not synthesized. As a result the photosynthetic unit size decreases, and this decrease has been found to be much more marked when measured on the optical cross section of PSII units ($t_{\frac{1}{2}}$ of variable fluorescence 3 to 4 times longer than in control) than when measured on overall photosynthetic units (only about 2 times less chlorophyll per evolved chlorophyll molecule or per PSII reaction center) (Dubertret and Lefort-Tran, 1978b; 1981). This confirms that chlorophylls associated with LHCP are mainly involved in harvesting light for system II reaction centers and justifies the correlation between the amount of LHCP and the size of the optical cross section of PSII units. This correlation can be extended to other organisms since it has been shown that in bean (Akoyunoglou, 1977) or in pea (Armond *et al.*, 1977), which contain larger amounts of LHCP than *Euglena,* greening under intermittent light and therefore absence of LHCP entails a 6 to 8 times extension of $t_{\frac{1}{2}}$ of variable fluorescence which, according to our interpretation, would be a 6 to 8 times decrease in the size of system II units. Moreover, measurement of optical cross sections of system II units in the mature chloroplast in the same experimental conditions provides higher values for organisms containing larger amounts of LHCP and chlorophyll *b* (spinach, *Chlorella*) than *Euglena* (G. Dubertret, unpublished data). These observations are consistent with the hypothesis proposed by Diner and Wollman (1979) that system II units are formed by the association of a basic PSII center–antennae complex, or core unit of small and constant size, with variable amounts of LHCP, depending on the organisms or on environmental conditions.

Figures 31 and 32 show that greening of *Euglena* under intermittent light induces the development of unpaired thylakoids, which thus appear to be isolated from each other. However, more significant to our interest concerning the structural organization of photosynthetic membranes themselves (Figs. 34 and 35) is that the decrease in chlorophyll *b* (and probably in LHCP) content and in optical cross section of the PSII unit induced by intermittent light is associated with a significant decrease of the size of EF freeze–fracture particles, from 130 Å to

---

Fig. 31. General view of a chloroplast fully developed for 4 days during greening under intermittent light (15 seconds of light/15 minutes of darkness). Thylakoids are either isolated (single arrow) or grouped by pairs (double arrows) or organized as grana-like structures (g.l.). In these last configurations, thylakoids, however, are not appressed. The triple layered structure of the chloroplast envelope (e) is clearly visible. Mitochondria (m). ×48,600.

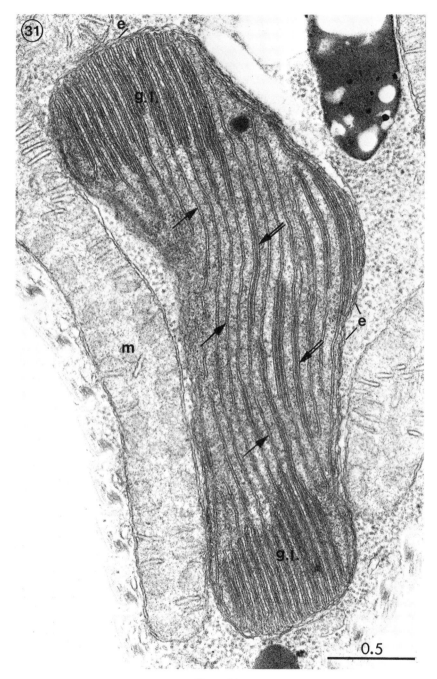

Figure 31

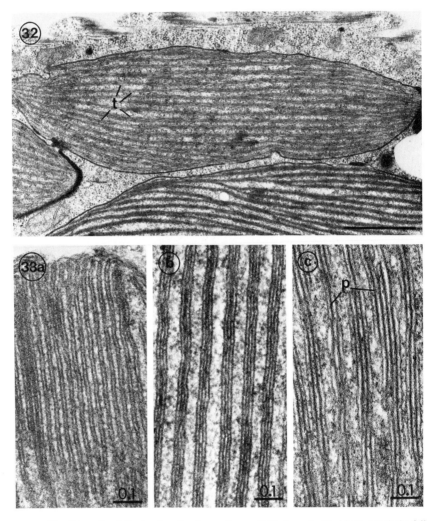

Fig. 32. Chloroplast fully developed under continuous light. Long thylakoids (t) stretch the full length of the chloroplast and appear in stacks of two or three. ×20,200.

Fig. 33. Details of thylakoid general organization in the chloroplast. (a and b): Chloroplasts developed under intermittent light; (c) chloroplast developed under continuous light. ×79,200; (a) Thylakoids are isolated and regularly spaced, (b) thylakoids are grouped by pairs without stacking and typical partitions cannot be observed; (c) organization in bands containing two or three thylakoids closely appressed with the formation of typical partitions (p).

70–80 Å (Fig. 36). Moreover, the same figure shows a small decrease of particle size in mixotrophically grown *Euglena* cells, in which nutritional conditions are responsible for a slight decrease in chlorophyll *b* content (chl *a*/chl *b* = 7–9 instead of 5–7). This confirms the direct correlation between the size of EF particles and the amount of LHCP observed by Henriques and Park (1976) in lettuce and by Armond *et al.* (1977) in pea and provides strong evidence in favor of the hypothesis that EF particles may correspond to the morphological expression of PSII units.

When such analyses are extended to different organisms, it appears that the minimal size of EF particles measured when LHCP is absent is similar in organisms as different as the primitive eukaryotic algae *Cyanidium* (Wollman, 1979), *Euglena*, and higher plants such as lettuce (Henriques and Park, 1976) or pea (Armond *et al.*, 1977). However, the maximal size of these particles appears to be related to the maximal amount of LHCP synthesized in these organisms. As a matter of fact, *Euglena*, which contains few LHCP (Brown *et al.*, 1975, Genge *et al.*, 1974), exhibits 110–130 Å EF particles, smaller than those of higher plants and green algae (∼150–170 Å). It thus seems that the general architecture of system II units would be similar in plants and algae, and that it would correspond, as proposed by Staehelin (1978), to the association of a basic core unit of similar size (80–100 Å) and of variable amounts of LHCP or other light-harvesting systems. Furthermore, recent analyses on *Chlamydomonas* and tobacco mutants led Wollman *et al.* (1979) and Miller and Cushman (1979) to propose that this core unit, which would correspond to the chlorophyll–protein complex of system II reaction centers, would be required for assembly of LHCP complements as large EF particles. Structural measurements of the size of EF particles and functional measurements of the optical cross section of system II light-harvesting antennae thus can well be correlated with biochemical measurement of qualitative and quantitative variations in chlorophyll–protein complexes in photosynthetic membranes and lead to a similar conclusion concerning the basic architecture of system II units.

The relation between the size of EF freeze–fracture particles and the light-harvesting system associated with PSII reaction centers does not answer the question whether each of these particles contains one system-II unit or more. Park and Biggins (1964) calculated that a lamellar subunit containing some 230 chlorophylls would be expected to have a diameter of ∼170 Å. More recently Williams (1977), taking into account the molecular weight, the chlorophyll content, and the estimated volume of each chlorophyll–protein of the two major complexes, calculated that a multimolecular complex corresponding to an overall photosynthetic unit containing 230 chlorophyll molecules would exhibit a diameter of 135–170 Å. Each of the EF particles then would contain only one system II unit. The question arises then as to how energy transfers, which require distances lower than 50 Å, can occur between system II units, since most of the EF

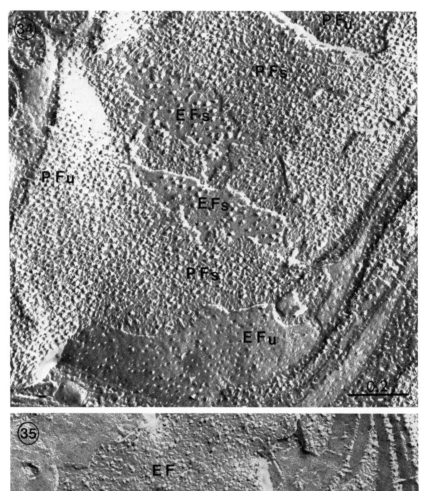

particles appear to be isolated from each other, even in stacked regions where their repartition density per area unit is increased. This contradiction is not well understood and is still open to debate (see Paillotin and Swenberg, 1979).

C. RELATIVE ORGANIZATION OF THYLAKOIDS IN THE CHLOROPLAST

Thylakoids of *Euglena* are appressed by two or three over their entire length (Buetow, 1968; Schiff, 1973) and therefore do not form grana stacks separated by stroma lamellae as they do in higher plants and most green algae (Fig. 32). However, the structure of the contact between two adjacent appressed thylakoids does not seem to differ basically from the partition between stacked membranes in grana of higher plants (Fig. 33).

In spite of extensive studies on the composition and structure of such membrane appressions in higher plant chloroplasts (for a review, see Anderson, 1975; Arntzen and Briantais, 1975; Arntzen, 1978; Boardman *et al.,* 1978), their physiological significance is still not well understood. Since most PSII activity and LHCP are concentrated in grana regions, it has been proposed that membrane stacking was necessary for PSII activity. This seemed to be confirmed by numerous observations relating the absence of PSII activity in chloroplasts containing unstacked thylakoids (see Arntzen and Briantais, 1975). Similar observations of a correlation between a lack of PSII activity and unpaired thylakoids in *Euglena* have been reported by Schwelitz *et al.* (1972), Polak-Charcon *et al.* (1975), and Ophir and Ben Shaul (1974) in mutants.

A complete lack of correlation between PSII activity and grana structure has been demonstrated in *Chlamydomonas* mutants ac-5 and ac-31 (Goodenough and Staehelin, 1971), in mutants of higher plants (see Arntzen and Briantais, 1975), and in pea greened under intermittent light (Armond *et al.,* 1977). *Euglena* does not differ from higher plants and green algae in this respect, since greening under intermittent light also leads to unpaired thylakoids (Figs. 31 and 33) with a high photosynthetic competence (Dubertret and Lefort-Tran, 1978b; 1981). Moreover, loss of grana structures obtained by washing isolated chloroplasts in a low salt medium does not affect photochemical activity (Izawa and Good, 1966).

On the other hand, studies of chloroplast fractionation of mutants on C_4 plants,

---

Fig. 34. Freeze–fracture of thylakoids in a chloroplast developed under continuous light. Four fracture faces can be recognized: EFs and PFs in stacked regions and EFu and PFu in unstacked regions. ×83,200.

Fig. 35. Freeze–fracture of unappressed thylakoids developed for 4 days under intermittent light exhibits only two fracture faces, EF and PF. The histograms of particle sizes (Fig. 36) show that at least EF faces are different from EFs and EFu faces found in fully greened chloroplasts. ×83,200.

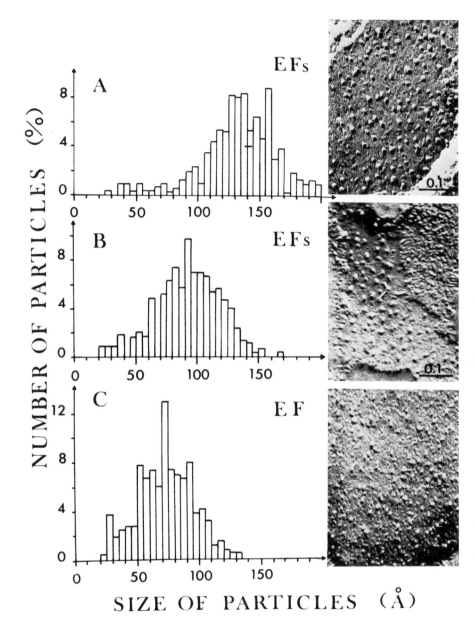

on algae, and on organisms greened under continuous or intermittent light offer several lines of evidence that LHCP could play a role in membrane stacking (for a review, see Anderson, 1975). Another supporting argument is that the general organization of chloroplast membranes in grana stacks depends on appropriate concentration of cations (Izawa and Good, 1966) and that the same concentration of cations induces the aggregation of isolated LHCP (Burke *et al.*, 1978). However, it should be kept in mind that a chlorophyll *b*-less barley mutant, which totally lacks LHCP (Miller *et al.*, 1976, Simpson, 1979), still exhibits some grana stacks, and this raises some questions about the role of LHCP in the formation of these grana stacks.

Thylakoid stacking thus is not required for PSII activity and seems to depend on the presence of LHCP. These observations then suggest that membrane stacking could ensure a higher yield for photochemical reactions. Senger *et al.* (1975) and Armond *et al.* (1976) observed during the greening of a *Scenedesmus* mutant and of pea an increase in quantum efficiency that correlates with the development of the stacking process. If this increase in quantum efficiency does not originate in decreasing proportions of unorganized chlorophylls as greening proceeds, one has to assume that it could result from an improved distribution of the absorbed energy among photosynthetic units when thylakoids are stacked (see Arntzen, 1978).

Energy distribution between photosynthetic units supposes energy transfers between these units. Most of these transfers occur between system II units or from system II to system I units (for a review, see Williams, 1977) and can easily be measured under conditions in which thylakoids are stacked or unstacked. As already stated, thylakoids are unstacked in some chlorophyll *b*-less mutants and in chloroplasts developed under intermittent light, or they can be experimentally unstacked by washing isolated mature chloroplasts with low-salt mediums. When measured in mature chloroplasts containing stacked membranes (*in vivo* or isolated in the presence of sufficient concentration of cations), energy transfer between system II units is efficient, whereas energy transfer from PSII to PSI (spillover) is reduced. Destacking thylakoids by incubating these mature chlorop-

---

Fig. 36. Histograms of particle sizes from exoplasmic fractured faces of stacked (EFs) and unstacked (EF) thylakoids. (a) EFs face of stacked thylakoids developed during greening under continuous light (72 hours, Chl *a*/Chl *b* =5-7). The particle density is 750 particles/μm^2. (b) EFs face of stacked thylakoids developed in green cells mixotrophically grown in the light for several generations (Chl *a*/Chl *b* = 7-9). The particle density is 973 particles/μm^2. (c) EF face of unstacked thylakoids developed under intermittent light (100 hours, Chl *a*/Chl *b* > 18). The particle density is 880 particles/μm^2. Depending on experimental conditions, particle size distributions are shifted to low values compared to the control greened under continuous light. This decrease in EF particle size is visible in micrographs on the right side of the figure. It can be correlated to decreasing relative amounts of chlorophyll *b* in the analyzed cells.

lasts in low-salt medium entails an important increase in spillover, concomitant with a decrease and disappearance of energy transfer between system II units (for a review, see Williams, 1977). Vernotte *et al.* (1976) showed that a reduced spillover exists in the *Su/su* tobacco mutant, which is deficient in chlorophyll *b*, LHCP, and grana structures. However, they did not report on the intersystem II energy transfer. When the development of unstacked LHCP-less thylakoids is induced by intermittent light in higher plants (Armond *et al.*, 1976; Akoyunoglou, 1977) or in *Euglena* (Dubertret and Lefort-Tran, 1978b; 1981), energy transfer does not occur between system II units. Moreover, Armond *et al.* (1976) showed that the cation-dependent regulation of energy transfer from system II to system I correlates with the appearance of LHCP and of grana during the final steps of membrane assembly under continuous light.

These observations confirm that energy distribution among photosynthetic units is related to the presence of LHCP and to the extent of stacking of photosynthetic membranes. In mature chloroplasts, spillover and efficient intersystem II energy transfers appear to be mutually exclusive, and depending on cation concentration, energy distribution in stacked membranes favors system II, whereas in unstacked thylakoids, more of the absorbed energy is funneled toward system I. Analyses of modified chloroplasts lacking grana stacks, in which intersystem II energy transfers do not occur systematically, should clarify the existence and the extent of the spillover.

However, Vandermeulen and Govindjee (1974) observed that the kinetics of spillover changes after removal of cations in the suspending medium of isolated chloroplasts was faster than changes in light scattering. These results were interpreted as reflecting macrostructural changes in membrane organization. Moreover, Telfer *et al.* (1976) observed that the effect of Mg^{2+} on stacking can be divorced from that of spillover. Membrane appression then could be only involved indirectly in the distribution of energy between photosynthetic units.

Basically, spillover or intersystem II energy transfers require that the concerned photosynthetic units be in contact or very close to each other. Changes in the efficiency of energy transfers may, *a priori*, be controlled by different types of mechanisms.

One possibility could be that some discrete, cation-induced conformational changes within photosynthetic units are sufficient to regulate energy transfers. Assuming that system I and system II units are included in a whole structure that cannot be dissociated, Seely (1973) proposed that energy absorbed by PSII pigments may be funneled to PSI reaction centers through a set of a few chlorophylls, whose relative orientation, depending on cation concentration, would regulate the spillover. In this case, the fact that the onset of spillover regulation during greening coincides with LHCP formation (Armond *et al.*, 1976) would suggest that these key chlorophylls are included in this chlorophyll–protein complex. However, this model does not explain how intersystem II transfers occur and are regulated.

Alternatively, another possibility could be that changes in energy transfers result from cation-induced, macrostructural changes affecting the relative disposition of individual system I and system II units in the photosynthetic membrane. In the framework of this hypothesis, one has to remember that changes in cation concentration are associated with changes in membrane stacking and in freeze–fracture particle distribution and density per unit area when LHCP is present in photosynthetic membranes (Ojakian and Satir, 1974; Staehelin, 1976). One can then consider that membrane stacking per se is a secondary effect of cation concentration and, therefore, that all types of energy transfers occur in the plane of each membrane. In this case, the dispersion of EF particles under low salt conditions would favour interactions between system I and system II units rather than interactions between system II units; the concentration of EF particles in stacked regions under high salt conditions would, on the contrary, reduce spillover and increase intersystem II energy transfers. However, since most EF particles appear to be isolated from each other, even when concentrated in stacked regions, one has to postulate that intersystem II energy transfers occur through pigment–protein complexes, which would remain associated with protoplasmic fracture faces (Wollman et al., 1980; Simpson, 1979).

In contrast, one can consider that membrane appression is directly involved in the nature of energy transfers between two adjacent membranes, and these transfers could therefore occur along the membrane or across the partition (Arntzen, 1978). In this case, direct contact of EF particles with those of adjacent membranes should allow the formation of domains, including several system II units in which intersystem II energy transfers would be favored. Low salt concentration, by unstacking thylakoids, would interrupt energy transfer between system II units and would allow interaction between system I and system II to be manifest.

The lack of LHCP in mutants or in chloroplasts developed under intermittent light and the subsequent lack either of EF particle concentration along each membrane or of contacts between EF particles across the partition would then be responsible for the absence of energy transfers between system II units. Concerning spillover, more analyses are required in order to determine whether or not it exists in unappressed photosynthetic membranes devoid of LHCP.

Such models of energy transfer regulation, which have been proposed for higher plants and green algae, seem to be valid in the case of *Euglena*, in which in addition to intersystem II energy transfers, spillover has been characterized by Jennings and Forti (1974a,b). However, these authors suggested that changes in spillover are not cation-dependent and are related to the amount of chlorophyll forms absorbing ~695 nm.

IV. Concluding Remarks

In "The Biology of Euglena" Cook (1968) wrote ". . . to the euglenaphiles, at least, the next decade should prove even more exciting than the last". Whether this was a premonition or an acute prediction, the present Volumes III and IV should demonstrate the soundness of this opinion. General chloroplast organization and function have entered the mainstream of molecular biology, and the *Euglena* system has been particularly rewarding because of its capacity to differentiate from a proplastid stage to a fully developed chloroplast under a variety of conditions.

Despite, or perhaps because of, our improved knowledge of *Euglena* chloroplast biology, many questions arise. Why, among Chlorophyta, are *Euglena* chloroplasts surrounded by three membranes instead of two? What is the nature of the third sheet? What is the functional and the evolutionary significance of such an organization? Concerning the first two questions, the isolation of intact *Euglena* chloroplasts coupled with studies on membrane fractions would lead to a better understanding of the nature of the three envelope membranes and their roles. The last question involves speculations about the evolutionary origins of the organelle. The *Euglena* chloroplast envelope may be a strong endorsement for the endosymbiotic hypothesis of chloroplast origin.

As regards the photosynthetic apparatus, *Euglena* differs from higher plants and green algae by its low content of chlorophyll *b* and LHCP and by the general organization of thylakoids, which are appressed by two or three over their entire length. A first, unanswered question concerns the determinism of such a type of thylakoid stacking. However, even if different in general organization, this appression of thylakoids seems to allow the same properties of energy transfer and, therefore, most likely the same functions as in higher plants.

Considered at the level of the light-harvesting system, the low amounts of chlorophyll *b* and LHCP seem to constitute only a quantitative difference between *Euglena* and higher plants. These low amounts account for a reduced size of system II light-harvesting antennae and of EF freeze-fracture particles, just as in higher plants, in which the concentration of LHCP has been experimentally decreased by intermittent light. Moreover, the absence of the light-harvesting system induces the same effects on the size of the system II unit and of EF particles, on energy transfers, and on the general organization of thylakoids, as its absence does in the chloroplasts of higher plants.

More specific to *Euglena* are system II reaction centers and their functional properties. In higher plants and green algae, the stable components of system II units appear to be the reaction centers: They are the first basic building blocks of system II units to be formed during greening and are later completed by the progressive development of light-harvesting antennae. Moreover, the adaptation to different light intensities consists in regulating the size of light-harvesting antennae, thereby adapting the number of excitations reaching system II reaction centers, for which there is little functional turnover. In contrast, PSII reaction centers seem to be more labile in *Euglena*. They are the last components of system II units to be formed during greening and after the light-harvesting antennae have been completed. Also they are more sensitive to the photodestroying effects of high light intensities than are light-harvesting antennae. In both cases, the low number of active reaction centers during the first stages of greening and their large optical cross section under high light intensities should constitute an unfavorable situation for *Euglena*. However, an adaptation of the turnover rate of functional PSII reaction centers in *Euglena* appears to be the mechanism that compensates for the specific differences in its case. Further studies should help to define the origins of the differences in the properties of PSII reaction centers of *Euglena* compared to those of higher plants.

Acknowledgments

We are grateful to M. Pouphile for her expert assistance in electron microscopy, to F. Ambard-Bretteville for her skillful help in functional measurements, and to R. Boyer for his helpful photographic work. We thank also E. Marnet and M. Montaggioni for their valuable and patient secretarial tasks. All electron microscopy figures were reproduced from the author's original micrographs.

References

Akoyunoglou, G. (1977). *Arch. Biochem. Biophys.* **183**, 571–580.

Anderson, J. M. (1975). *Biochim. Biophys. Acta* **416**, 191–235.

Anderson, J. M., Waldron, J. C., and Thorne, S. W. (1978). *FEBS Lett.* **92**, 227–233.

Armond, P. A., Arntzen, C. J., Briantais, J. M., and Vernotte, C. (1976). *Arch. Biochem. Biophys.* **175**, 54–63.

Armond, P. A., Staehelin, L. A., and Arntzen, C. J. (1977). *J. Cell Biol.* **73**, 400–418.

Arntzen, C. J. (1978). *Curr. Top. Bioenerg.* **8**, 111–160.

Arntzen, C. J., and Briantais, J. M. (1975). *In* "Bioenergetics of Photosynthesis" (Govindjee, ed.), pp. 51–113. Academic Press, New York.

Arntzen, C. J., Dilley, R. A., Peters, G. A., and Shaw, E. R. (1972). *Biochim. Biophys. Acta* **256**, 85–107.

Baker, N. R., and Leech, R. M. (1977). *Plant Physiol.* **60**, 640–644.

Bennoun, P., and Chua, N. H. (1976). *In* "Genetics and Biogenesis of Chloroplasts and Mitochondria" (T. H. Büchner *et al.*, eds.), pp. 33–39. Elsevier/North-Holland Biomedical Press, Amsterdam.

Ben-Shaul, Y., Silman, R., and Ophir, I. (1972). *Physiol. Veg.* **10**, 255–268.

Billecocq, A. (1975). *Ann. Immunol. (Paris)* **126C**, 337–352.

Bisalputra, T. (1974). *In* "Algal Physiology and Biochemistry" (W. D. P. Stewart, ed.), Vol. 10, p. 124–160. Blackwell, Oxford.

Bisalputra, T., and Bailey, A. (1973). *Protoplasma* **76**, 443–454.

Blair, G. E., and Ellis, R. J. (1973). *Biochim. Biophys. Acta* **319**, 223–234.

Boardman, N. K., Anderson, J. M., and Goodchild, D. J. (1978). *Curr. Top. Bioenerg.* **8**, 35–109.

Bogorad, L. (1975). *In* "Membrane Biogenesis" (A. Tzagoloff, ed.), pp. 201–245. Plenum, New York.

Bottomley, W., Spencer, D., and Whitfeld, P. R. (1974). *Arch. Biochem. Biophys.* **164**, 106–117.

Bovarnick, J. G., Chang, S. W., Schiff, J. A., and Scwartzbach, S. D. (1974). *J. Gen. Microbiol.* **83**, 51–62.

Bradbeer, J. W., and Börner, T. (1978). In "Chloroplast Development" (G. Akoyunoglou et al., eds.), pp. 727-732. Elsevier/North-Holland Biomedical Press, Amsterdam.

Bradbeer, J. W., Atkinson, Y. E., Börner, T., and Hagemann, R. (1979). Nature (London) 279, No. 5716, 816-817.

Branton, D., Bullivant, S., Gilula, N. B., Karnovsky, M. J., Moor, H., Mühlethaler, K., Northcote, D. H., Packer, L., Satir, P., Speth, V., Staehelin, L. A., Steere, R. L., and Weinstein, R. S. (1975). Science 190, 54-56.

Branton, P., and Park, R. B. (1967). J. Ultrastruct. Res. 19, 283-303.

Brown, J. S., Alberte, R. S., and Thornber, J. P. (1975). Proc. Int. Congr. Photosynth., 3rd, 1974 pp. 1951-1962.

Buetow, D. E. (1968). In "The Biology of Euglena" (D. E. Buetow, ed.), Vol. I, pp. 109-184. Academic Press, New York.

Buetow, D. E. (1976). J. Protozool. 23 (1), 41-47.

Burke, J. J., Ditto, C. L., and Arntzen, C. J. (1978). Arch. Biochem. Biophys. 187, 252-263.

Butler, W. L. (1965). Biochim. Biophys. Acta 102, 1-8.

Butler, W. L., and Kitajima, M. (1975). Proc. Int. Congr. Photosynth., 3rd, 1974 pp. 13-24.

Butler, W. L., and Kitajima, M. (1975). Biochim. Biophys. Acta 376, 116-125.

Cahen, D., Malkin, S., Shochat, S., and Ohad, I. (1976). Plant Physiol. 58, 257-267.

Cahen, D., Malkin, S., Gurevitz, M., and Ohad, I. (1978). Plant Physiol. 62, 1-5.

Chan, P. H., and Wildman, S. G. (1972). Biochim. Biophys. Acta 277, 677-680.

Chua, N. H., and Schmidt, G. W. (1978). In "Photosynthetic Carbon Assimilation" (H. W. Siegelman and G. Hind, eds.), pp. 325-347. Plenum, New York.

Chua, N. H., and Schmidt, G. W. (1979). J. Cell Biol. 81, 461-483.

Cobb, A. A., and Wellburn, A. R. (1974). Planta 121, 273-282.

Cobb, A. A., and Wellburn, A. R. (1976). Planta 129, 127-131.

Cohen, D., and Schiff, J. A. (1976). Arch. Biochem. Biophys. 177, 201-216.

Delepelaire, P., and Chua, N. H. (1979). Proc. Natl. Acad. Sci. U.S.A. 76, 111-115.

Dempsey, G. P., Bullivant, S., and Watkins, W. B. (1973). Science 179, 190.

Diamond, J. (1976). Planta 130, 145-149.

Diner, B. A., and Mauzerall, D. (1973). Biochim. Biophys. Acta 292, 285-290.

Diner, B. A., and Wollman, F. A. (1979). Plant Physiol. 63, 20-25.

Dobberstein, B., Blöbel, G., and Chua, N. H. (1977). Proc. Natl. Acad. Sci. U.S.A. 74, 1082-1085.

Dodge, J. D. (1974). In "The Fine Structure of Algal Cells" (J. D. Dodge, ed.), pp. 81-103. Academic Press, New York.

Douce, R., and Joyard, J. (1978). In "Chloroplast Development" (G. Akoyunoglou et al., eds.), pp. 283-296. Elsevier/North Holland Biomedical Press, Amsterdam.

Douce, R., and Joyard, J. (1979). Adv. Bot. Res. 7, 1-116.

Dubertret, G. (1981a). Plant Physiol. 67, 47-53.

Dubertret, G. (1981b). Plant Physiol. 67, 54-58.

Dubertret, G., and Joliot, P. (1974). Biochim. Biophys. Acta 357, 399-411.

Dubertret, G., and Lefort-Tran, M. (1978a). Biochim. Biophys. Acta 503, 316-332.

Dubertret, G., and Lefort-Tran, M. (1978b). In "Chloroplast Development" (G. Akoyunoglou et al., eds.), pp. 419-425. Elsevier/North-Holland Biomedical Press, Amsterdam.

Dubertret, G., and Lefort-Tran, M. (1981). Biochim. Biophys. Acta 634, 52-69.

Egneus, H., Reftel, S., and Sellden, G. (1972). Physiol. Plant 27, 48-55.

Ellis, R. J. (1977). Biochim. Biophys. Acta 463, 185-215.

Emerson, R., and Arnold, A. (1932). J. Gen. Physiol. 16, 191-205.

Fehlmann, M., Bellemare, G., and Godin, C. (1975). Biochim. Biophys. Acta 378, 119-124.

Fenna, R. E., and Matthews, B. W. (1977). Brookhaven Symp. Biol. 28, 170-182.

Freyssinet, G. (1976). *Plant Physiol.* **57**, 831–835.

Freyssinet, G. (1977a). *Biochimie* **59**, 597–610.

Freyssinet, G. (1977b). *In* Doctoral Thesis, University Claude Bernard, Lyon I. Lyon.

Freyssinet, G. (1978). *Exp. Cell Res.* **115**, 207–219.

Gaffron, A., and Wohl, K. (1936). *Naturwissenschaften* **24**, 81–103.

Genge, S., Pilger, D., and Hiller, R. G. (1974). *Biochim. Biophys. Acta* **347**, 22–30.

Gibbs, S. P. (1962). *J. Cell Biol.* **14**, 433–444.

Gibbs, S. P. (1970). *Ann. N.Y. Acad. Sci.* **175**, 454–473.

Gibbs, S. P. (1978). *Can. J. Bot.* **56**, 2883–2889.

Goodchild, D. J., and Park, R. B. (1971). *Biochim. Biophys. Acta* **226**, 393–399.

Goodenough, U. W., and Staehelin, L. A. (1971). *J. Cell Biol.* **48**, 594–619.

Gray, J. C., and Kekwick, R. G. O. (1974a). *Eur. J. Biochem.* **44**, 481–489.

Gray, J. C., and Kekwick, R. G. O. (1974b). *Eur. J. Biochem.* **44**, 491–500.

Gray, J. C., Kung, S. D., and Wildman, S. G. (1978). *Arch. Biochem. Biophys.* **185**, 272–281.

Gulik-Krzywicki, T., and Costello, M. J. (1977). *Proc.—Annu. Meet., Electron Microsc. Soc. Am.* **35**, 330–333.

Gurevitz, M., Kratz, H., and Ohad, I. (1977). *Biochim. Biophys. Acta* **461**, 475–488.

Haan, G. A., Warden, J. T., and Duysen, L. N. M. (1973). *Biochim. Biophys. Acta* **325**, 120–125.

Hackenbrock, C. R., and Miller, K. J. (1975). *J. Cell Biol.* **65**, 615–630.

Hayden, D. B., and Hopkins, W. G. (1977). *Can. J. Bot.* **55**, 2525–2529.

Heber, U., and Krause, G. H. (1971). *In* "Photosynthesis and Photorespiration" (M. D. Hatch *et al.*, eds.), pp. 218–225. Wiley (Interscience), New York.

Heber, U., and Walker, D. A. (1979). *Trends Biochem. Sci.* **4**, 252–256.

Heizmann, P. (1974). *Biochim. Biophys. Acta* **353**, 301–312.

Heizmann, P., Trabuchet, G., Verdier, G., and Nigon, V. (1972). *Biochim. Biophys. Acta* **277**, 149–160.

Heldt, H. W. (1969). *FEBS Lett.* **5**, 11–14.

Heldt, H. W., and Rappley, Y. L. (1970). *FEBS Lett.* **10**, 143–148.

Heldt, H. W., and Sauer, F. (1971). *Biochim. Biophys. Acta* **234**, 83–91.

Henningsen, K. W., and Boardman, N. K. (1973). *Plant Physiol.* **51**, 1117–1126.

Henriques, F., and Park, R. B. (1976). *Proc. Natl. Acad. Sci. U.S.A.* **73**, 4560–4564.

Henriques, F., and Park, R. B. (1978). *Plant Physiol.* **62**, 856–860.

Herron, H. A., and Mauzerall, D. (1972). *Plant Physiol.* **51**, 141–148.

Holowinsky, A. W., and Schiff, J. A. (1970). *Plant Physiol.* **45**, 339–347.

Holt, S. C., and Stern, A. I. (1970). *Plant Physiol.* **45**, 475–483.

Howell, S. H., and Moudrianakis, E. N. (1967a). *J. Mol. Biol.* **27**, 323–333.

Howell, S. H., and Moudrianakis, E. N. (1967b). *Proc. Natl. Acad. Sci. U.S.A.* **58**, 1261–1268.

Izawa, S., and Good, N. E. (1966). *Plant Physiol.* **41**, 544–553.

Jennings, R. C., and Forti, G. (1974a). *Plant Sci. Lett.* **3**, 29–33.

Jennings, R. C., and Forti, G. (1974b). *Biochim. Biophys. Acta* **347**, 299–310.

Joliot, A., and Joliot, P. (1964). *C.R. Hebd. Seances Acad. Sci.* **258**, 4622–4625.

Joyard, J., and Douce, R. (1976). *Physiol. Veg.* **14**, 31–48.

Junge, W. (1977). *Encycl. Plant Physiol., New Series,* **5**, 59–93.

Karakashian, S. J., Karakashian, M. W., and Rudzinska, M. A. (1968). *J. Protozool.* **15**, 113–128.

Kawashima, N., and Wildman, S. G. (1971). *Biochim. Biophys. Acta* **229**, 749–760.

Kawashima, N., and Wildman, S. G. (1972). *Biochim. Biophys. Acta* **262**, 42–49.

Kirk, J. T. O., and Tilney-Basset, R. A. E. (1967). *In* "The Plastids" (J. T. O. Kirk *et al.*, eds.), pp. 572–583. Freeman, San Francisco, California.

Kivic, P. A., and Vesk, M. (1974). *Can. J. Bot.* **52**, 695–699.

Laetsch, W. M. (1968). *Am. J. Bot.* **55**, 875–883.

310 G. Dubertret and M. Lefort-Tran

Laetsch, W. M. (1971). In "Photosynthesis and Photorespiration" (M. D. Hatch et al., eds.), pp. 323-349. Wiley (Interscience), New York.

Laetsch, W. M., and Rice, J. (1969). Am. J. Bot. 56, 77-87.

Laulhere, J. P., and Dorme, A. M. (1977). Plant Sci. Lett. 8, 251-256.

Lavorel, J. (1976). Courr. du CNRS 22, 19-24.

Leedale, G. F. (1967). In "Euglenoid Flagellates" (G. F. Leedale, ed.), pp. 169-184. Prentice-Hall, Englewood Cliffs, New Jersey.

Leedale, G. F. (1968). In "The Biology of Euglena" (D. E. Buetow, ed.), Vol. 1, pp. 185-242. Academic Press, New York.

Lefort, M. (1959). Rev. Gen. Bot. 66, 1-13.

Lefort, M. (1965). C. R. Hebd. Seances Acad. Sci. 261, 233-236.

Lefort-Tran, M., Gulik, T., Plattner, H., Beisson, J., and Wiessner, W. (1978). Electron Micros. Proc. Int. Congr., 9th, 1978 Vol. II, pp. 146.

Lefort-Tran, M., Pouphile, M., Feyssinet, G., and Pineau, B. (1980). J. Ultrastruct. Res. 73, 43-63.

Machold, O., Simpson, D. J., and Moller, B. L. (1979). Carlsberg Res. Commun. 44, 235-254.

Malkin, P., and Kok, B. (1966). Biochim. Biophys. Acta 126, 413-432.

Malnoë, P., Rochaix, J. D., Chua, N. H., and Spahr, P. H. (1979). J. Mol. Biol. 133, 417-434.

Markwell, J. P., Thornber, J. P., and Boggs, R. T. (1979). Proc. Natl. Acad. Sci. U.S.A. 76, 1233-1235.

Meier, R., Reisser, W., Wiessner, W., and Lefort-Tran, M. (1980). Z. Naturforsch., C. Biosci. 35c, 1107-1110.

Melis, A., and Akoyunoglou, G. (1977). Plant Physiol. 59, 1156-1160.

Melkonian, M., and Robenek, H. (1979). Protoplasma 100, 183-197.

Menke, W. (1960). Z. Naturforsch. B: Anorg. Chem., Org. Chem., Biochem., Biophys., Biol. 15B, 479-482.

Menke, W. (1962). Annu. Rev. Plant Physiol. 13, 27-44.

Mereschkowsky, C. (1905). Biol. Zentralbl. 25, 593-604.

Miller, K. R., and Cushman, R. A. (1979). Biochim. Biophys. Acta 546, 481-497.

Miller, K. R., and Staehelin, L. A. (1973). Protoplasma 77, 55-78.

Miller, K. R., Miller, G. J., and McIntyre, K. R. (1976). J. Cell Biol. 71, 624-638.

Miller, K. R., Miller, G. J., and McIntyre, K. R. (1977). Biochim. Biophys. Acta 459, 145-156.

Mitchell, P. (1966). Biol. Rev. Cambridge Philos. Soc. 41, 445-502.

Mühlethaler, K., and Frey-Wyssling, A. (1959). J. Biophys. Biochem. Cytol. 6, 507-512.

Myers, J., and Graham, J. R. (1971). Plant Physiol. 48, 282-286.

Neuburger, M., Joyard, J., and Douce, R. (1977). Plant Physiol. 59, 1178-1181.

Neville, D. M., Jr., and Chang, T. M. (1978). Curr. Top. Membr. Transp. 1, 65-150.

Ohad, I., Siekevitz, P., and Palade, G. E. (1967). J. Cell. Biol. 35, 521-551.

Ojakian, G. K., and Satir, P. (1974). Proc. Natl. Acad. Sci. U.S.A. 21, 2052-2056.

Ophir, I., and Ben Shaul, Y. (1974). Protoplasma 80, 109-127.

Ortiz, W., and Stutz, E. (1980). FEBS Lett. 116, 298-302.

Paillotin, G., and Swenberg, C. E. (1979). Ciba Found. Symp. [N.S.] 61, 201-215.

Park, R. B., and Biggins, J. (1964). Science, 144, 1009-1011.

Park, R. B., and Branton, D. (1966). Broohaven Symp. Biol. 19, 341-352.

Park, R. B., and Pfeifhofer, A. O. (1968). Proc. Natl. Acad. Sci. U.S.A. 60, 337-343.

Park, R. B., and Pon, N. P. (1963). J. Mol. Biol. 6, 105-114.

Park, R. B., and Sane, P. V. (1971). Annu. Rev. Plant Physiol. 22, 395-430.

Philippovitch, I. I., Tongur, A. M., Alina, B. A., and Oparin, A. I. (1970). Exp. Cell Res. 62, 399-406.

Phung Nhu Hung, P., Lacourly, A., and Sarda, C. (1970). Z. Pflantzenphysiol. 62, 1-16.

Pineau, B. (1980). Thése de doctorat d'Etat. Paris XI.

Pineau, B., and Douce, R. (1974). *FEBS Lett.* **47**, 255-259.

Pineau, B., Ledoigt, G., Maillefer, C., and Lefort-Tran, M. (1979). *Plant Sci. Lett.* **15**, 331-343.

Plattner, H., Schmitt-Fumian, W. W., and Bachmann, L. (1973). *In* "Freeze-etching Techniques and Applications" (E. L. Benedetti and P. Favard, eds.), pp. 81-100. S.F.M.E., Paris.

Polak-Charcon, S., Porat, N., Shneyour, A., and Ben Shaul, Y. (1975). *Proc. Int. Congr. Photosynth., 3rd, 1974* Vol. 3, pp. 1913-1923.

Priestley, D. A. (1977). Ph.D. Thesis, University of Leeds, England.

Remy, R. (1969). *C.R. Hebd. Seances Acad. Sci., Ser. D,* 268, 3057-3060.

Remy, R. (1971). *FEBS Lett.* **13**, 313-317.

Remy, R., and Hoarau, J. (1978). *In* "Chloroplast Development" (G. Akoyunoglou *et al.*, eds.), pp. 235-240. Elsevier/North-Holland Biomedical Press, Amsterdam.

Rosado-Alberio, J. T., Weier, T. E., and Stocking, C. R. (1968). *Plant Physiol.* **43**, 1325-1331.

Roy, H., Patterson, R., and Jagendorf, A. T. (1976). *Arch. Biochem. Biophys.* **172**, 64-73.

Salvador, G., Lefort-Tran, M., Nigon, V., and Jourdan, F. (1971). *Exp. Cell Res.* **64**, 458-462.

Sane, P. V., Goodchild, D. J., and Park, R. B. (1970). *Biochim. Biophys. Acta* **216**, 162-178.

Satir, P., and Satir, B. (1974). *Cold Spring Harbor Conf. Cell Proliferation* **1**, 233-249.

Schiff, J. A. (1970). *Symp. Soc. Exp. Biol.* **24**, 227-302.

Schiff, J. A. (1973). *Adv. Morphog.* **10**, 265-312.

Schimper, A. F. W. (1885). *Jahrb. Wiss. Bot.* **16**, 1-247.

Schmid, G. H., and Gaffron, A. (1971). *Photochem. Photobiol.* **14**, 451-464.

Schnepf, E. (1966). *In* "Problem des Biologischen keduplikation" (P. Sitte, ed.), pp. 372-393. Springer-Verlag, Berlin and New York.

Schwartzbach, S. D., Schiff, J. A., and Klein, P. (1976). *Planta* **131**, 1-9.

Schwelitz, F. D., Dilley, R. A., and Crane, F. L. (1972). *Plant Physiol.* **50**, 166-170.

Seely, G. R. (1973). *J. Theor. Biol.* **40**, 189-199.

Senger, H., Bishop, N. I., Wehrmeyer, W., and Kulandaivelu, J. (1975). *Proc. Int. Congr. Photosynth., 3rd, 1974* Vol. 3, pp. 1913-1923.

Shumway, L. K., and Weier, T. (1967). *Am. J. Bot.* **54**, 773-780.

Simpson, D. J. (1979). *Carlsberg Res. Commun.* **44**, 305-336.

Sprey, B., and Laetsch, W. M. (1976a). *Z. Pflanzenphysiol.* **78**, 146-163.

Sprey, B., and Laetsch, W. M. (1976b). *Z. Pflanzenphysiol.* **78**, 360-371.

Sprey, B., and Laetsch, W. M. (1978). *Z. Pflanzenphysiol.* **87**, 37-53.

Staehelin, L. A. (1976). *J. Cell Biol.* **71**, 136-158.

Staehelin, L. A. (1978). *In* "Light Transducing Membranes" (D. Deamer, ed.), pp. 335-355. Academic Press, New York.

Stern, A. I., Schiff, J. A., and Epstein, H. T. (1964). *Plant Physiol.* **39**, 220-226.

Strotmann, H., and Berger, S. (1969). *Biochem. Biophys. Res. Commun.* **35**, 20-26.

Taylor, F. J. R. (1976). *J. Protozool.* **23**, No. 1, 28-41.

Telfer, A., Nicolson, J., and Barber, J. (1976). *FEBS Lett.* **65**, 77-83.

Thornber, J. P. (1975). *Annu. Rev. Plant Physiol.* **26**, 127-158.

Thornber, J. P., and Highkin, H. R. (1974). *Eur. J. Biochem.* **41**, 109-116.

Thornber, J. P., Gregory, R. P. F., Smith, C. A., and Leggett-Bailey, J. (1967). *Biochemistry* **6**, 391-396.

Thornber, J. P., Alberte, R. S., Hunter, F. A., Shiozawa, J. A., and Kan, K. S. (1977). *Brookhaven Symp. Biol.* **28**, 132-148.

Vandermeulen, D. L., and Govindjee (1974). *Biochim. Biophys. Acta* **368**, 61-70.

Vasconcelos, A. C. (1976). *Plant Physiol.* **58**, 719-721.

Vasconcelos, A. C., Mendiola-Morgenthaler, R., Floyd, G. L., and Salisbury, J. R. (1976). *Plant Physiol.* **58**, 87-90.

Vernotte, C., Briantais, J. M., and Remy, R. (1976). *Plant Sci. Lett.* **6,** 135–141.

Virgin, H., Kahn, A., and von Wettsein, D. (1963). *Photochem. Photobiol.* **2,** 83.

Vivier, E., Petiprez, A., and Chive, A. E. (1967). *Protistologica* **3,** 325–335.

von Wettstein, D. (1959). *Brookhaven Symp. Biol.* **16,** 123–160.

Wild, A. (1969). *Prog. Photosynth. Res., Proc. Int. Congr. [1st], 1968* pp. 871–876.

Williams, W. P. (1977). *In* "Primary Processes of Photosynthesis" (J. Barber, ed.), pp. 99–147. Elsevier/North-Holland Biomedical Press, Amsterdam.

Wollman, F. A. (1979). *Plant Physiol.* **63,** 375–381.

Wollman, F. A., Olive, J., Bennoun, P., and Recouvreur, M. (1980). *J. Cell Biol.* **87,** 728–735.

CHAPTER 8

PHOTOCONTROL OF CHLOROPLAST DEVELOPMENT IN *EUGLENA*

Jerome A. Schiff and Steven D. Schwartzbach

I. Introduction

Evolution, through adaptation and selection, has resulted in the complex cellular process of photosynthesis that traps light energy and makes it available to living systems. Evolution has also selected systems in which light serves as a means of controlling the development and replication of the chloroplast. In order to be adaptive, such control systems must respond to the same wavelengths of light that are effective for photosynthesis since there would be little evolutionary advantage in a system which brought about the formation of a chloroplast in

THE BIOLOGY OF *EUGLENA*, VOL. 3

response to wavelengths of light which are ineffective; for this reason, blue and red light are the most effective regions of the spectrum in controlling plastid development in green organisms such as *Euglena*.

The control system shows many resemblances to those already familiar to us from substrate induction in bacterial systems. In *Escherichia coli,* for example, if organisms grown on glucose are presented with, say, lactose, several enzyme activities are induced that enable the cells to utilize lactose. In the case of photosynthesis, however, the substrate to be utilized is light, and in order to do this, the induction of many enzymes, structural organization in the form of membranes, pigment complexes, etc., must be formed, all being properly coordinated in time. It is this series of coordinated substrate–induced enzyme inductions that we call chloroplast development and the end result is the formation of an organelle, rather than just a few enzymes, to enable the organism to utilize light for energy storage through photosynthesis (Schiff, 1980a).

This review will be mainly concerned with the photocontrol of plastid development in *Euglena*. It will not be possible to provide an exhaustive survey of everything that has been written about chloroplast development in this organism. However, since the field has been reviewed at intervals, most of this material can be found in the following publications: Nigon and Heizmann (1978), Schmidt and Lyman (1976), Schiff (1973, 1975, 1978, 1980a,b), Leedale (1967), Buetow (1968), Buetow and Wood (1978), Buetow *et al.* (1980), Parthier (1981), and in the other chapters of this treatise.

II. Light in the Hierarchy of Inducing Substrates

As far-reaching as the effects of light are on chloroplast development, light is only one of several substrates to which the cell can respond. If we are to understand light induction, we must also understand the role of light in relation to regulation by other substrates. Free-living organisms often use a wide variety of organic compounds as sources of energy and carbon. For example, cells of *Euglena gracilis* Klebs var. *bacillaris* Cori and strain Z Pringsheim can grow phototrophically using carbon dioxide and light or organotrophically on carbohydrates, tricarboxylic acid cycle intermediates, or a series of organic acids and alcohols beginning with two carbon compounds (Hutner and Provasoli, 1951; 1955). Photoorganotrophic growth is also possible on organic substrates in the presence of light. Although it is advantageous for a cell to be able to utilize those organic compounds that might occur in its environment, it is energetically wasteful to synthesize enzymes which are unnecessary (gratuitous) for growth in any specific environment. For this reason, regulatory systems have evolved that allow the cell to optimize its enzyme complement for the utilization of readily available nutrients while retaining the flexibility to rapidly adapt its metabolic machinery to utilize other nutrients, which become available from time to time. These systems have been best studied in prokaryotes in which a considerable amount is known about the metabolic means of control (Goldberger, 1979).

Enzymes that are required for growth on all substrates are synthesized constitutively (i.e., they are always present in growing cells). Enzymes involved in catabolism of novel substrates are induced by their substrates; anabolic enzymes which catalyze biosynthetic reactions are repressed by the end products of their pathways. In growing *Euglena* cells, the enzymes required for growth on glutamate, malate, or succinate are synthesized constitutively since they are part of the tricarboxylic acid cycle or its associated reactions. Enzymes of the glyoxylate cycle, on the other hand, are induced when ethanol or acetate are present in the growth medium (Collins and Merrett, 1975; Horrum and Schwartzbach, 1981; Woodcock and Merrett, 1980). Since light is an energy substrate for photosynthesis that allows carbon dioxide to be used as a carbon source, the photocontrol of chloroplast development can be viewed as a complex example of induction by a novel substrate and light must take its place in the hierarchy of inducing substrates available to the cell.

The control of gene expression by substrate availability prevents the synthesis of enzymes that are gratuitous for growth only if a single source of carbon and energy is available in the environment. When the environment contains a number of compounds, any of which can serve as a sole source of carbon and energy, it is far more efficient for the cell to establish priorities of utilization. Maximum utilization of available nutrients for growth is obtained when the cell uses a single source for which it contains the metabolic machinery and when this source has been fully utilized, adapts to utilize another source. This phenomenon is called diauxic growth. Since adaptation itself requires energy, diauxie ensures that the cell will maximally utilize a substrate that does not require adaptation before expending energy on the formation of the machinery to utilize a substrate that requires enzyme induction. Catabolite repression is an example of a well-characterized bacterial regulatory system that coordinates the expression of substrate-inducible genes to ensure that only those enzymes required to utilize the most energetically favorable carbon source are induced (Goldberger, 1979).

Light-induced chloroplast development in *Euglena* is repressed in the presence of certain utilizable carbon sources (App and Jagendorf, 1963; Buetow, 1967; Garlaschi *et al.*, 1974; Harris and Kirk, 1969; Horrum and Schwartzbach, 1980a; Schwelitz *et al.*, 1978). Ethanol, a carbon source whose metabolism requires inducible enzymes, specifically inhibits the synthesis of a number of light-induced chloroplast-localized enzymes in nondividing cells (Horrum and Schwartzbach, 1980a,c). When cells are grown under photo-organotrophic conditions, the synthesis of light-induced chloroplast constituents is low compared with levels found in phototrophic cells (Nicolas *et al.*, 1980). The synthesis of chloroplast-localized macromolecules increases when cells enter the stationary phase due to the depletion of exogenous carbon. Thus, when alternative sources of carbon and energy are available that require induction, chloroplast development is repressed (Horrum and Schwartzbach, 1980a). Since the development of a chloroplast, which is necessary for utilization of light and carbon dioxide, requires a significantly larger expenditure of energy than the synthesis of the few enzymes necessary to utilize any other carbon substrate, it is not surprising that a control mechanism has evolved in *Euglena,* similar to catabolite repression in *E. coli,* to repress chloroplast development when other sources of carbon and energy are available whose utilization requires a less elaborate induction. Thus the hierarchy of control in *Euglena* seems to ensure that constitutive substrates such as glutamate, malate, and succinate, which require no induction of enzymes for utilization, are utilized first and that these do not repress light-induced chloroplast development. Substrates such as ethanol and acetate, which require enzyme induction, must compete with light to obtain the wherewithall for induction. In this competition, ethanol and acetate take precedence over light, and the use of these two carbon sources is observed at the expense of chloroplast development.

Since light-induced chloroplast development in *Euglena* is frequently studied in cells which are not dividing due to starvation for carbon and/or nitrogen (Stern *et al.*, 1964a), it is important to understand the effect of nutrient limitation on the control processes we have just discussed. The cells respond to this nutritional down-shifting by lowering their rates of protein and RNA synthesis, indicating a coordinated response to the absence of a carbon source. Provision of a utilizable carbon source, a shift-up, increases the rate of RNA and protein synthesis, and the cells eventually divide. Exposure of dark-grown nondividing *Euglena* to light is analogous to providing a carbon source, because light brings about paramylum break-down leading to a readily utilized internal source of carbon unavailable from outside the cells. Turnover of constituents, as in starving bacterial cells, allows the *Euglena* cell to sacrifice components not immediately required to provide the precursors for materials of higher priority.

This brings us at once to the first photoreceptor system active in chloroplast development in *Euglena,* the blue light system thought to control the central metabolism and other events external to the developing chloroplast to provide constituents for the developing plastid (Schiff, 1980b). Light, acting through this photoreceptor system in nondividing cells, induces the breakdown of storage carbohydrates (Schwartzbach *et al.*, 1975; Horrum and Schwartzbach, 1980b), increases the rate of cellular respiration (Schiff, 1963; Fong and Schiff, 1977), increases the rate of RNA synthesis including the transcription of cytoplasmic rRNA's (cyt rRNA) (D. Cohen and Schiff, 1976), increases the rate of protein synthesis (Schwartzbach and Schiff, 1979), and raises the level of the constitutive mitochondrial enzyme activities of fumarase and succinate dehydrogenase

(Horrum and Schwartzbach, 1980b). Light, ethanol, or malate does not induce fumarase in growing cells indicating that the induction of this enzyme represents a shift-up response in nondividing cells (Horrum and Schwartzbach, 1982). This generalized response external to the developing chloroplast ensures that sufficient carbon and energy will be made available for the synthesis of the specific machinery to utilize the inducing substrate whether this is the induction of chloroplast development by light or the induction of glyoxysomal machinery, for example, by ethanol. Without the mobilization of such internal reserves by an inducing substrate, cells might starve in the midst of plenty, being unable to form the enzymatic or organellar systems necessary to use the inducing substrate.

There is another photoreceptor system absorbing both blue and red light that is also involved in chloroplast development (Schiff, 1980b). This system appears to induce the synthesis of constituents required specifically for the utilization of light as a substrate for photosynthesis: the plastid rRNAs (pl rRNA) (D. Cohen and Schiff, 1976) and the plastid proteins, whether coded in nuclear DNA (nDNA) or plDNA (Egan *et al.*, 1975; Freyssinet *et al.*, 1979; Bingham and Schiff, 1979a,b; Russel *et al.*, 1978; Schmidt and Lyman, 1975); the red–blue receptor is also responsible for the reduction of protochlorophyll(ide) to chlorophyll(ide) during chlorophyll biosynthesis (Egan *et al.*, 1975; Egan and Schiff, 1974; Holowinsky and Schiff, 1970).

The result is two photoreceptor systems acting in tandem, the blue light system controlling the availability of molecules and energy and the blue and red–blue systems coordinately directing what specific chloroplast constituents shall be made using these resources.

III. Arrested Development of the Plastid in Darkness and Chloroplast Development in the Light

Light is indispensible for plastid development in *E. gracilis* var. *bacillaris* and the Z strain. During growth in darkness, plastid development is arrested at the stage of a small proplastid (Fig. 1) (Klein *et al.*, 1972; Osafune *et al.*, 1980a). On exposure to light, the proplastid in either dividing or nondividing cells is induced to develop into a chloroplast (Stern *et al.*, 1964a,b). Cells dividing in darkness maintain about ten proplastids and those dividing in the light, about ten chloroplasts (Klein *et al.*, 1972) through plastid division. Light-grown cells placed in darkness on a complete medium divide and the chloroplasts become reduced to proplastids (Ben-Shaul *et al.*, 1965; Pellegrini, 1980b). If the cells are placed on a resting medium lacking carbon and nitrogen, they cease division and the chloroplasts remain much as they were; they are not reduced to proplastids in the absence of cell division (Ben-Shaul *et al.*, 1965). In both dividing and nondividing cells one of the first steps in chlorophyll loss appears to be the

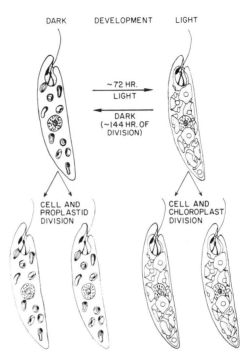

Fig. 1. Plastid replication and development in *E. gracilis.* (From Schiff, 1973; light-grown cell after Leedale, 1967.)

loss of magnesium from chlorophyll *a* to form pheophytin *a;* this is followed by loss of the carboxymethyl group on the isocyclic ring to form pyropheophytin *a* (Greenblatt and Schiff, 1959; Schoch *et al.*, 1980; 1981). A selective loss of one type of antenna pigment appears to occur, judging from the absorption spectra of the intact cells; this also occurs when the cells are placed in the inhibitor of photosynthesis DCMU (Calvayrac *et al.*, 1979). Perhaps darkness and DCMU lead to the same end result: lack of proton pumping and consequent accumulation of hydrogen ions in close enough proximity to the chlorophyll to cause pheophytin formation. Although our information on loss of plastid structure and plastid pigments during the return of the chloroplast to the proplastid in darkness is rather meager, there is considerable information on the development of the proplastid into the chloroplast in the light.

A. THE DEVELOPMENTAL SYSTEM

Although light-induced chloroplast development from the proplastid occurs in dividing cells on a complete medium, many investigators have chosen to work with nondividing cells maintained on a resting medium devoid of utilizable carbon and nitrogen (e.g., see Stern *et al.*, 1964a). While these conditions provide a developmental system in which variables due to cell division are absent, the cells are starved for essential metabolites, and for this reason metabolic turnover is high. The lack of external substrates has important consequences for the response of the cells to light and other inducing substrates.

The proplastid of *Euglena* (Fig. 2) in dark-grown resting cells is rather small and undifferentiated, like the proplastids of dark-grown *Ochromonas danica* (Gibbs, 1962) and the younger etiolated leaves of higher plants, when compared with the etioplasts of older angiosperm leaves (Klein and Schiff, 1972). The highly crystalline prolamellar bodies and extensive prothylakoids characteristic of etioplasts are not present, but what it lacks in quantity of internal membranes the *Euglena* proplastid more than makes up in the three-dimensional complexity of its form. Based on reconstructions from serial sections (Osafune *et al.*, 1980), the *Euglena* proplastid resembles a polyp with many protuberances extending out into the surrounding cytoplasm; at the extremity of each of these protuberances is a noncrystalline prolamellar body connected with prothylakoids that extend through the plastid (Fig. 3). Each proplastid contains several prolamellar bodies, and this probably explains why several red-fluorescing centers were seen in proplastids by fluorescence microscopy (Klein *et al.*, 1972). The proplastid may be somewhat plastic in extent and form, since it has been reported that plastids at various stages in the life cycle of *Euglena* may come together or separate to form larger or smaller structures (Lefort-Tran, 1975; Pellegrini, 1980a). The proplastid is closely associated with other cellular organelles such as mitochondria, Golgi, and microbody-like organelles (Schiff, 1973; Osafune *et al.*, 1980). Also evident are membrane whorls, which have been seen in many systems and in *Euglena* appear to connect the developing proplastid to other organelles. The extended form of the proplastid results in an extensive surface in contact with the surrounding cytoplasm explaining the many patches of cytoplasm seen within random sections of proplastids; these patches of cytoplasm are outside the proplastid and are separated from it by the proplastid outer membranes, but one must be careful when interpreting random sections to distinguish what is plastid stroma and what is cytoplasm. The different sizes of plastid and cytoplasmic ribosomes (Schwartzbach *et al.*, 1974; Avadhani and Buetow, 1972) are frequently helpful in deciding, the larger cytoplasmic ribosomes serving as markers for the cytoplasmic intrusions into the proplastid.

On exposure to light (Osafune *et al.*, 1980; Klein *et al.*, 1972), the proplastid expands somewhat and becomes more globular; the expansion of the arms of the proplastid reduces the areas of intrusive cytoplasm to smaller patches. Mitochondria, Golgi, and microbody-like objects are seen in very close association with the developing plastid, and membrane whorls are much in evidence sometimes as connections to other organelles. During this period membrane whorls appear as

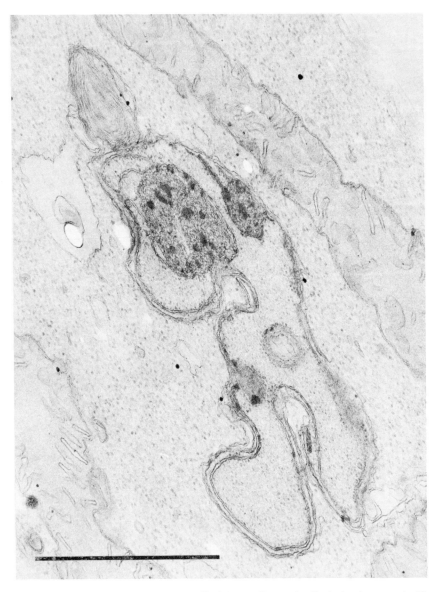

Fig. 2. Section through a dark-grown cell of *E. gracilis* var. *bacillaris* showing a proplastid. Marker indicates 1.0 μm.

transient structures in the interiors of the prolamellar bodies; these are soon replaced by elongated straight tublues which in turn disappear to leave a cavity within the prolamellar body. As development proceeds there is a progressive pairing of thylakoids with the deposition of a dense matrix between them. By 24 hours of light exposure, paired thylakoids are the rule, and the plastid has expanded considerably (Fig. 4). This expansion continues concommitantly with the differentiation of the pyrenoid. The pyrenoid in *Euglena*, as in other algae, may contain ribulosebisphosphate (RuBP) carboxylase and perhaps other enzymes of the Calvin cycle in condensed form (Schiff, 1973, 1980a). Since the paramylum is formed outside the chloroplast membrane surrounding the pyrenoid region, perhaps the pyrenoid serves as a center for efficient utilization of assimilatory power (NADPH and ATP) formed in the thylakoids for carbon dioxide fixation. It might be remembered, however, that pyrenoids are not indispensible for carbon dioxide fixation, since many species of algae lack them, and they may be absent or indistinct in otherwise normal chloroplasts in photosynthetic cells of *Euglena* grown under various conditions.

The fully developed chloroplast at 72 – 96 hours of development (Fig. 5) still has appressed mitochondria but has expanded considerably in its development from the proplastid. There has been an extensive formation of thylakoids, which are now organized into stacks. The pyrenoid and paramylum sheath are clearly seen (Klein *et al.*, 1972; Ben-Shaul *et al.*, 1964). The development of better ways to break *Euglena* cells and to isolate chloroplasts using Percoll gradients have yielded chloroplasts that show high rates of light-dependent protein synthesis and carbon dioxide fixation (Ortiz *et al.*, 1980; Tokunaga *et al.*, 1976; Shigeoka *et al.*, 1980). These chloroplasts are very similar to those in the intact cells (Fig. 6) when viewed in the electron microscope.

The morphological development of the chloroplast is correlated with a series of biochemical changes that occur concomitantly. Both soluble and membrane proteins are made *de novo* or increase during chloroplast development; the same is true of lipids including the plastid thylakoid pigments (Fig. 7), nucleic acids, and small molecules (see reviews cited in Section I). These constituents are synthesized within the developing chloroplast or are made outside the organelle and are transported in. Although certain soluble plastid enzymes, such as the NADP-triosephosphate dehydrogenase, which are synthesized outside the plastid, appear without a lag during light-induced development and reach their peak concentrations after about 24–36 hours (Schiff, 1973; Bovarnick *et al.*, 1974b)), many other constituents such as chlorophyll and cytochrome c_{552} (Freyssinet *et al.*, 1979) show a lag in development and only reach their maximum levels after 72–96 hours of development. Thus, there are reasons to suppose that there are early and late events in plastid development, but of course these may overlap to form a sequence of inductions that favor the proper assembly of the organelle (Fig. 8). The length of the lag period in nondividing cells induced by light to

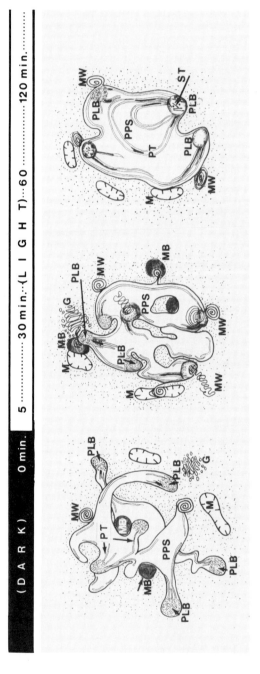

Fig. 3. The structure of the proplastid in dark-grown nondividing cells of *E. gracilis* var. *bacillaris* and after exposure of the cells to light for up to 2 hours, as reconstructed from serial sections. G, Golgi; M, mitochondrion; MB, microbody-like object; MW, membrane whorl; PLB, prolamellar body; PPS, proplastid stroma; PT, prothylakoid; ST, straight tubule. (From Osafune *et al.*, 1980.)

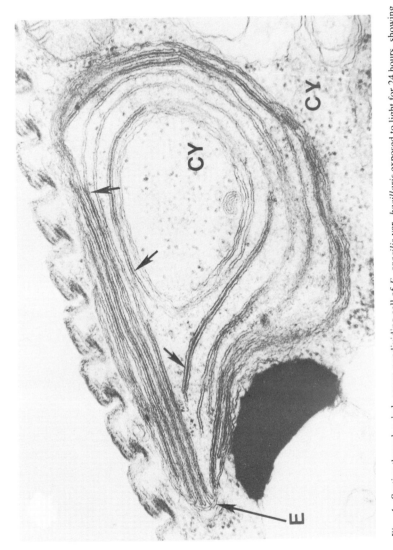

Fig. 4. Section through a dark-grown nondividing cell of *E. gracilis* var. *bacillaris* exposed to light for 24 hours, showing a developing chloroplast. C, cytoplasm; E, plastid envelope; arrows designate stacked regions of thylakoid with dense matrix between paired thylakoids. (From Osafune *et al.*, 1980.)

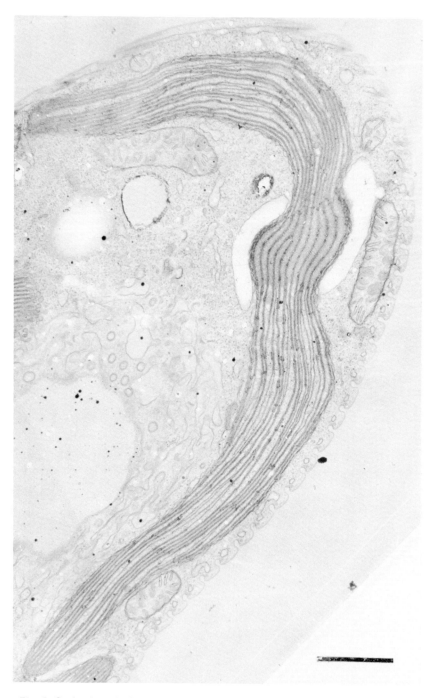

Fig. 5. Section through a light-grown cell of *E. gracilis* var. *bacillaris* showing a fully developed chloroplast with central pyrenoid bearing paramylum sheaths. Marker indicates 1.0 μm.

form chloroplasts is dependent on the previous nutritional history of the cells. Cells grown on a medium containing a high carbon/nitrogen ratio show a greater accumulation of paramylum and a shorter lag period compared with those grown on a lower carbon/nitrogen ratio (Freyssinet, 1976a,b). Net photosynthesis makes its appearance at about 3 hours of development (Stern *et al.*, 1964a), and

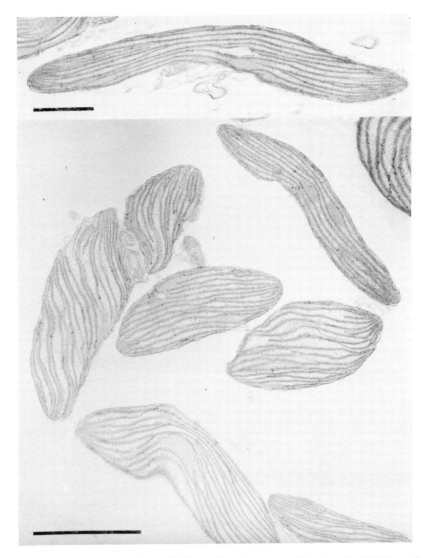

Fig. 6. Section through highly purified intact chloroplasts prepared by the method of Ortiz *et al.* (1980) from cells of *E. gracilis* var. *bacillaris* grown in light on low vitamin B_{12}. Marker indicates 1.0 μm (B. Gomez-Silva and J. A. Schiff, 1981).

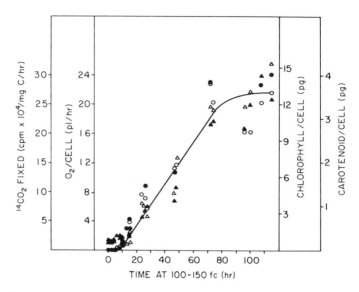

Fig. 7. Pigments and photosynthesis during chloroplast development on exposure of dark-grown nondividing cells of *E. gracilis* var. *bacillaris* to light for various periods of time. Key: CO$_2$(●); O$_2$ (○); chlorophyll (△); carotenoid (▲). (From Schiff, 1973.)

studies of the incoming of photosystem II have been conducted (Cohen *et al.*, 1978). The formation of the various plastid constituents are highly coordinated through tight control mechanisms, which undoubtedly serve to insure that the various structures are assembled correctly. This is reflected in the close correlation of chlorophyll and carotenoid synthesis and the incoming of photosynthetic function (Fig. 7) as well as in the tight mutual control of formation of plastid lipids, plastid thylakoid polypeptides, and other plastid thylakoid constituents (to be discussed later) that lead to the proper assembly of the photosynthetic membranes.

There is a concomitant activation of synthesis outside the developing chloroplast as well. For example, microbody enzymes such as glycolate dehydrogenase increase after 24 hours of light-induced chloroplast development and are thought to be the consequence of induction by products of photosynthesis such as glycolate (Lor and Cossins, 1978; Merrett and Lord, 1973; Horrum and Schwartzbach, 1981). In nondividing cells undergoing light-induced chloroplast development, mitochondrial enzymes such as fumarase and succinate dehydrogenase (Horrum and Schwartzbach, 1980b; 1982) are observed to increase transiently; maximum activity is seen at the end of the lag period of development. Similar transients are observed in the activities of enzymes of early carbohydrate metabolism such as β-(1,3)-glucan phosphorylase (Dwyer and Smillie, 1970)

FIRST 12 HOURS OF LIGHT
(LAG PERIOD)

> 12 HOURS OF LIGHT
(MAJOR DEVELOPMENTAL PERIOD)

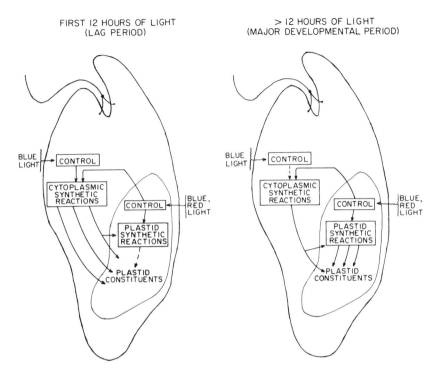

Fig. 8. Model for events of the lag period and the period of active development. (From Schiff, 1973.)

and hydrolase. Enzymes of glycolysis such as phosphofructokinase, NAD-glyceraldehyde-3-phosphate dehydrogenase, and aldolase, which are presumably free in the cytoplasm, decrease during light-induced chloroplast development (Dockerty and Merrett, 1979; Karlan and Russel, 1976; Bovarnick *et al.,* 1974b). These changes are selective in that other enzymes in the same compartments do not change in activity such as the hydroxypyruvate reductase of the microbodies and serine hydroxymethyltransferase of mitochondria and cytoplasm (Horrum and Schwartzbach, 1980d). It is likely that those nonplastid enzymes, which increase in activity during light-induced chloroplast development, are concerned with the general step up in cellular activity already mentioned, whereas those which decrease may undergo metabolic turnover to provide the amino acids for the synthesis of enzymes necessary for biosynthesis and function of the developing chloroplast. [Protease inhibitors that block chloroplast development in *Euglena* may block hydrolases that make amino acids available through turnover (Zeldin *et al.,* 1973).] Those activities which do not change are probably necessary for general cellular housekeeping, whether or not the chloroplast is being formed (i.e., they are constitutive). Further work should uncover the detailed metabolic relations among these various groups of proteins. The extensive turnover observed in cyt rRNAs (D. Cohen and Schiff, 1976) during light-induced chloroplast development, which is under control of the blue-light receptor external to the developing chloroplast, is another example of the means available to a carbon-limited cell for the reutilization of constituents for the synthesis of macromolecules of high priority.

B. ORIGIN AND PHOTOCONTROL OF ENERGY AND METABOLITES FOR CHLOROPLAST DEVELOPMENT

There are several lines of evidence which indicate that the developing chloroplast need not obtain energy and essential metabolites from its own photosynthetic functions. The light intensity which is optimal for chloroplast development is lower than that required to saturate photosynthesis and photosynthetic activity and is low during early development when many new constituents are being synthesized (Stern et al., 1964a,b). Concentrations of inhibitors of photosynthesis, such as DCMU, which block carbon dioxide fixation completely, have little effect on chloroplast development in otherwise nutritionally sufficient cells (Schiff et al., 1967; Dwyer and Smillie, 1971; Horrum and Schwartzbach, 1981). On the other hand, inhibitors of mitochondrial respiration such as azide (Fong and Schiff, 1977) and anaerobiosis (Klein et al., 1972), and of mitochondrial phosphorylation such as dinitrophenol (Evans, 1971) are extremely effective inhibitors of chloroplast development; an increase in oxygen uptake occurs on light induction of chloroplast development (Schiff, 1963; Fong and Schiff, 1977). Also, mutants of various photosynthetic organisms including *Euglena*, blocked in some essential step in photosynthesis, make otherwise normal chloroplasts (see Schiff, 1980a; Schiff et al., 1971; Schneyour and Avron, 1975). Many organisms make chlorophyll and chloroplasts in the dark and, therefore, do not require photosynthesis for the formation of their photosynthetic machinery (see Schiff, 1980a). All of this evidence indicates that photosynthetic metabolism is gratuitous for chloroplast development—it can be dispensed with as long as the rest of the cell can supply the energy, reducing power, and metabolites required. The close association of the developing plastid with other cellular organelles such as mitochondria, microbodies, and Golgi, and with the surrounding cytoplasm (Osafune et al., 1980) already mentioned is undoubtedly part of the means by which this feeding of the developing plastid is accomplished. We have also mentioned that one important area of control of chloroplast development by light is the general step-up induction of the cell to provide the energy and constituents for chloroplast development. This activation of central metabolism external to the developing chloroplast by light is one means of providing the necessary metabolites to the developing organelle. It also insures that the organism is able to respond to substrate induction by light even under starvation conditions. The ability of a chloroplast to utilize light energy when it becomes available, even though no carbon or nitrogen is available from the environment, would be strongly selected for during evolution to insure that the cell always has enough internal reserves to make the machinery to use the light before it starves to death.

1. *Endogenous Sources of Energy and Metabolites*

Although turnover of proteins and nucleic acids in nondividing cells of *Euglena* provides a source of constituents for light-induced chloroplast development (Bovarnick *et al.*, 1974b; Horrum and Schwartzbach, 1980a; D. Cohen and Schiff, 1976), carbon storage compounds such as paramylum (Schwartzbach *et al.*, 1975) and wax (Rosenberg and Pecker, 1964) are also mobilized on exposure of the dark-grown cells to light; wax can also be utilized for limited development in darkness (Shihira-Ishikawa *et al.*, 1977). Since light-induced paramylum breakdown takes place at a lower rate in dark-grown bleached mutants such as W_3BUL and $W_{10}BSmL$ (Schwartzbach *et al.*, 1975; Horrum and Schwartzbach, 1980b) compared with dark-grown wild-type cells (Schwartzbach *et al.*, 1975; Freyssinet, 1976a; Freyssinet *et al.*, 1972; Dwyer and Smillie, 1970, 1971), this process may be controlled, at least in part, by a non-plastid photoreceptor system. The higher rate in wild-type cells may be due to control of paramylum degradation by both the nonplastid and plastid photoreceptors, or because the presence in wild-type cells of a proplastid capable of development provides a metabolic sink for the products formed from paramylum.

Cycloheximide, a specific inhibitor of cytoplasmic protein synthesis in *Euglena* (Kirk, 1970), or levulinic acid, a competitive inhibitor of δ amino-levulinic acid (ALA) utilization for chlorophyll synthesis (Richard and Nigon, 1973), can replace light as an inducer of paramylum degradation (Schwartzbach *et al.*, 1975; Horrum and Schwartzbach, 1980b). A model based on negative transcriptional control has been proposed which suggests that levulinic acid may act as an analog of a normal corepressor (ALA) in controlling the formation of a protein on cytoplasmic ribosomes, which inhibits paramylum degradation (Schwartzbach *et al.*, 1975; Schiff, 1978). Another model, based on light-induced changes in membrane transport, is also applicable. Although cyclo-heximide does not inhibit respiration in *Euglena* (Kirk, 1970), it does appear to inhibit membrane transport in a manner unrelated to its action on protein synthesis (Evans, 1971; Parthier, 1974). Evans suggested that cycloheximide produces a redistribution of intracellular ions, which causes a decrease in the pH gradient between the inside and the outside of the cell; this altered pH gradient is thought to be responsible for the inhibition of transport of various compounds into the cells. Levulinic acid might act in a similar manner, since it is a weak acid which is taken up by the cells. Thus light might induce paramylum breakdown in *Euglena* by altering membrane transport in such a manner as to change the intracellular pH. A similar model has been offered to explain light-induced starch degradation in *Scenedesmus* (see G. Brinkmann and H. Senger, in Senger, 1980).

Light, acting through the nonchloroplast photoreceptor, decreases the rate of amino acid transport into dark-grown *Euglena* cells (Schwartzbach and Schiff, 1979). However, since paramylum degradation begins immediately on illumination and the inhibition of amino acid transport occurs 3–4 hours later, the two processes may be independently regulated.

2. Influence of Exogenous Sources of Energy and Metabolites

A number of utilizable carbon sources have been reported to repress light-induced chloroplast development (App and Jagendorf, 1963; Buetow, 1967; Garlaschi *et al.*, 1974; Harris and Kirk, 1969; Horrum and Schwartzbach, 1980a; Schwelitz *et al.*, 1978). In some cases, repression occurs even when cells grown in darkness on the repressing carbon source are exposed to light in the absence of the carbon source.

The presence of a utilizable nitrogen source can reverse the apparent repression of chloroplast development due to growth on a particular carbon source (Harris and Kirk, 1969; Schwelitz *et al.*, 1978). Since the amount of chlorophyll formed on reaching the stationary phase is proportional to the nitrogen content of the medium (Horrum and Schwartzbach, 1980a), the repression of chloroplast development observed during growth of the cells on media containing a high carbon/nitrogen ratio probably results from an inability of the nitrogen-deficient cells to synthesize chlorophyll rather than from a specific repression by a particular carbon source.

By using nitrogen-sufficient dark-grown resting cells supplemented with various carbon sources at the time of light exposure, it has been shown that ethanol or acetate specifically represses light-induced chloroplast development (Horrum and Schwartzbach, 1980a). Ethanol specifically represses the light-induced synthesis of chlorophyll, NADP-glyceraldhyde-3-phosphate dehydrogenase, glycolate dehydrogenase, phosphoglycolate phosphatase, and chloroplast-localized valyl-tRNA synthetase (Horrum and Schwartzbach, 1980a, 1981; D. D. McCarthy and S. D. Schwartzbach, unpublished). The addition of ethanol to dark-grown resting cells induces the formation of malate synthase and fumarase, promotes cell division, and increases total cellular protein synthesis indicating that the repression of chloroplast development is not due to nitrogen deficiency (Horrum and Schwartzbach, 1980a,b; 1981; 1982). Since chloroplast development is specifically repressed by acetate or ethanol but not by malate or succinate (Garlaschi *et al.*, 1974; Horrum and Schwartzbach, 1980a), glyoxylate might be the actual repressing compound.

Light, ethanol, and/or malate induce fumarase, succinic dehydrogenase, and phosphoglycerate phosphatase in dark-grown resting cells (Horrum and Schwartzbach, 1980b; James and Schwartzbach, unpublished), but light, ethanol, and/or malate do not induce fumarase in cells maintained in balanced growth (Horrum and Schwartzbach, 1982). Enzymes induced by light or organic carbon are required for growth on all carbon sources suggesting that their induction represents a general step-up response, a consequence of providing a starving cell with a utilizable carbon source. Enzymes induced by light and repressed by ethanol, however, are specifically required for phototrophic growth. The catabolite repression of light-induced chloroplast development by those organic substrates which require enzyme induction for their utilization indicates the inferior position of light in the hierarchy of inducing substrates.

C. Origin and Photocontrol of Genetic Information for Chloroplast Development

1. Mutants Blocked in Chloroplast Development

Mutants of *Euglena* can be obtained that are blocked in chloroplast development in various ways (Schiff *et al.*, 1971, 1980; Bingham and Schiff, 1979a). These are thought to be mutants in plDNA because the nucleus appears to be $2N$ or higher in ploidy, and nuclear mutants would not be expressed unless dominant. Mitochondrial mutants affecting function also are unlikely since *Euglena* is an obligate aerobe, and such mutants would be lethal.

The bleached mutants of *Euglena* (many of these are designated W for white) seem to lack much or all of their plDNA. These mutants are produced by treatment with bleaching agents such as uv, temperatures above 34° (the chloroplast appears to be a heat-sensitive organelle in a more heat-resistant cell), and antibiotics such as streptomycin (see references in Schiff, 1980a; Schiff *et al.*, 1971, 1980; also reviews cited in Section I). As the cells divide during or after treatment with these agents, replication of plDNA appears to cease, and the remaining plDNA is diluted among the progeny; in some cases it may be more or less destroyed. On plating these cells in the light, white colonies are recovered, which after cloning never revert to green (Srinivas and Lyman, 1980; Goldstein *et al.*, 1974; Nicolas *et al.*, 1977).

In one of the best-studied of these mutants, W_3BUL, plDNA is undetectable by analytic ultracentrifugation in cesium chloride (Edelman *et al.*, 1965), separation on columns (E. Stutz, personal communication) and by fluorescence microscopy using a DNA-specific fluorochrome (Coleman, 1979). It has been suggested that some of these bleached mutants may contain very small amounts of residual plDNA limited perhaps to some ribosomal cistrons or somewhat more (P. Heizmann *et al.*, 1981). This is clearly an area that needs more work using the highly sensitive techniques of molecular biology now available. A complication has recently appeared that could affect investigations of DNA in *Euglena*. It has been reported that most of our laboratory strains of *Euglena* are infected with *Mycoplasma* (Riggs, 1979; W. A. C. Riggs, personal communication). The base compositions of *Mycoplasma* DNAs are very close to those of pl and mtDNA's of *Euglena* (A + T = 60–70 moles %: Buchanan and Gibbons, 1974). This should be considered when interpreting experimental findings, particularly those dealing with minor DNA components.

One puzzling observation has been the varying phenotypes of bleached mutants. Mutants can be obtained, for example, which have reduced carotenoid levels or an accumulation of different carotenoids when compared with wild-type cells or with each other; W_3BUL belongs in this group. Others lack colored carotenoids completely, such as $W_{10}BSmL$. The amount of stigma material also varies among these mutant and wild-type cells (see Osafune and Schiff, 1980b, and references therein), and stigma differentiation is under control by light. Similarly, carotenoid biosynthesis is under light control in some bleached mutants but not in others (Dolphin, 1970; Fong and Schiff, 1979; Steinitz *et al.*, 1980). The bleached mutants W_3 and W_{10} both show increased paramylum break-down and changes in rates of leucine incorporation on illumination, although W_3 shows a transient increase in fumarase activity on illumination; W_{10} does not (Horrum and Schwartzbach, 1980b; Schwartzbach *et al.*, 1975; Schwartzbach and Schiff, 1979). Are these differences among bleached strains due to retention of different trace fragments of residual plDNA, or are they due to different regulatory situations established in the cells when the plDNA was lost?

Although wild-type cells can synthesize colored carotenoids in darkness and do not accumulate ζ-carotene, certain bleached mutants accumulate this intermediate of carotenoid biosynthesis in the *cis* form. Illumination of the cells with blue light causes a *cis* to *trans* photoisomerization of the ζ-carotene. There may be two means of isomerizing ζ-carotene in wild-type cells, one a direct photochemical reaction, the other a dark enzymatic reaction (Fong and Schiff, 1979; Steinitz *et al.*, 1980; Schiff *et al.*, 1982).

Dark-grown cells of the bleached mutant W_3BUL contain a proplastid remnant which is reduced to a thin leaflet lacking plastid ribosomes and most of the stroma and prothylakoids (Fig. 9) (Osafune and Schiff, 1980a; Parthier and Neumann, 1977; Schiff and Epstein, 1968; Palisano and Walne, 1976). As in wild-type, close associations with other cellular organelles such as mitochondria are seen. When the cells are exposed to light, a limited expansion of the plastid occurs with the formation of a large vacuole and a crystalline prolammelar body, but plastid development does not proceed beyond this point. These bodies are of two types, but only one type is found in any one cell; this points again to a heterogeneity among cells of even highly cloned bleached mutants. Since no plastid ribosomes are seen and no plrRNA is detected (D. Cohen and Schiff, 1976) in W_3BUL, if any plastid DNA remnants are present, it is unlikely that they provide

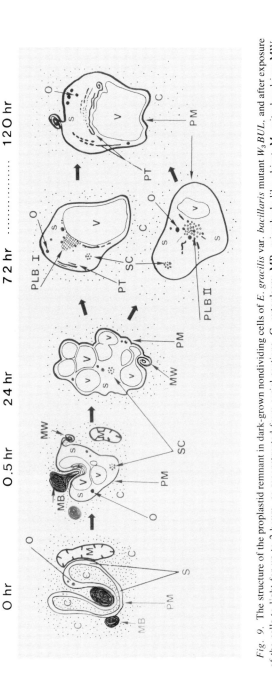

Fig. 9. The structure of the proplastid remnant in dark-grown nondividing cells of *E. gracilis* var. *bacillaris* mutant W_3BUL, and after exposure of the cells to light for up to 2 hours, as reconstructed from serial sections. C, cytoplasm; MB, microbody-like object; M, mitochondrion; MW, membrane whorl; O, osmiophilic granule; PLB I, prolamellar body type I (crystalline); PLB II, prolamellar body type II (noncrystalline); PM, plastid membrane; PT, prothylakoid; S, proplastid stroma; SC, stroma center; V, vacuole. (From Osafune and Schiff, 1980a).

translatable information for the formation of plastid proteins. Therefore, it is likely that the prothylakoids and other structures seen in this proplastid remnant are nuclear-coded, and that the vacuole represents an area which would ordinarily be filled with materials formed if plDNA were present and active, as in wild-type cells.

Various properties of bleached mutant $W_3 BUL$ relevant to chloroplast development are summarized in Table I (Schiff, 1980b). In general, chloroplast components are low in concentration (are present at repressed levels comparable to those in dark-grown wild-type cells) or are undetectable. Since plDNA is undetectable in W_3 by conventional methods, this mutant has been useful as an indicator of where various plastid components may be coded. Those present in W_3 (even at repressed levels) are thought to be nuclear-coded, since there is no conclusive evidence that the limited mitochondrial genome contributes genetic information for plastid development. Those plastid components that are not detectable in W_3 may be coded in plDNA, but it should be recalled that a structural gene for a protein may be elsewhere than in plDNA with a regulatory gene in plDNA regulating its expression.

Various photoresponses associated with light-induced chloroplast development, particularly those thought to be localized outside the plastid, persist in W_3. Examples are the induction of transcription of cyt rRNA from nDNA by blue light (D. Cohen and Schiff, 1976), light-induced paramylum breakdown (Schwartzbach et al., 1975), induction of the photoreactivating enzyme (Diamond et al., 1975), induction of the chloramphenicol-reducing system (Vaisberg et al., 1976; Schiff, 1975), and a light-induced increase in cellular respiration (Schiff, 1963; Fong and Schiff, 1977). As a general rule, W_3 appears to retain (at least partially) those blue light photoresponses connected with the general step-up of the central metabolism of the cell necessary for feeding the developing chloroplast. This mutant, however, appears to lack the photoresponses that are concerned with the specific induction of various plastid components such as those enzymes which remain low or undetectable in illuminated cells of W_3 (Table I).

2. Sources of Genetic Information for the Formation of Plastid Constituents

There are three known potential sources of genetic information* in *Euglena* cells (Edelman et al., 1965). Nuclear DNA constitutes about 90% of the cell DNA with mtDNA and plDNA making up the total with about 5% each, although the actual amounts vary with the growth conditions (Chelm et al., 1977b; Rawson and Boerma, 1976b). The nuclear genome is thought to be diploid, based on renaturation kinetics (Rawson, 1975; Rawson et al., 1979), but octaploidy has been suggested, based on target theory (Hill et al., 1966; see Schiff et al.,

---

*This topic is covered in detail in Volume IV, Chapters 2, 3, and 6–9.

1980). Mitochondrial DNA is thought to have a molecular weight of about 40×10^6 daltons based on renaturation experiments and hybridization with mt rRNA (Crouse *et al.*, 1974; Talen *et al.*, 1974). Plastid DNA is circular, with a molecular weight of 92×10^6 daltons (Manning and Richards, 1972; Manning *et al.*, 1971; Slavik and Hershberger, 1975). Hybridization experiments indicate that cyt rRNA is coded in nDNA (Gruol and Haselkorn, 1976) and that mt rRNA is coded in mtDNA (Crouse *et al.*, 1974). The pl rRNAs are coded in plDNA and are present in three copies arranged as tandem repeats (Gray and Hallick, 1978; Helling *et al.*, 1979; Kopecka *et al.*, 1977; Rawson *et al.*, 1978) with interspersed tRNA cistrons (El-Geweley *et al.*, 1981; Hallick *et al.*, 1979; Keller *et al.*, 1980), although the exact arrangement may vary from one strain to another (Wurtz and Buetow, 1981). The plDNA codes for 26–30 tRNAs (Gruol and Haselkorn, 1976; McCrea and Hershberger, 1976; Mubumbila *et al.*, 1980; Schwartzbach *et al.*, 1976a), which are scattered throughout the plDNA genome (El Geweley *et al.*, 1981; Hallick *et al.*, 1979); the cyt tRNAs are coded in nDNA (Gruol and Haselkorn, 1976).

Estimates based on hybridization and ultracentrifugation indicate a doubling of plDNA during chloroplast development in nondividing cells (Chelm *et al.*, 1977b; Rawson and Boerma, 1976b; Uzzo and Lyman, 1972). The significance of this increase is not clear; under some conditions of normal development there is no net DNA synthesis (Stolarsky *et al.*, 1976; Walfield and Hershberger, 1978).

Light-induced chloroplast development is accompanied by qualitative and quantitative changes in the transcription of the nuclear and chloroplast genomes. The fraction of single-copy nDNA (i.e., the nonrepetitive sequences) transcribed in dark-grown cells is higher than in light-grown cells (Curtis and Rawson, 1979). During light-induced chloroplast development, this fraction increases to a maximum 12 hours after light exposure and then declines to the value characteristic of light-grown cells. Some of the sequences present 24 hours after the beginning of light exposure were not present prior to light exposure, indicating the transcription of new sequences during chloroplast development (Verdier, 1979a). These patterns of transcription are consistent with changes already noted in levels of various enzyme proteins, thought to be nuclear-coded, during chloroplast development.

It has been estimated that a maximum of 57% of the chloroplast genome is expressed during light-induced chloroplast development (Chelm and Hallick, 1976; Rawson and Boerma, 1976a; Verdier, 1979b). A significant fraction of the chloroplast genome is transcribed in dark-grown cells, and new sequences are expressed during chloroplast development (Chelm and Hallick, 1976; Chelm *et al.*, 1978, 1979; Rawson and Boerma, 1976a, 1979). Most of the sequences expressed in the dark-grown cells continue to be expressed during light-induced chloroplast development. Some sequences that are expressed in the dark are not expressed early in development but reappear toward the end (Chelm *et al.*, 1979). Sequences unique to light-grown cells appear at various stages in development, but the order and extent of their expression is not related to their positions in the genome (Chelm *et al.*, 1978, 1979; Rawson and Boerma, 1979). In general, various soluble and membrane proteins show comparable patterns of expression during light-induced chloroplast development.

Consistent with earlier genetic experiments in various organisms (see Schiff, 1980a), genetic information for the formation of plastid proteins originates in both the plastid and nuclear genomes. Plastid proteins that are present in bleached mutants in which plDNA is undetectable are thought to be nuclear-coded; those whose synthesis is also sensitive to cycloheximide but insensitive to streptomycin and chloramphenicol and thought to be translated on cytoplasmic ribosomes. These include: NADP-glyceraldehyde-3-phosphate dehydrogenase (Bovarnick *et al.*, 1974b), photoreactivating enzyme (Diamond *et al.*, 1975), chloroplast-specific aminoacyl-tRNA synthetases (Hecker *et al.*, 1974); chloroplast ribosomal proteins (Freyssinet, 1977; 1978), chloroplast elongation factors G and EF-TS (Breitenberger *et al.*, 1979; Breitenberger and Spremulli, 1980; Fox *et al.*, 1980), alkaline DNase (Egan and Carrell, 1972), class I aldolases (Karlan and Russel, 1976; Russel *et al.*, 1978; Schmidt and Lyman, 1975), phosphoglycerate kinase (Schmidt and Lyman, 1975; Russel *et al.*, 1978); phosphoglycolate phosphatase (Horrum and Schwartzbach, 1980c), and a number of thylakoid polypeptides (Bingham and Schiff, 1979a,b). The enzymes of carotenoid biosynthesis (Vaisberg and Schiff, 1976) and of sulfolipid formation (Bingham and Schiff, 1979a) and of sulfate activation and reduction (Brunold and Schiff, 1976) also appear to belong to this group.

Studies with specific inhibitors of chloroplast translation such as streptomycin and chloramphenicol (Avadhani and Buetow, 1972; Schwartzbach and Schiff, 1974) and labeling of various proteins in isolated chloroplasts supplied with radioactive amino acids have served to identify proteins that are translated (and might be coded) in the chloroplast. Absence of a protein from a mutant in which plDNA is undetectable may be an indication of chloroplast coding, but further evidence is necessary since the structural gene could be in nDNA (or mDNA) and be regulated by a gene product of plDNA. One fairly unambiguous case is the large subunit of ribulosebisphosphate carboxylase. This polypeptide is undetectable in mutants in which plDNA is undetectable (Sagher *et al.*, 1976), the increase in enzyme activity during chloroplast development is inhibited by streptomycin (Bovarnick *et al.*, 1974b), the mRNA for the polypeptide is found in the non-poly (A) fraction of plRNA (Sagher *et al.*, 1976), the polypeptide is synthesized in isolated chloroplasts (Vasconcelos, 1976), and the gene that codes for it has been tentatively localized to a site in the plDNA genome (Hallick *et al.*, 1979). The following plastid enzymes may be chloroplast-coded and translated (but more evidence is necessary to be certain): various thylakoid membrane proteins (Bingham and Schiff, 1979a,b; Vasconcelos, 1976), a number of plastid ribosomal proteins (Freyssinet, 1977; 1978) and various unidentified soluble plastid proteins (Vasconcelos, 1976).

Although the first enzyme(s) of chlorophyll synthesis is made in the cytoplasm (Benney and Nigon, 1975; Richard and Nigon, 1973), it is not known with certainty where the other enzymes of chlorophyll biosynthesis are coded and translated. A long pathway of this sort would be expected to have enzymes from both the nuclear–cytoplasmic group and the chloroplast group and would explain why chlorophyll synthesis is inhibited both by inhibitors of cytoplasmic and chloroplastic translation (see Freyssinet et al., 1979). It should be remembered, however, that this could also be a consequence of the tight coregulation of the formation of one thylakoid constituent by another (Bingham and Schiff, 1979a,b).

D. PHOTOCONTROL OF FORMATION OF THYLAKOID MEMBRANE CONSTITUENTS

Here we consider the influence of light on the formation of those plastid constituents associated with the thylakoid membranes. The work conducted so far points to an extremely tight control of the formation of each thylakoid membrane constituent by the others. Unusual light regimes have been used to defeat this control in some instances (Gurevitz et al., 1977), but under normal continuous illumination the formation of thylakoid polypeptides, lipids, and pigments appear to be closely coupled, ensuring that constituents will be present at the right time and in the correct amount to achieve an orderly assembly of the membranes (Bingham and Schiff, 1979a,b).

1. Chlorophyll Synthesis and the Consequences of Preillumination

As already indicated, chlorophyll synthesis begins slowly on illumination of dark-grown cells and after a lag of some 12 hours enters the period of rapid synthesis (Stern et al., 1964a). Chlorophyll synthesis during the lag period does not require chloroplast protein synthesis, but cytoplasmic protein synthesis is necessary (Bovarnick et al., 1974a; Schwartzbach et al., 1976b). Protein synthesis in both compartments is necessary during the subsequent period of rapid synthesis of chlorophyll. There is also a lag in the synthesis of thylakoids and their polypeptides, a period during which newly synthesized chlorophyll appears to fill already-existing membrane sites (Bingham and Schiff, 1979b; Klein et al., 1972; Bovarnick et al., 1974a).

The enzyme(s) producing the first committed precursor of chlorophyll, δ-aminolevulinic acid (ALA), is synthesized on cytoplasmic ribosomes upon light induction (Beney and Nigon, 1975; Richard and Nigon, 1973). Light is required for maintenance of the enzyme level, which falls if cytoplasmic protein synthesis is inhibited or the cells are darkened. Synthesis of ALA is required for chlorophyll synthesis during both the lag and the period of rapid synthesis of chlorophyll (Schwartzbach et al., 1976b), but conditions which eliminate the lag period do not produce an accumulation of ALA or increase the capacity of the cell for ALA synthesis.

The length of the lag period in chlorophyll synthesis depends on the previous nutritional history of the cells. Cells exposed to light in the presence of ethanol or acetate have an indefinite lag period and never enter the period of rapid chlorophyll synthesis (Horrum and Schwartzbach, 1980a). The length of the lag period in cells grown on other substrates (Freyssinet, 1976a) is inversely related to their paramylum content and consumption. Cells differing in paramylum content undoubtably differ in their content of other macromolecules and various reserves of carbon, nitrogen, and phosphorous (Freyssinet et al., 1972); paramylum content may be only one indicator of the general nutritional state of the cells. Nutrient-rich cells would, in general, be expected to synthesize the machinery required for rapid chlorophyll formation more quickly than nutrient-poor cells.

The lag period in chlorophyll synthesis can be eliminated by exposing the dark-grown cells to a 2-hour preillumination period followed by a 12-hour dark period prior to placing them in continuous light for chloroplast development (Holowinsky and Schiff, 1970) (see Fig. 10). Synthesis of proteins on chloroplastic and cytoplasmic ribosomes during the preillumination period and during the subsequent dark period are required for lag elimination (Schwartzbach et al., 1976b).

The optimal length of the dark period required for lag elimination is approximately equal to the length of the lag period (Freyssinet, 1976b; Holowinsky and Schiff, 1970); consequently, cells with higher paramylum contents require shorter dark periods. Although the degradation of paramylum is strictly light-dependent (Freyssinet et al., 1972; Schwartzbach et al., 1975), the optimal length of the preillumination period for lag elimination is not related to the paramylum content of the cells. Cells that have received a preillumination period followed by a dark period and cells not so treated show similar rates of paramylum degradation when subsequently exposed to continuous light to allow chloroplast development to take place. Thus, it appears that the preillumination period provides the carbon and energy for the synthesis of proteins in the subsequent dark period that are required for rapid chlorophyll synthesis; the same events must occur during the normal lag period in nonpreilluminated cells, since the end result in both cases is rapid chlorophyll synthesis without a further lag.

Other early strictly light-dependent events of chloroplast development that are not influenced by preillumination include the kinetics of NADP-glyceraldehyde 3-phosphate dehydrogenase synthesis (Bovarnick, 1969), alkaline DNase synthesis (J. M. Egan and J. A. Schiff, unpublished experiments), fumarase syn-

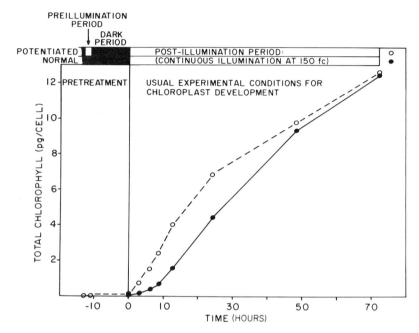

Fig. 10. Time course of chlorophyll accumulation in control (●) and potentiated (○) cells. Zero time is taken as the beginning of the postillumination period. Cells from a 3-day dark-grown resting culture were exposed to 90 minutes of preillumination (white light, 150 ft-c) starting at -12 hours and to a dark period (potentiated cells) before exposure to continuous illumination with white light (postillumination). Control cells experienced an uninterrupted dark period until they were exposed to continuous illumination at 0 hour. (From Holowinsky and Schiff, 1970.)

thesis (Horrum and Schwartzbach, 1980), and cyt rRNA synthesis (D. Cohen and Schiff, 1976). As might be expected from the observation that carotenoid levels do not change significantly during the lag period of normal development (Stern *et al.*, 1964a; Bovarnick *et al.*, 1974a), preillumination does not eliminate the lag in carotenoid synthesis (Bovarnick, 1969); the lag period in appearance of carbon dioxide fixation is also not eliminated by preillumination (Schwartzbach *et al.*, 1976b).

Preillumination does increase the capacity of the developing chloroplast to synthesize proteins. Thus preillumination induces the synthesis of pl rRNA, and this continues through the dark period (D. Cohen and Schiff, 1976). Since rRNA and ribosomal protein synthesis are usually coregulated, plastid ribosomes may be formed in the dark period; plastid ribosomes require proteins synthesized in the cytoplasm and in the plastid (Freyssinet, 1977; 1978). Preillumination increases the rate of cytochrome c_{552} synthesis on subsequent illumination (Freyssinet *et al.*, 1979) and shortens the lag period for thylakoid membrane formation

(Klein *et al.*, 1972). Perillumination also increases the rate of formation of cytoplasmic polysomes (Freyssinet, 1977).

Both photoreceptor systems known to participate in the regulation of chloroplast development are involved in the regulation of chlorophyll biosynthesis. The action spectrum for chlorophyll synthesis in preilluminated or nonpreilluminated cells exposed to continuous light indicates that a protochlorophyll(ide) holochrome is the light-receptor molecule (Egan *et al.*, 1975; Egan and Schiff, 1974); blue and red light are both effective. The preillumination action spectrum indicates the participation of a blue-light receptor system augmented to some extent by the red-blue protochlorophyll(ide) system (Egan *et al.*, 1975; Holowinsky and Schiff, 1970).

The experiments just described lend themselves to the following interpretation. The blue light system controls the central metabolism of the cell and other machinery largely external to the developing plastid; this system is thought to be predominantly involved in the step-up response that provides the materials and energy for feeding the developing chloroplast. For this reason (Fig. 8) this system should be the first to be activated. The lag period in plastid development appears to be the interval in which these externally supplied constituents are entering the developing chloroplast from the rest of the cell. The end of the lag period would signal the accumulation of sufficient imports to allow the machinery within the plastid (Fig. 8) to operate at full capacity, and development enters the rapid period of synthesis and expansion (Fig. 7) in which induction of formation of specific plastid constituents by the blue-red receptor becomes the predominant feature.

2. *Protochlorophyll(ide) and Related Pigments*

The proplastid thylakoids of *Euglena* contain protochlorophyll and protochlorophyll(ide), which are transformable to chlorophyll(ide) in the light (C. Cohen and Schiff, 1976; Kindman *et al.*, 1978). Large amounts of protopheophytin and protopheophorbide are found as well (C. Cohen and Schiff, 1976), but it is not known for certain whether these are normal constituents or are formed during extraction and purification of the pigments. Control experiments and the constancy of protopheophorbide/protopheophytin levels, however, indicate that they may not be wholly artifacts. Studies using fluorescence spectroscopy indicate the possibility of magnesium insertion into protopheophytin/ protopheophrobide to form protochlorophyll(ide) (Frey *et al.*, 1979). The growing evidence that pheophytins of chlorophyll and bacteriochlorophyll are normal functional constituents of photosynthetic reaction centers (Straley *et al.*, 1973; Klimov *et al.*, 1980; Rutherford *et al.*, 1981) should keep us alert to the possibility that not all of the protopheophytin, protopheophorbide, pheophytin, and bacteriopheophytin found on extraction of cells and photosynthetic membranes is artifactual, and that biosynthetic pathways may exist for the formation of these magnesium-free pigments.

The absorption spectrum of dark-grown cells of *Euglena* indicates that they contain only a protochlorophyll(ide) species absorbing at 635 nm but lack the form of protochlorophyll(ide) absorbing at 650 nm (Kindman *et al.*, 1978). On illumination, the 635 nm absorption decreases, and a new absorption due to chlorophyll(ide) appears at 676 nm. This absorption very rapidly shifts a few nenometers to shorter wavelengths to lie at about 673 nm. In darkness, after illumination, protochlorophyll(ide) absorbing at 635 nm is resynthesized to the same levels it was before illumination, indicating a tight control of its synthesis. When dividing cells of *Euglena* are placed under nondividing conditions on a medium lacking available carbon and nitrogen, protochlorophyll(ide) is progressively lost from the cells, undoubtedly being subject to turnover into other cellular constituents needed for survival under starvation conditions (Alhadeff *et al.*, 1979, 1980). However, on illumination of the cells or addition of a carbon source such as glutamate or malate, the protochlorophyll(ide) is rapidly regenerated. Since light is known to induce paramylum breakdown (Schwartzbach *et al.*, 1975), light or an exogenous carbon source appears to provide needed carbon for protochlorophyll(ide) biosynthesis and, for a general step-up in metabolism, signalling that nutrients are available and that the turnover and loss of certain constituents can cease.

On the whole, the proplastids and protopigments of *Euglena* are more like those of younger etiolated seedlings than they are like the better-studied older etiolated seedlings (Lancer *et al.*, 1976; Klein and Schiff, 1972). Thus *Euglena* cells have proplastids rather than etioplasts, lack crystalline prolamellar bodies, contain protochlorophyll(ide) absorbing at 635 nm rather than 650 nm, etc. The smaller, less highly developed proplastid system of *Euglena* (like that in *Ochromonas* (Gibbs, 1962)) may be the more normal route of chloroplast development; the more highly developed etioplasts of older higher plants appear to be an adaptation to prolonged etiolation when faster development on illumination is an asset to allow rapid formation of photosynthesis before seed reserves are exhausted (Schiff, 1980a).

3. *Plastid Thylakoid Polypeptides, Sulfolipid, and Carotenoids*

Since the thylakoid sulfolipid is present in wild-type cells and in bleached mutants such as W_3BUL, it is likely that this thylakoid constituent is synthesized by enzymes coded in a DNA other than plDNA, very likely in nDNA (see Bingham and Schiff, 1979a). Since it moves with chlorophyll on gradients designed to resolve thylakoid membranes of wild-type cells and chloroplasts, the sulfolipid is a convenient marker for thylakoid membranes from mutants and from cells undergoing light-induced chloroplast development. Like similar preparations from other organisms, plastid thylakoid membranes from chloroplasts or from wild-type light-grown cells of *E. gracilis* var. *bacillaris* or strain Z display 30–40 polypeptides on SDS polyacrylamide gel electrophoresis (Bingham and Schiff, 1979a). Some of the more prominent bands among them (those thought to be associated with the pigment–protein complexes) are not detectable in similar preparations from dark-grown wild-type cells. The patterns from thylakoid membranes of wild-type dark-grown cells are identical to those obtained from either light or dark-grown bleached mutants such as W_3BUL indicating that most, if not all, of the thylakoid polypeptides of the proplastid may be coded elsewhere than in plDNA, probably in nDNA. Certain plastid proteins are present at low levels in dark-grown cells of the wild-type but are undetectable in bleached mutants such as W_3BUL. Such molecules, at least in part, are candidates for coding in plDNA; some of these polypeptides can be found among those labeled in isolated chloroplasts supplied with radioactive amino acids (Bingham and Schiff, 1979a; Vasconcelos, 1976).

When dark-grown wild-type cells are exposed to light, some new plastid thylakoid polypeptides appear, but all plastid thylakoid polypeptides increase concomitantly with development of the plastid membranes as seen by electron microscopy (Bingham and Schiff, 1979a,b; Klein *et al.*, 1972; Ben-Shaul *et al.*, 1964).

The thylakoid lipids, among them the sulfolipid and the polyunsaturated fatty acids, increase during chloroplast development (Rosenberg and Pecker, 1964; Erwin, 1968). The carotenoids are at low levels in the dark-grown cells and also increase during light-induced chloroplast development (Stern *et al.*, 1964a). The selective inhibitor of carotenoid biosynthesis SAN 9789, which blocks the pathway just after phytoene formation, has been useful in studying the developmental consequences of carotenoid depletion. In SAN-treated cells, the levels of plastid thylakoid polypeptides, chlorophyll, and sulfolipid formed during light-induced chloroplast development are closely related to the amount of carotenoid formed, another example of the coregulation of synthesis of plastid thylakoid components that insures correct assembly of the thylakoid structure (Vaisberg and Schiff, 1976). Lack of synthesis of chlorophyll and other constituents when carotenoid synthesis is blocked would have been highly selected during evolution, since carotenoids are known to protect the cells from deleterious photoxidations sensitized, in some cases, by chlorophyll. In higher plants, for some reason, this control is not as stringent; higher plants make normal proplastids and protochlorophyll(ide) in the absence of carotenoids (Pardo and Schiff, 1980) and, in some cases, can even undergo a limited amount of light-induced chloroplast development in dim light.

Since the plastid thylakoid polypeptide patterns of dark-grown wild-type cells and of bleached mutants are the same, the absence of detectable protochlorophyll(ide) in the mutants does not appear to prevent the formation of these polypeptides in darkness (Bingham and Schiff, 1979a). The addition of inhibitors of translation on either 87 S cytoplasmic ribosomes of *Euglena* (such as cycloheximide) or on 70 S chloroplast ribosomes (such as streptomycin or chloramphenicol) blocks the incoming of all plastid thylakoid polypeptides during light-induced chloroplast development (Bingham and Schiff, 1979b); the same patterns of inhibition is observed for cytochrome c_{552} and chlorophyll (Freyssinet *et al.*, 1979). Cytochrome c_{552} formation is also correlated with the presence of protochlorophyll(ide) (Russel *et al.*, 1978). The uniformity of these responses point once again to a high degree of coregulation of the biosynthesis of various membrane constituents.

IV. Photoreceptors and Levels of Control

It is apparent from the foregoing discussion that at least two photoreceptors control chloroplast development in *Euglena*. Although action spectra are not easy to measure or to interpret, they constitute the best evidence, at present, for the nature of the molecules involved. Mutants blocked in each of the photoreceptors or in the expression of their function (Russel *et al.*, 1978; Schmidt and Lyman, 1975) should also help in their identification and in sorting out the developmental systems that they control.

A. THE RED–BLUE PHOTORECEPTOR SYSTEM

The action spectrum for chlorophyll biosynthesis during continuous greening is very similar to the absorption spectrum of the protochlorophyll(ide) holochrome from beans, within the precision of the measurements (Egan *et al.*, 1975; Egan and Schiff, 1974). Thus the action spectrum shows a ratio of blue/red peaks of about 5:1. The action spectrum is in accord with the information we have on the photoconversion of protochlorophyll(ide) to chlorophyll(ide) in intact dark-grown cells (Kindman *et al.*, 1978). Detailed action spectra during greening for the formation of various plastid enzymes thought to be nuclear-coded and cytoplasmically synthesized, such as alkaline DNase and NADP-triosephosphate dehydrogenase, also resemble the absorption of protochlorophyll(ide) holochrome (Egan *et al.*, 1975). Action spectra exist for the formation of plastid rRNA and for the postillumination period of potentiation, which are consistent with the same type of photoreceptor (Egan *et al.*, 1975; Holowinsky and Schiff, 1970; D. Cohen and Schiff, 1976). It should be noted, however, that the data do not rule out the participation of other photoreceptors as well. For example, the action spectra are frequently rather broad, the dose response curves are complex enough to suggest that more than one photoreceptor system is operating, and red and far-red light are sometimes additive. It is not conclusively shown that all red–blue phenomena are the same in *Euglena*. For example, although protochlorophyll(ide) photoreduction to chlorophyll(ide) is an obligatory step in chlorophyll biosynthesis, it is not known whether protochlorophyll(ide) itself or other pigments which mimic its absorption are active in the other red–blue phenomena, such as photocontrol of protein and nucleic acid synthesis (D. Cohen and Schiff, 1976).

B. THE BLUE PHOTORECEPTOR SYSTEM

In *Euglena*, this photoreceptor makes itself known as an unusually high blue peak in action spectra for the preillumination which brings about the elimination of the lag in chlorophyll synthesis (Egan *et al.*, 1975; Holowinsky and Schiff, 1970); the blue/red ratio is about 40:1. It appears to be unmixed with the red/blue receptor in broad band action spectra for the formation of cytoplasmic rRNAs during chloroplast development (D. Cohen and Schiff, 1976) and in mutants lacking the red/blue response (Schmidt and Lyman, 1974; Russel *et al.*, 1978). Since the same action spectrum can be measured in mutant W_3BUL, in which plastid DNA, plastid ribosomes, and protochlorophyll(ide) are undetectable, and since it appears to control the transcription of cyt rRNA genes in nDNA, this photoreceptor probably occurs externally to the plastid. Whether this receptor has peaks in regions of the spectrum other than the blue is not known with certainty; the absorption of this system in the near ultraviolet appears to be quite low (Egan *et al.*, 1975). This blue light system, therefore, joins a long list of other similar blue receptors of unknown composition referred to collectively as cryptochrome (Senger, 1980; Schiff, 1980b). Blue light receptors are known to be of importance in controlling chloroplast development in other organisms, such as the light-requiring mutants of *Scenedesmus* and *Chlorella* (see Senger, 1980), organisms that ordinarily produce chlorophyll and chloroplasts in darkness.

C. COREGULATION BY THE TWO PHOTORECEPTOR SYSTEMS

Although phenomena controlled by each of the two photoreceptor systems can be identified, there is also evidence for coordinate control by both systems. For example, a number of plastid enzymes thought to be nuclear-coded and cytoplasmically translated are repressed in dark-grown wild type cells and are induced to increase by light. The action spectra indicate that the red/blue photoreceptor controls these inductions (Egan *et al.*, 1975), and consistent with this (Table I), mutant W_3BUL that lacks protochlorophyll(ide) and the red–blue induction of pl rRNA synthesis shows no derepression of these enzymes on illumination (Bovarnick *et al.*, 1974b). However, another mutant (Y_9ZNAlL), lacking protochlorophyll(ide) but thought to retain plDNA, shows a derepression of these enzymes by light absorbed by the blue receptor (Russel *et al.*, 1978; Schmidt and Lyman, 1975). This evidence indicates a coordinate control of the induction of

Table I

SELECTED PROPERTIES OF W_3BUL COMPARED WITH DARK- AND LIGHT-GROWN WILD-TYPE
Euglena CELLS

Parameter	W_3BUL	Wild-Type Dark-Grown	Light-Grown
Plastid DNA	Not detectable	Present	Present
Chlorophyll	Not detectable	Not detectable	Present
Carotenoids	Low	Low	High
Photosynthesis	Not detectable	Not detectable	Present
Various plastid enzymes thought to be nuclear-coded[a]	Low	Low	High
Various plastid enzymes thought to be plastid-coded, at least in part[b]	Not detectable	Not detectable or low	High
Non-plastid enzymes[c]	High	High	High
Plastid rRNAs, ribosomes, and tRNAs	Not detectable	Low	High
Cytoplasmic rRNAs, ribosomes, and tRNAs	High	High	High

[a] For example, NADP-triosephosphate dehydrogenase, alkaline DNase, leucyl-tRNA synthetase, enzymes forming sulfolipid, proplastid thylakoid polypeptides.
[b] For example, RuBP-carboxylase, cytochrome c_{552}.
[c] For example, NAD-triosephosphate dehydrogenase.

these enzymes in wild-type cells by the blue and red/blue photoreceptors, as in lag elimination in chlorophyll synthesis (Holowinsky and Schiff, 1970). Consistent with this interpretation, red light alone can bring about normal chloroplast development in *Euglena* (Bovarnick *et al.*, 1969; Bingham and Schiff, 1979b). If we assume that the blue system has no red absorption whatsoever, the absorption of light by the red–blue system must somehow generate a signal that induces the blue-controlled system in the absence of blue light. This coregulation may be related to the coregulation of one plastid thylakoid constituent by another, already discussed.

D. LEVELS OF CONTROLS

Since rather low levels of light are effective in inducing chloroplast development in *Euglena* (Stern *et al.*, 1964b; Holowinsky and Schiff, 1970), some sort of cellular amplification must occur to convert the photoresponse into an effective metabolic signal. By analogy to other prokaryotic and eukaryotic systems, one might expect control at any level of cellular function, including information processing (transcription, translation, and posttranslational modification) and modulation of membrane transport. So far, in *Euglena* evidence exists only for transcriptional control by light [e.g., the control of rRNA transcription from nDNA and plDNA (D. Cohen and Schiff, 1976)]. A detailed model has been offered based on control of prokaryotic transcription in which pigmented repressors are modified by light in such a manner as to derepress the operons they control (Schiff, 1981a,b). Since light is an inducer of the formation of light-utilizing machinery for photosynthesis resulting in chloroplast development, this is analogous to substrate induction by organic molecules in prokaryotes.

We should not overlook the possibility of control at the level of membrane transport by either of the two photoreceptors. The control of movement of ions or substrates by photoreceptors localized in membranes could achieve the amplification needed for control (see Brinkmann and Senger, in Senger, 1980). Although the photoreduction of protochlorophyll(ide) to chlorophyll(ide) serves as a control point in chlorophyll biosynthesis, it is not known whether the other manifestations of control by the red–blue receptor are due to protochlorophyll(ide) itself or to another molecule with similar absorption properties. Since protochlorophyll(ide) is membrane-localized and requires two hydrogens for reduction, it could play a strategic role in the photocontrol of processes such as transcription of plDNA by modulating the levels of hydrogen and other ions within the chloroplast. This would appear to be the way for bulky and often insoluble molecules to control processes some distance away from their resident sites.

V. Conclusions

The picture of chloroplast development in *Euglena* which we have tried to assemble is shown in Fig. 11. We consider light-induced chloroplast development to be a substrate induction by light, where an entire organelle is constructed to utilize the inducing substrate. This substrate induction has at least two components: a blue light system and a red–blue system. Illumination of the blue receptor, thought to be localized outside the chloroplast, produces a step-up in the central metabolism of the cell to provide the energy and metabolites necessary to feed the developing chloroplast, particularly during the lag period early in development. It also controls transcription of nuclear DNA. In higher plants, phytochrome usually assumes these functions; phytochrome has not been detected in *Euglena* (Holowinsky and Schiff, 1970; Boutin and Klein, 1972). The red–blue system is probably localized in the plastid and is thought to control the formation of chlorophyll and the transcription in *Euglena* of plDNA. The blue light and red–blue systems appear to act coordinately in the induction of plastid proteins thought to be coded in nDNA. In this way, the specific substrate induction of individual chloroplast proteins is achieved.

We end where we began. These various control systems, whatever their

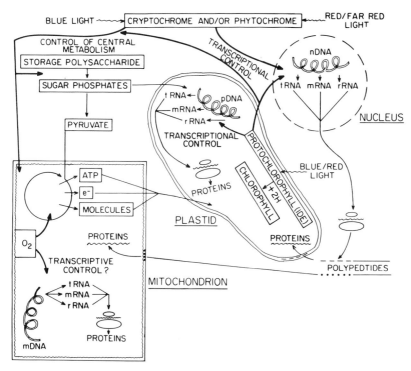

Fig. 11. Summary of nutritional and informational interactions among organelles during chloroplast development in *Euglena* and other organisms. The DNAs of either the mitochondrion or the chloroplast code for each organelle's tRNAs and rRNAs and for certain proteins that are translated on the ribosomes of each organelle and remain within it. Nuclear DNA codes for rRNAs and tRNAs of the cytoplasm and for proteins that are translated on cytoplasmic ribosomes. Some of these remain in the cytoplasm; others bearing signal polypeptides recognize and enter the appropriate organelle losing the signal portion enroute. Light-induced breakdown of storage carbohydrate via glycolysis feeds mitochondrial respiration, which supplies small molecules, energy, and reducing power to the developing chloroplast. Blue light via cryptochrome (in *Euglena*) or red/far red light via phytochrome (in higher plants) is thought to control the non-chloroplast aspects of plastid development including the central metabolism of the cell and nuclear transcription of genes coding for plastid proteins. Blue and red light [a protochlorophyll(ide)-type of photoreceptor] is thought to control plastid transcription and chlorophyll(ide) formation. Since those plastid enzymes that are nuclear-coded show protochlorophyll(ide)-type action spectra for induction, it is suggested that illumination of the plastid receptor results in a regulatory signal which passes to the nucleus and cytoplasm to activate processes ordinarily activated by the blue light receptor. Oxygen apparently controls mitochondrial development, particularly in yeast. (From Schiff, 1981a.)

chemistry or level of control, absorb in the blue and red regions of the spectrum because the substrates to be utilized by the fully-induced chloroplast for photosynthesis are blue and red light. There would have been a strong selection during evolution for control pigments which absorbed in the same regions of the spectrum as the sensitizers of photosynthesis to ensure that cellular resources would be expended to form chloroplasts only when light of the appropriate quality for photosynthesis was available.

Acknowledgment

This study was based on work supported by grants from the National Institutes of Health GM 14595 (JAS) and GM 26994 (SDS).

References

Alhadeff, M., Coronado, R., Figueroa, N., and Schiff, J. A. (1979). *Plant Physiol.* **63–S,** 98.

Alhadeff, M., Coronado, R., Figueroa, N., and Schiff, J. A. (1980). *In* ''Photoreceptors and Plant Development'' (J. deGreef, ed.), pp. 175–177 Antwerpen Univ. Press.

App, A. A., and Jagendorf, A. T. (1963). *J. Protozool.* **10,** 340–343.

Avadhani, N. G., and Buetow, D. E. (1972). *Biochem J.* **128,** 353–365.

Beney, G., and Nigon V. (1975). *Proc. Int. Congr. Photosynth., 3rd, 1974,* pp. 1801–1808.

Ben-Shaul, Y., Schiff, J. A., and Epstein, H. T. (1964). *Plant Physiol.* **39,** 231–240.

Ben-Shaul, Y., Epstein, H. T., and Schiff, J. A. (1965). *Can. J. Bot.* **43,** 129–136.

Bingham, S., and Schiff, J. A. (1979a). *Biochim. Biophys. Acta* **547,** 512–530.

Bingham, S., and Schiff, J. A. (1979b). *Biochim. Biophys. Acta* **547,** 531–543.

Boutin, M. E., and Klein, R. M. (1972). *Plant Physiol.* **49,** 656–657.

Bovarnick, J. G. (1969). Dissertation, Brandeis University, Waltham, Massachusetts.

Bovarnick, J. G., Zeldin, M. H., and Schiff, J. A. (1969). *Dev. Biol.* **19,**321–340.

Bovarnick, J. G., Chang, S. W., Schiff, J. A., and Schwartzbach, S. D. (1974a). *J. Gen. Microbiol.* **83,** 51–62.

Bovarnick, J. G., Schiff, J. A., Freedman, Z., and Egan, J. M., Jr. (1974b). *J. Gen. Microbiol.* **83,** 63–71.

Breitenberger, C. A., and Spremulli, L. L. (1980). *J. Biol. Chem.* **255,** 9814–9820.

Breitenberger, C. A., Graves, M. C., and Spremulli, L. L. (1979). *Arch. Biochem. Biophys.* **194,** 265–270.

Brunold, C., and Schiff, J. A. (1976). *Plant Physiol.* **57,** 430–436.

Buchanan, R. E., and Gibbons, N. E., eds. (1974). ''Bergey's Manual of Determinative Bacteriology,'' 8th ed., p. 932. Williams & Wilkins, Baltimore, Maryland.

Buetow, D. E. (1967). *Nature (London)* **213,** 1127–1128.

Buetow, D. E., ed. (1968). ''The Biology of *Euglena,*'' Vols. 1 and 2. Academic Press, New York.

Buetow, D. E., and Wood, W. M. (1978). *Subcell. Biochem.* **5,** 1–85.

Buetow, D. E., Wurtz, E. A., and Gallagher, T. (1980). *In* ''Nuclear–Cytoplasmic Interactions in the Cell Cycle'' pp. 9–55. Academic Press, New York.

Calvayrac, R., Laval-Martin, D., Dubertret, G., and Bomsel, (1979). *Plant Physiol.* **63,** 866–872.

Chelm, B. K., and Hallick, R. B. (1976). *Biochemistry* **15,** 593–599.

Chelm, B. K., Hoben, P. J., and Hallick, R. B. (1977a). *Biochemistry* **16,** 776–782.

Chelm, B. K., Hoben, P. J., and Hallick, R. B. (1977b). *Biochemistry* **16,** 782–785.

Chelm, B. K., Gray, P. W., and Hallick, R. B. (1978). *Biochemistry* **17,** 4239–4244.

Chelm, B. K., Hallick, R. B., and Gray, P. W. (1979). *Proc. Natl. Acad. Sci. U.S.A.* **76**, 2258-2262.

Cohen, C., and Schiff, J. A. (1976). *Photochem. Photobiol.* **24**, 555-566.

Cohen, D., and Schiff, J. A. (1976). *Arch. Biochem. Biophys.* **177**, 201-216.

Cohen, D., Malkin, S., Gurevitz, M., and Ohad, I. (1978). *Plant Physiol.* **62**, 1-5.

Coleman, A. W. (1979). *J. Cell Biol.* **82**, 299-305.

Collins, N., and Merrett, M. J. (1975). *Plant Physiol.* **55**, 1018-1022.

Crouse, E. J., Vandrey, J. P., and Stutz, E. (1974). *FEBS Lett.* **42**, 262-266.

Curtis, S. E., and Rawson, J. R. Y. (1979). *Biochemistry* **18**, 5299-5304.

Diamond, J. S., Schiff, J. A., and Kelner, A. (1975). *Arch. Biochem. Biophys.* **167**, 603-614.

Dockerty, A., and Merrett, M. J. (1979). *Plant Physiol.* **63**, 468-473.

Dolphin, W. D. (1970). *Plant Physiol.* **46**, 685-691.

Dwyer, M. R., and Smillie, R. M. (1970). *Biochim. Biophys. Acta* **216**, 392-401.

Dwyer, M. R., and Smillie, R. M. (1971). *Aust. J. Biol. Sci.* **24**, 15-22.

Edelman, M., Schiff, J. A., and Epstein, H. (1965). *J. Mol. Biol.* **11**, 769-774.

Egan, J., and Schiff, J. A. (1974). *Plant Sci. Lett.* **3**, 101-105.

Egan J., Dorsky, D., and Schiff, J. A. (1975). *Plant Physiol.* **56**, 318-323.

Egan, J. M., Jr., and Carrell, E. F. (1972). *Plant Physiol.* **50**, 391-395.

El-Gewely, M. R., Lomax, M. I., Lau, E. T., Helling, R. B., Farmerie, W., and Barnett, W. E. (1981). *Mol. Gen. Genet.* **181**, 296-305.

Erwin, J. A. (1968). In ''The Biology of Euglena'' (D. E. Buetow, ed.), Vol. 2, pp. 133-148. Academic Press, New York.

Evans, W. R. (1971). *J. Biol. Chem.* **245**, 6144-6151.

Fong, F., and Schiff, J. A. (1977). *Plant Physiol.* **59**, 5-92.

Fong, F., and Schiff, J. A. (1979). *Planta* **146**, 119-127.

Fox, L., Erion, J., Tarnowski, J., Spremulli, L., Brot, N., and Weissbach, H. (1980). *J. Biol. Chem.* **255**, 6018-6019.

Frey, M. A., Alberte, R. S., and Schiff, J. A. (1979). *Biol. Bull. (Woods Hole, Mass.)* **157**, 368-369.

Freyssinet, G. (1976a). *Plant Physiol.* **57**, 824-830.

Freyssinet, G. (1976b). *Plant Physiol.* **57**, 831-835.

Freyssinet, G. (1977). *Physiol. Veg.* **15**, 519-550.

Freyssinet, G. (1978). *Exp. Cell Res.* **115**, 207-219.

Freyssinet, G., Heizmann, P., Verdier, G., Trabuchet, G., and Nigon, V. (1972). *Physiol. Veg.* **10**, 421-442.

Freyssinet, G., Harris, G. C., Nasitir, M., and Schiff, J. A. (1979). *Plant Physiol.* **63**, 908-915.

Garlaschi, F. M., Garlaschi, A. M., Lombardi, A., and Forti, G. (1974). *Plant Sci. Lett.* **2**, 29-39.

Gibbs, S. P. (1962). *J. Cell Biol.* **14**, 433-444.

Goldberger, R. F., ed. (1979). ''Biological Regulation and Development,'' Vol. 1. Plenum, New York.

Goldstein, N. H., Schwartzbach, S.D., and Schiff, J. A. (1974). *J. Protozool.* **21**, 443.

Gomez-Silva, B., and Schiff, J. A. (1981). *Plant Physiol.* **67** (suppl.) 32.

Gray, P. W., and Hallick, R. B. (1978). *Biochemistry* **17**, 284-289.

Greenblatt, C., and Schiff, J. A. (1959). *J. Protozool.* **6**, 23-28.

Gruol, D. J., and Haselkorn, R. (1976). *Biochim. Biophys. Acta* **447**, 82-95.

Gurevitz, M., Kratz, H., Ohad, I. (1977). *Biochim. Biophys. Acta* **461**, 475-488.

Hallick, R. B., Rushlow, K. E., Orozco, E. M., Jr., Stiegler, G. L., and Gray, P. W. (1979). *ICN-UCLA Symp. Mol. Cell. Biol.* **15**, 127-141.

Harris, R. C., and Kirk, J. T. O. (1969). *Biochem. J.* **113**, 195-205.

Hecker, L. I., Egan, J., Reynolds, R. J., Nix, C. E., Schiff, J. A., and Barnett, W. E. (1974). *Proc. Natl. Acad. Sci. U.S.A.* **71**, 1910-1944.

Heizmann, P., Doly, J., Hussein, Y., Nicolas, P., Nigon, V., and Bernardi, G. (1981). *Biochim. Biophys. Acta* **653**, 412–415.

Helling, R. B., El-Gewely, M. R., Lomax, M. I., Baumgartner, J. E., Schwartzbach, S. D., and Barnett, W. E. (1979). *Mol. Gen. Genet.* **174**, 1–10.

Hill, H. Z., Schiff, J. A., and Epstein, H. T. (1966). *Biophys. J.* **6**, 125–133.

Holowinsky, A. W., and Schiff, J. A. (1970). *Plant Physiol.* **45**, 339–347.

Horrum, M. A., and Schwartzbach, S. D. (1980a). *Plant Physiol.* **65**, 382–386.

Horrum, M. A., and Schwartzbach, S. D. (1980b). *Planta* **149**, 376–383.

Horrum, M. A., and Schwartzbach, S. D. (1980c). *Plant Physiol.* **65**, S-19.

Horrum, M. A., and Schwartzbach, S. D. (1980d). *Plant Sci. Lett.* **20**, 133–139.

Horrum, M. A., and Schwartzbach, S. D. (1981). *Plant Physiol.* **68**, 430–434.

Horrum, M. A., and Schwartzbach, S. D. (1982). *Biochim. Biophys. Acta* (in press).

Hutner, S., and Provasoli, L. (1951). *Biochem. Physiol. Protozoa* **1**, 29–121.

Hutner, S., and Provasoli, L. (1955). *Biochem. Physiol. Protozoa* **2**, 17–41.

Karlan, A. W., and Russel, G. K. (1976). *J. Protozool.* **23**, 176–179.

Keller, M., Burkard, G., Bohnert, H. J., Mubumbila, M., Gordon, K., Steinmetz, A., Heiser, N., Crouse, E. J., and Weil, J. H. (1980). *Biochem. Biophys. Res. Commun.* **95**, 47–54.

Kindman, L. A., Cohen, C. E., Zeldin, M. H., Ben-Shaul, Y., and Schiff, J. A. (1978). *Photochem. Photobiol.* **27**, 787–794.

Kirk, J. T. O. (1970). *Nature (London)* **226**, 182.

Klein, S., Schiff, J. A., and Holowinsky, A. W. (1972). *Dev. Biol.* **28**, 253–273.

Klein, S. H., and Schiff, J. A. (1972). *Plant Physiol.* **49**, 619–626.

Klimov, V. V., Dolan, E., and Ke, B. (1980). *FEBS Lett.* **112**, 97–100.

Kopecka, H., Crouse, E. J., and Stutz, E. (1977). *Eur. J. Biochem.* **72**, 525–535.

Lancer, J., Cohen, C. E., and Schiff, J. A. (1976). *Plant Physiol.* **57**, 369–374.

Leedale, G. F. (1967). "Euglenoid Flagellates." Prentice-Hall, Englewood Cliffs, New Jersey.

Lefort-Tran, M. (1975). *In* "Les cycles cellulaires et leur blockage chez plusieurs protistes," pp. 297–308. CNRS, Paris.

Lor, K. L., and Cossins, E. A. (1978). *Phytochemistry* **17**, 659–665.

McCrea, J. M., and Hershberger, C. L. (1976). *Nucleic Acids Res.* **3**, 2005–2018.

Manning, J. E., and Richards, O. C. (1972). *Biochim. Biophys. Acta* **259**, 285–296.

Manning, J. E., Wolstenholme, D. R., Ryan, R. S., Hunter, J. A., and Richards, O. C. (1971). *Proc. Natl. Acad. Sci. U.S.A.* **68**, 1169–1173.

Merrett, M. J., and Lord, J. M. (1973). *New Phytol.* **72**, 751–767.

Mubumbila, M., Byrkard, G., Keller, M., Steinmetz, A., Crouse, E., and Weil, J. H. (1980). *Biochim. Biophys. Acta* **609**, 31–39.

Nicolas, P., Innocent, J. P., and Nigon, V. (1977). *Mol. Gen. Genet.* **155**, 123–129.

Nicolas, P., Freyssinet, G., and Nigon, V. (1980). *Plant Physiol.* **65**, 631–634.

Nigon, V., and Heizmann, P. (1978). *Int. Rev. Cytol.* **53**, 212–290.

Ortiz, W., Reardon, E. M., and Price, C. A. (1980). *Plant Physiol.* **66**, 291–294.

Osafune, T., and Schiff, J. A. (1980a). *J. Ultrastruct. Res.* **73**, 64–76.

Osafune, T., and Schiff, J. A. (1980b). *J. Ultrastruct. Res.* **73**, 336–349.

Osafune, T., Klein, S. H., and Schiff, J. A. (1980). *J. Ultrastruct.* **73**, 77–90.

Palisano, J. R., and Walne, P. L. (1976). *Nova Hedwigia* **27**, 455–481.

Pardo, A., and Schiff, J. A. (1980). *Can. J. Bot.* **58**, 25–35.

Parthier, B. (1974). *Biochem. Physiol. Pflanz.* **166**, 555–560.

Parthier, B., and Neumann, D. (1977). *Biochem. Physiol Pflanz.* **171**, 547–562.

Parthier, B. (1981). *In* "Biochemistry and Physiology of Protozoa," 2nd ed., Vol. 4, pp. 261–300. Academic Press, New York.

Pellegrini, M. (1980a). *J. Cell Sci.* **43**, 137–166.

Pellegrini, M. (1980b). *J. Cell Sci.* **46**, 313–340.

Rawson, J. R. Y. (1975). *Biochim. Biophys. Acta* **402**, 171–178.

Rawson, J. R. Y., and Boerma, C. L. (1976a). *Biochemistry* **15**, 588–592.

Rawson, J. R. Y., and Boerma, C. L. (1976b). *Proc. Natl. Acad. Sci. U.S.A.* **73**, 2401–2404.

Rawson, J. R. Y., and Boerma, C. L. (1979). *Biochem. Biophys. Res. Commun.* **89**, 743–749.

Rawson, J. R. Y., Kushner, S. R., Vapnek, D., Alton, N. K., and Boerma, C. L. (1978). *Gene* **3**, 191–209.

Rawson, J. R. Y., Eckenrode, V. K., Boerma, C. L., and Curtis, S. (1979). *Biochim. Biophys. Acta* **563**, 1–16.

Richard, F., and Nigon, V. (1973). *Biochim. Biophys. Acta* **313**, 130–149.

Riggs, W. A. C. (1979). *J. Protozool.* **26**, 7a.

Rosenberg, A., and Pecker, M. (1964). *Biochemistry* **3**, 254–258.

Russel, G. K., Draffan, A. G., Schmidt, G. W., and Lyman, H. (1978). *Plant Physiol.* **62**, 678–682.

Rutherford, A. W., Mullet, J. E., Paterson, D. R., Robinson, H. H., Arntzen, C. J., and Crofts, A. R. (1981). *Proc. Int. Congr. Photosynth., 5th, 1980* (in press).

Sagher, D., Grosfeld, H., and Edelman, M. (1976). *Proc. Natl. Acad. Sci. U.S.A.* **73**, 722–726.

Schiff, J. A. (1963). *Year Book—Carnegie Inst. Washington* **62**, 375–378.

Schiff, J. A. (1970). *Symp. Soc. Exp. Biol.* **24**, 277–301.

Schiff, J. A. (1973). *Adv. Morphog.* **10**, 265–312.

Schiff, J. A. (1975). *Proc. Int. Congr. Photosynth., 3rd, 1974* pp. 1691–1717.

Schiff, J. A. (1978). *In* "Photocontrol of Chloroplast Development in Euglena" (G. Akoyounouglou, *et al.*, eds.), pp. 747–767. Elsevier, Amsterdam.

Schiff, J. A. (1980a). *In* "The Biochemistry of Plants" (P. Stumpf and E. Conn, eds.), Vol. 1, pp. 209–272. Academic Press, New York.

Schiff, J. A. (1980b). *In* "The Blue Light Syndrome" (H. Senger, ed.), pp. 495–511. Springer-Verlag, Berlin and New York.

Schiff, J. A. (1981a). *Ann. N.Y. Acad. Sci.* **361**, 166–192.

Schiff, J. A. (1981b). *BioSystems* **14**, 123–147.

Schiff, J. A., and Epstein, H. T. (1968). *In* "The Biology of *Euglena*" (D. E. Buetow, ed.), Vol. 2, p. 285. Academic Press, New York.

Schiff, J. A., Zeldin, M. H., and Rubman, J. (1967). *Plant Physiol.* **42**, 1716–1725.

Schiff, J. A., Lyman H., and Russel, G. K. (1971). *In* "Methods in Enzymology" (A. San Pietro, ed.), Vol. 23, pp. 143–162. Academic Press, New York.

Schiff, J. A., Lyman, H., and Russel, G. K. (1980). *In* "Methods in Enzymology" (A. San Pietro, ed.), Vol. 69, Part C pp. 23–29. Academic Press, New York.

Schiff, J. A., Cunningham, F. X., and Green, M. S. (1982). *In* "6th Int. Symp. on Carotenoids, Liverpool." Pergamon Press (in press).

Schmidt, G., and Lyman, H. (1975). *Proc. Int. Congr. Photosynth., 3rd, 1974* pp. 1755–1764.

Schmidt, G., and Lyman, H. (1976). *In* "Genetics of Algae" (R. A. Lewin, ed.), pp. 257–299. Univ. of California Press, Berkeley.

Schneyour, A., and Avron, M. (1975). *Plant Physiol.* **55**, 137–141.

Schoch, S., Sheer, H., Rüdiger, W., Schiff, J. A., and Siegelman, H. W. (1980). *Plant Physiol.* **65-S**, 19.

Schoch, S., Sheer, H., Rüdiger, W., Schiff, J. A., and Siegelman, H. W. (1981). *Z. Naturforsch.* **36C**, 827–833.

Schwartzbach, S. D., and Schiff, J. A. (1974). *J. Bacteriol.* **120**, 334–341.

Schwartzbach, S. D., and Schiff, J. A. (1979). *Plant Cell Physiol.* **20**, 837–838.

Schwartzbach, S. D., Freyssinet, G., and Schiff, J. A. (1974). *Plant Physiol.* **53**, 533–542.

Schwartzbach, S. D., Schiff, J. A., and Goldstein, N. H. (1975). *Plant Physiol.* **56**, 313–317.

Schwartzbach, S. D., Hecker, L. I., and Barnett, W. E. (1976a). *Proc. Natl. Acad. Sci. U.S.A.* **73**, 1984–1988.

Schwartzbach, S., Schiff, J. A., and Klein, S. (1976b). *Planta* **131**, 1-9.

Schwelitz, F. D., Cisneros, P. L., Jagielo, J. A., Comer, J. L., and Butterfield, K. A. (1978). *J. Protozool.* **25**, 257-261.

Senger, H. ed. (1980). "The Blue Light Syndrome." Springer-Verlag, Berlin and New York.

Shigeoka, S., Yokota, A., Nakano, Y., and Kitaoka, S. (1980). *Bull. Univ. Osaka Prefect., Ser. B* **32**, 37-41.

Shihira-Ishikawa, I., Osafune, T., Ehara, T., Ohkuro, I., and Hase, E. (1977). *Plant Cell Physiol., Spec. Issue* pp. 445-454.

Slavik, N. S., and Hershberger, C. L. (1975). *FEBS Lett.* **52**, 171-174.

Srinivas, Y., and Lyman, H. (1980). *Plant Physiol.* **66**, 295-301.

Steinitz, Y. L., Schiff, J. A., Osafune, T., and Green, M. S. (1980). *In* "The Blue Light Syndrome" (H. Senger, ed.), pp. 269-280. Springer-Verlag, Berlin and New York.

Stern, A. I., Schiff, J. A., and Epstein, H. T. (1964a). *Plant Physiol.* **39**, 220-226.

Stern, A. I., Epstein, H. T., and Schiff, J. A. (1964b). *Plant Physiol.* **39**, 226-231.

Stolarsky, L., Walfield, A., Birch, R., and Hershberger, C. (1976). *Biochim. Biophys. Acta* **425**, 438-450.

Straley, S. C., Parson, W. W., Mauzerall, D., and Clayton, R. K. (1973). *Biochim. Biophys. Acta* **305**, 597-609.

Talen, J. L., Sanders, J. P. M., and Flavell, R. A. (1974). *Biochim. Biophys. Acta* **374**, 129-135.

Tokunaga, M., Nakano, Y., and Kitaoka, S. (1976). *Agric. Biol. Chem.* **40**, 1439-1440.

Uzzo, A., and Lyman, H. (1972). *Photosynth., Two Centuries After Its Discovery Joseph Priestley Proc. Int. Congr. Photosynth. Res., 2nd, 1971* pp. 2585-2599.

Vaisberg, A. J., and Schiff, J. A. (1976). *Plant Physiol.* **57**, 260-269.

Vaisberg, A. J., Schiff, J. A., Li, L., and Freedman, Z. (1976). *Plant Physiol.* **57**, 594-601.

Vasconcelos, A. C. (1976). *Plant Physiol.* **58**, 719-721.

Verdier, G. (1979a). *Eur. J. Biochem.* **93**, 573-580.

Verdier, G. (1979b). *Eur. J. Biochem.* **93**, 581-586.

Walfield, A. M., and Hershberger, C. L. (1978). *J. Bacteriol.* **133**, 1437-1443.

Woodcock, E., and Merrett, M. J. (1980). *Arch. Microbiol.* **124**, 33-38.

Wurtz, E. A., and Buetow, D. E. (1981). *Curr. Genetics* **3**, 181-187.

Zeldin, M. H., Skea, W., and Matteson, O. (1973). *Biochem. Biophys. Res. Commun.* **32**, 544-549.

INDEX